Beginning MATLAB and Simulink

From Beginner to Pro

Second Edition

Sulaymon Eshkabilov

APress®

Beginning MATLAB and Simulink: From Beginner to Pro

Sulaymon Eshkabilov
Agricultural and Biosystems Engineering Department, North Dakota State University,
Fargo, ND, USA

ISBN-13 (pbk): 978-1-4842-8747-7   ISBN-13 (electronic): 978-1-4842-8748-4
https://doi.org/10.1007/978-1-4842-8748-4

Managing Director, Apress Media LLC: Welmoed Spahr
Acquisitions Editor: Steve Anglin
Development Editor: James Markham
Coordinating Editor: Mark Powers
Copy Editor: Kim Wimpsett

Cover designed by eStudioCalamar

Cover image by Bradley Jasper Ybanez on Unsplash (www.unsplash.com)

Distributed to the book trade worldwide by Apress Media, LLC, 1 New York Plaza, New York, NY 10004, U.S.A. Phone 1-800-SPRINGER, fax (201) 348-4505, e-mail orders-ny@springer-sbm.com, or visit www. springeronline.com. Apress Media, LLC is a California LLC and the sole member (owner) is Springer Science + Business Media Finance Inc (SSBM Finance Inc). SSBM Finance Inc is a **Delaware** corporation.

For information on translations, please e-mail booktranslations@springernature.com; for reprint, paperback, or audio rights, please e-mail bookpermissions@springernature.com.

Apress titles may be purchased in bulk for academic, corporate, or promotional use. eBook versions and licenses are also available for most titles. For more information, reference our Print and eBook Bulk Sales web page at www.apress.com/bulk-sales.

Any source code or other supplementary material referenced by the author in this book is available to readers on GitHub (https://github.com/Apress). For more detailed information, please visit www.apress. com/source-code.

Printed on acid-free paper

To the memory of my father.
To my mother.
To my wife, Nigora, after 29 wonderful years together.

Table of Contents

About the Author

 **Dr. Sulaymon Eshkabilov** is an assistant professor in the Department of Agricultural and Biosystems Engineering at North Dakota State University. He obtained a Master of Engineering degree from Tashkent Automobile Road Institute; a Master of Sciences from Rochester Institute of Technology, NY; and a PhD from the Cybernetics Institute of Academy Sciences of Uzbekistan in 1994, 2001, and 2005, respectively. He was an associate professor at Tashkent Automobile Road Institute from December 2006 through January 2017. He held visiting professor and researcher positions at Ohio universities, from 2010 to 2011, and at Johannes Kepler University, from January through September 2017. He teaches a number of courses, including "Instrumentation and Measurement," "System Modelling with MATLAB," "Machine Design Analysis," "Agricultural Power," "Numerical Methods," "Introduction to Finite Element Modelling," and "Advanced MATLAB/Simulink Modelling" for undergraduate and graduate students.

His research interests are image processing, machine learning applications, mechanical vibrations, micro-electromechanical systems, mechatronic system design, and simulation and model development of dynamic systems. He has developed simulation and data analysis models for various image data, additive manufacturing process optimization, vibrating systems, autonomous vehicle control, and studies of mechanical properties of bones. He is the author of five books devoted to MATLAB/Simulink applications for mechanical engineering students and numerical analysis. From 2009 through 2022 he worked as an external academic expert for the European Commission to assess academic projects.

About the Technical Reviewers

Dr. Sanjarbek Ruzimov specializes in vehicle dynamics and energy management. He works as an associate professor in the Department of Mechanical and Aerospace Engineering at Turin Polytechnic University in Tashkent. He obtained his doctoral degree in Mechatronics from Politecnico di Torino, Italy. His active research interests include automotive mechatronic systems optimization, energy optimization of hybrid electric vehicles, and implementation of intelligent transportation systems to improve the vehicle performance and efficiency. He teaches courses about machine design, motor vehicle design, vehicle component design, HEV and EV design and modeling, and scientific computing with MATLAB and Microsoft Excel, where MATLAB and Simulink are the main tools used during the teaching process.

Dr. Joseph Mueller specializes in control systems and trajectory optimization. For his doctoral thesis, he developed optimal ascent trajectories for stratospheric airships. His active research interests include robust optimal control, adaptive control, applied optimization and planning for decision support systems, and intelligent systems to enable autonomous operations of robotic vehicles. Prior to joining SIFT in early 2014, Dr. Mueller worked at Princeton Satellite Systems for 13 years. In that time, he served as the principal investigator for eight Small Business Innovative Research contracts for NASA, Air Force, Navy, and MDA. He has developed algorithms for optimal guidance and control of both formation flying spacecraft and high altitude airships, and he developed a course of action planning tool for DoD communication satellites. In support of a research study for NASA Goddard Space Flight Center in 2005, Dr. Mueller developed the Formation Flying Toolbox for MATLAB, which is now used at NASA, ESA, and several universities and aerospace companies around the world. In 2006, Dr. Mueller developed the safe orbit guidance mode algorithms and software for the Swedish Prisma mission, which has successfully flown a two-spacecraft formation flying mission since its launch in 2010. Dr. Mueller also serves as an adjunct professor in the Aerospace Engineering & Mechanics Department at the University of Minnesota, Twin Cities, campus.

Acknowledgments

I would like to express my special gratitude to the two technical reviewers of this book. First, thank you to Dr. Sanjarbek Ruzimov for his very thorough work while reviewing the context and MATLAB code and Simulink models demonstrated in this the book. Without his critical insights, remarkable suggestions, and corrections of many points along the way, the book would not be as good as it is now. Second, thank you to Dr. Joseph Mueller for his incredible suggestions and comments while reviewing this book. Because of his critical remarks and advice, the book became richer and more educationally appealing. I would also like to express my special gratitude to Dr.Mark Powers for his timely and well-planned correspondence throughout this book project.

My cordial gratitude goes to my mother for her limitless support and love. She kept checking on my progress throughout the entire book progress.

I would like to thank my wife, Nigora. Without her support, I would not have been able to take up the challenging task of writing a second edition of this book. I have spent so much time over the weekends in my office writing and editing the book. In addition, I would like to thank our children—Anbara, Durdona, and Dovud—for being such delightful people and for being the inspiration for writing this book.

Introduction

This book is aimed at beginner-level learners of MATLAB/Simulink packages. It covers the essential, hands-on tools and functions of the MATLAB and Simulink packages and explains them via interactive examples and case studies. The main principle of the book is "learning by doing," and it progresses from simple to complex. It contains dozens of solutions and simulation models via M/MLX files/scripts and Simulink models, which help you learn the programming and modeling essentials. Moreover, there are many recommendations for avoiding pitfalls related to programming and modeling aspects of the MATLAB/Simulink packages. This second edition of the book has been updated with many interesting examples using the MATLAB 2022b version. Moreover, any errors that slipped into the first edition have been corrected.

Beginning MATLAB and Simulink explains various practical issues of programming and modeling in parallel by comparing the programming tools of MATLAB and blocks of Simulink. After studying this book, you'll be proficient at using the MATLAB/Simulink packages. You can apply the source code and models from the book's examples as templates to your own projects in data science, numerical analysis, modeling and simulation, or engineering.

Essential learning outcomes of the book include the following:

- Getting started using MATLAB and Simulink

- Performing data analysis and visualization with MATLAB

- Programming essentials of MATLAB and core modeling aspects of Simulink and how to associate scripts and models of the MATLAB and Simulink packages

- Developing GUI models and stand-alone applications in MATLAB

- Working with integration and numerical root-finding methods in MATLAB and Simulink

- Solving differential equations in MATLAB and Simulink

- Applying MATLAB for data analysis and data science projects

The book contains eight logically interlinked chapters.

- Chapter 1 is dedicated to introducing the MATLAB environment and creating MATLAB recognized data types, including numeric, cell, structure, character, logical, and table arrays and function handles.

- Chapter 2 covers most of the essential programming tools and functions, such as `for ... end` and `while ... end` loop operators, `if...elseif...else...end` condition operators, symbol referencing, and most common errors and warnings. It also covers M-file debugging tools and options in MATLAB. Moreover, this chapter addresses MATLAB-specific programming tools, including function files, function handles, symbol references, and display operators.

- Chapter 3 covers GUI development and how to write and edit GUI model callback functions via several simple but appealing examples, such as solving and plotting quadratic equations, computing, and plotting the sine cardinal function. This chapter also shows how to use and adjust various GUI dialog boxes.

- Chapter 4 addresses the issues of how to develop MEX files, C/C++ code, and stand-alone applications from the existing M-files and code written in C.

- Chapter 5 is dedicated to Simulink modeling essentials. The examples in this chapter cover matrix operations, computing function values, modeling mechanical engineering examples, and solving ordinary differential equations. Moreover, it covers issues around how to associate Simulink models with MATLAB scripts.

- Chapter 6 is devoted to data visualization issues, such as building 2D and 3D plots and animated plots in MATLAB. The chapter examples and code show how to build various appealing 2D and 3D plots using symbolic expressions and numerical values using plot, bar, errorbar, pie, pie3, mesh, contour, fplot, fsurf, etc. Moreover, there are several examples showing how to create animated 2D and 3D graphs.

- Chapter 7 is dedicated to matrix algebra and array operations. It addresses solving systems of linear equations, eigen-value problems, matrix decompositions, matrix and vector operations, and conversions of arrays and strings via examples in MATLAB and Simulink in parallel. Moreover, it addresses a dozen different ways to solve system linear equations and polynomials symbolically and numerically with MATLAB and Simulink. It also shows a few different interesting examples of how to use the least squares method.

- Chapter 8 covers some essential aspects of solving ordinary differential equations analytically and numerically and using MATLAB's built-in functions and commands, as well as Simulink modeling essentials in association with MATLAB. This chapter covers MATLAB functions such as `dsolve`, `laplace`, `ilaplace`, `ode23`, `ode45`, and `ode113`. And it demonstrates how to use the Simulink functions `simset`, `sim`, etc.

All of the source code (scripts, M/MLX/MAT files, Simulink models, SLX/MDL files, C code, MEX-files, and installation `*.exe` files) discussed in the book are available to readers via GitHub (`github.com/apress/beginning-matlab-simulink-2e`).

Note The scripts in the book may not always be the best solutions to the given problems, but this is done intentionally to emphasize methods used to improve them. In some other cases, it is found to be the most appropriate solution. Should I spot better alternative solutions to exercises, I will publish them via MathWorks' MATLAB Central User Community's file exchange, via my file exchange link there (under my name).

No matter how hard we worked to proofread this book, it is inevitable that there will be some typographical errors that might slip through and appear in print. My apologies.

Sulaymon Eshkabilov

October 2022

CHAPTER 1

Introduction to MATLAB

The MATLAB package is employed in a wide range of engineering and scientific computing applications and is associated with the dynamic system simulation package Simulink. MATLAB has remarkable strengths, such as user-friendly and intuitive programming syntaxes, high-quality numerical algorithms for various numerical analyses, powerful and easy-to-use graphics, simple command syntax to perform computations, and many add-ons such as toolboxes and real and complex vectors and matrices, including sparse matrices as fundamental data types.

MATLAB is used in many diverse areas, including simulation of various systems such as vehicle performance, mapping of the human genome, financial analysis in emerging economies, and image analysis and processing applications. In addition, it is used in microbiology applications for the diagnosis and treatment of small organisms, in dynamic simulations of large ships in down-scaled laboratory models, in simulations of next-generation network audio products, in teaching computer programming for undergraduates with real-time laboratory tests and measurements, and in image processing for underwater archaeology and geology.

In this chapter, we discuss some essential features of the graphical user interface (GUI) of MATLAB, including how to use the help tools and library sources, how to adjust the format options and accuracy settings, how to create various variables and variable structures, and how to employ M/MLX-editors to write and edit scripts and programs.

© Sulaymon Eshkabilov 2022
S. Eshkabilov, *Beginning MATLAB and Simulink*, https://doi.org/10.1007/978-1-4842-8748-4_1

Menu Panel and Help

You can launch the MATLAB application in the Windows operating system by clicking the icon/shortcut on the desktop or via Start ⮭ All Programs ⮭ MATLAB. As MATLAB loads, the user's last preserved files, entries, and commands appear, along with the menu bar and tools with the latest preferences. MATLAB's GUI tools and windows are customizable. Users can easily change the preferences of the package according to their needs. Figure 1-1 shows the default main window of MATLAB 2022a. Note that the package's GUI menu and tools have been changed over the years to make the package more user friendly and the tools more intuitive. The default window shown in Figure 1-1 contains the main menu tools (1), the current directory indicator (2), the Command window (3), the Workspace window (4), the Current Folder window (5), and the Command History window (not shown here). These windows can be docked/undocked, opened in a separate window, closed, or removed from the main window, and they can be dragged from one pane to another and maximized or minimized.

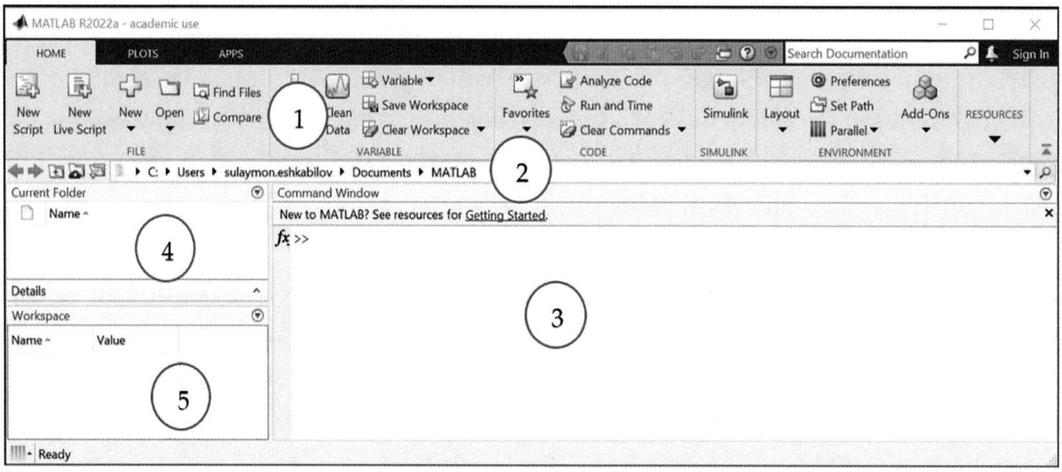

Figure 1-1. *Default MATLAB desktop window, MATLAB (R2022a)*

The main components of the package's GUI tools are as follows:

- The menus and toolbars are grouped into three tabs: HOME, PLOTS, and APPS (Figure 1-2). The HOME tab (1) shown in Figure 1-1 contains all the main tools, for creating new files and variables, importing data, analyzing code, and more.

- The Current Directory window (5) shown in Figure 1-1 is in the left pane by default. This window displays all the files in the current directory and folder directory.

- The Command window (3) shown in Figure 1-1 is in the central pane by default. All commands and (small) scripts/code can be entered directly after $fx \gg$. By clicking $fx \gg$ (see Figure 1-4) in the Command window, you can view all the built-in functions of the package and the installed MATLAB toolboxes. This option is available starting from the MATLAB R2008a version. All installed toolboxes of the package can be also viewed or accessed by clicking the APPS tab (see Figure 1-2), which is available only in later versions of MATLAB starting from the R2010a version.

- In the Workspace pane (4) in Figure 1-1 of the default desktop window, MATLAB shows all the current entries and saved variables during the session. These entries will be saved temporarily until the MATLAB application is closed. All essential attributes and properties of entries/variables (variable names, values, types) are displayed in the workspace.

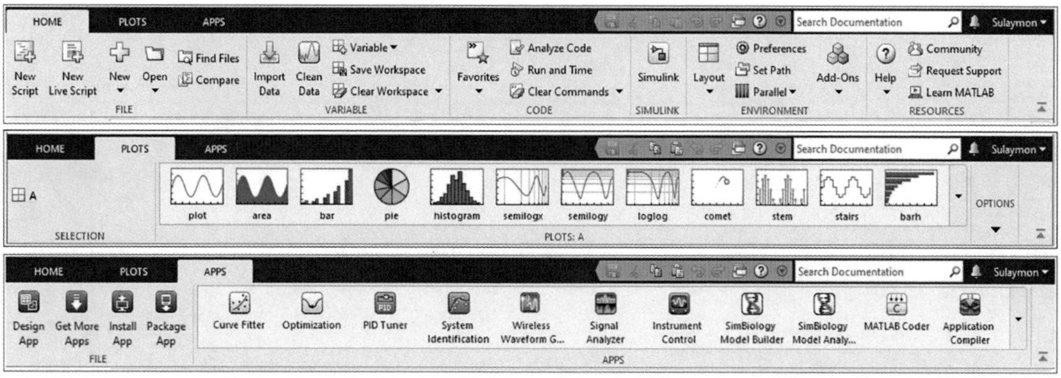

Figure 1-2. *MATLAB main menu: HOME, PLOTS, APPS (MATLAB R2022a)*

It should be noted that the GUI and tools of MATLAB are updated with new releases of the package. For example, a new GUI tool called Clean Data, Clean Data, which is used for to import timetable data, was introduced in MATLAB R2022a (Figure 1-2).

The help documentation, help libraries, and help tools are one of the strengths of the program. They can be accessed by using GUI tools and entries from the Command window. It is always useful to start using the package by exploring the options of the Getting Started documentation by clicking Getting Started above the Command window. This will lead you to the help library documentation of the package, as shown in Figure 1-3, that contains the most essential documentation of the package and its help tools, such as examples and syntaxes of the functions.

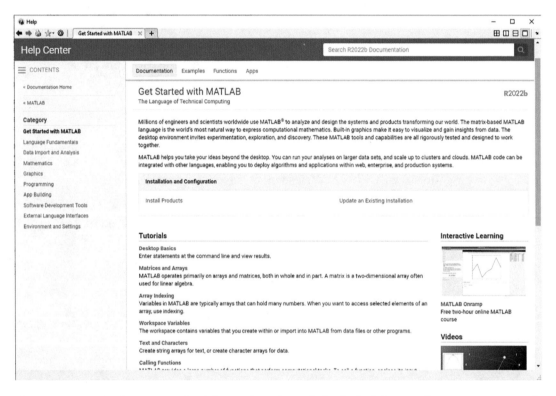

Figure 1-3. *Getting started with MATLAB and the help library documentation of MATLAB2022b*

Note   Before discussing the help options, it is worth highlighting one important point concerning comments. In MATLAB, users can write all necessary hints and help remarks as comments within the M/MLX files and in the Command window as well. Comments need to start with a % sign. There are some other options to add comments that we will discuss later while writing M/MLX files.

There are a few other hands-on ways to obtain help. For example, to get quick help on how to use MATLAB's built-in function tools and commands including the user-generated functions, in the Command window a user should type the following:

```
>> doc size;          % extended help on the command SIZE
>> helpwin size;      % help shown in a separate window on the command SIZE
>> help clear         % quick help on how to use the command CLEAR
>> help matrix;       % quick help how to use the command ISMATRIX
>> help +             % quick help on "+"
>> help size;         % quick help on SIZE
>> lookfor size       % extensive search for a list of functions and files
                        containing the command SIZE.
```

Figure 1-4 and Figure 1-5 show some of the results of the quick and extensive help. In addition, the application offers broad library resources, product descriptions, video tutorials, and open public forums on the MATHWORKS website.

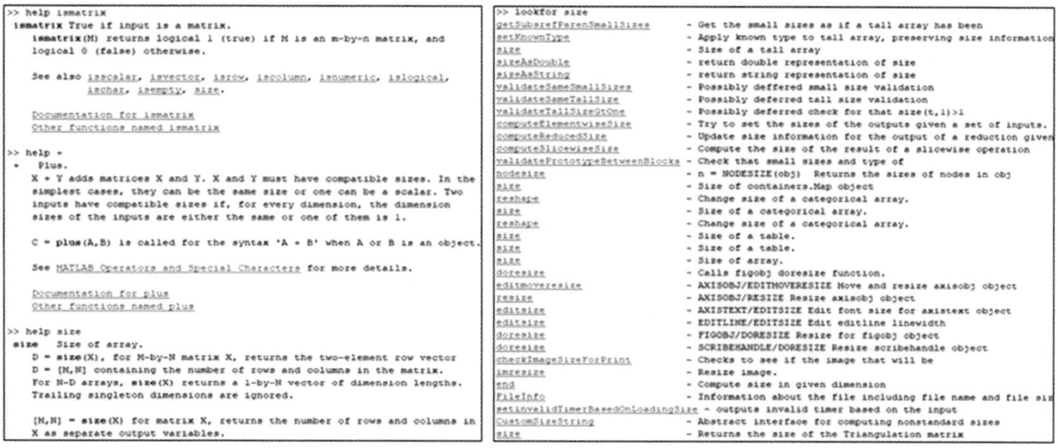

Figure 1-4. *Getting help*

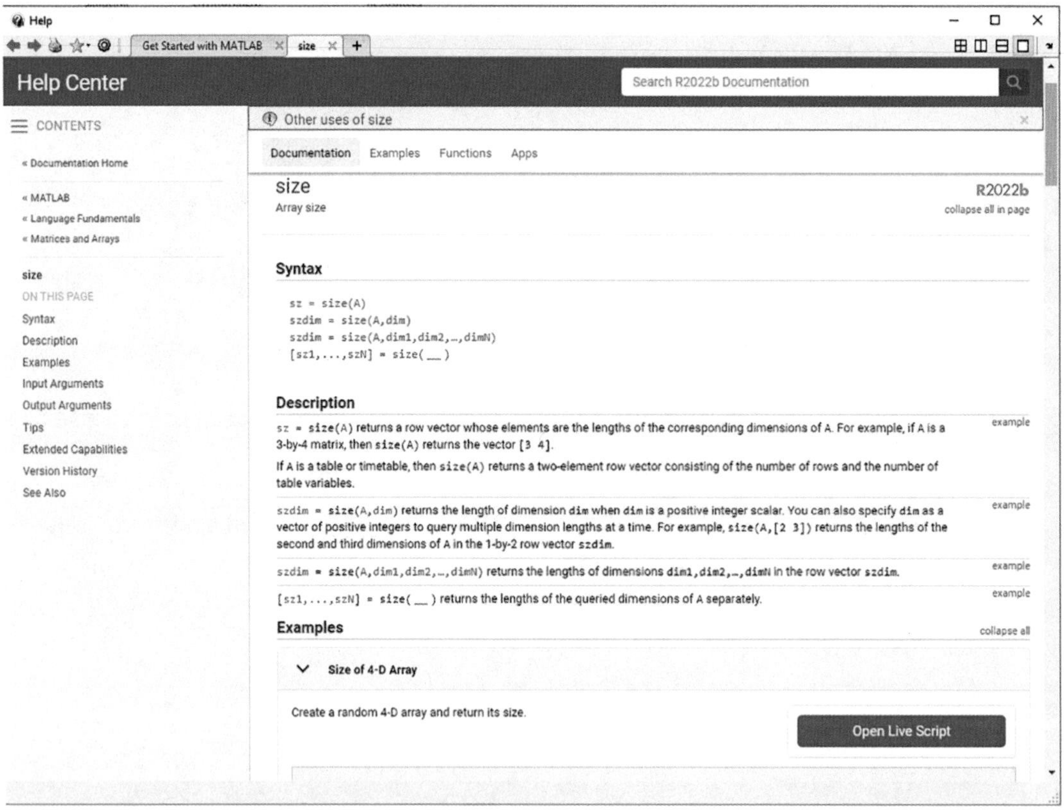

Figure 1-5. *Extensive help obtained from the >> doc size command*

You can search for help in MATLAB using the Command window and using the `help`, `lookfor`, `doc`, `docsearch`, and `helpwin` commands, for example.

- Quick help can be obtained from the Command window with the `help` command. Note that in this case, help hints not only from MATLAB's built-in commands/functions but also within a user's created/developed function files are displayed. This is a quick step to obtain help. Here's an example to get help with a clock:

  ```
  >> help clock
  ```

- Extensive help with examples will open in the help library window with the next commands only if such a function (e.g., `clock`) file exists:

```
>> doc clock
>> docsearch clock
```

- An extended list of M-files containing a searched keyword can be seen in the Command window with the next help command. Note that this option is much slower than the other two search options due to an exhaustive search for a keyword.

```
>> lookfor clock
```

- A function file explanation can be visualized in the help library by using the following command only if such a function file (e.g., clock) exists:

```
>> helpwin clock
```

- All extended help tips, examples, and command syntaxes can be viewed from the Help Library (the help browser displayed in Figure 1-3) that can be accessed by clicking Help menu options.

- The F1 functional key on the keyboard can be used to open the help browser and help documentation.

- By clicking the Help menu in the Main Menu panel, a user can get access to various help resources from MathWorks, such as Help Library resources, web resources, demo examples, updates, trials, and so forth.

There are numerous hands-on help resources available online, such as the MathWorks website, academia and the user community's published scripts and file exchanges [1], and the MATLAB answers forum [2], where users and developers post their questions and seek answers or conversely post their answers to posted questions. Moreover, there are function files, Simulink models, online forums, tutorials of numerous universities [3], and personal web pages of professors and researchers [4], just to name a few.

The MATLAB Environment

Let's start working in the MATLAB environment by making some changes to its layout and preferences. We'll use the Layout, Preferences, and Set Path tools located on the HOME tab. To make changes in the layout (Figure 1-6) from the HOME tab's main menu, a user has to click the Layout drop-down option (1), and a whole range of options of using different windows will be available to choose from. The Desktop window consists of Command, Command History, Current Directory, and Workspace windows if there are check marks before those window names. You can separate or drag any of these windows by clicking the title bar and dragging the window to the new location.

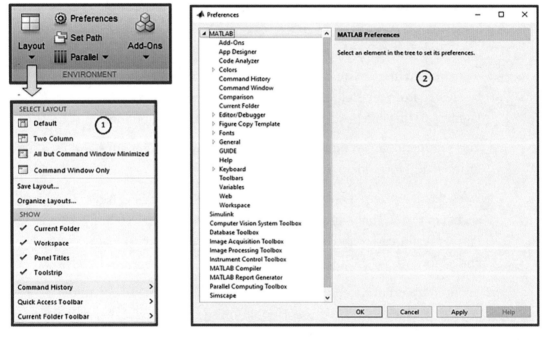

Figure 1-6. *Changes in Layout (1) and Preferences (2)*

To make changes to Preferences, you either click ⚙ Preferences on the main menu or type the following command in the Command window and press Enter:

```
>> preferences
```

Subsequently, the Preferences window (2) shown in Figure 1-6 will pop up. The directories/paths to the current directory can be altered, and new paths (3) can be added using the 🗁 Set Path GUI button in the HOME pane, as shown in Figure 1-7.

You can modify many options and tools including the GUI quick access tools and Editor/Command window displays from the Preferences window. For instance, you can adjust the fonts (size, type, color) of the Editor, Command, Workspace, Figure, and Command History windows and set up keyboard shortcuts, programming tools, and many more. To display the data tips and highlight the current line in the Editor window, you choose Preferences ➤ Editor/Debugger ➤ Display and then select Enable Data-Tips in Edit mode. In addition, many tools can be added to the main menu as shortcuts. Some of these key customizations in Preferences can also be attained by issuing commands from the Command window.

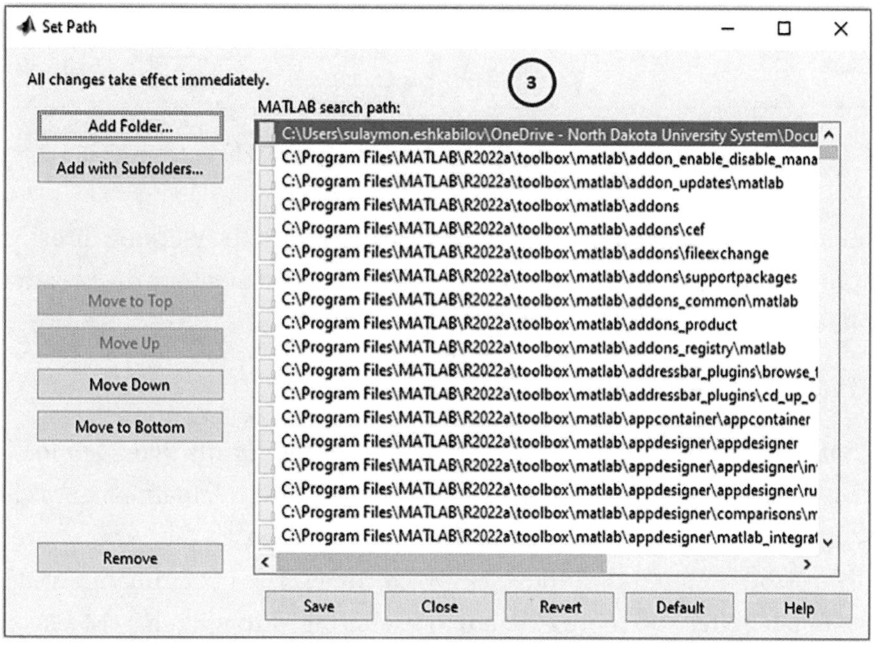

Figure 1-7. Setting a path (3), adding a folder, and removing a path

Working in the Command Window

Your work in MATLAB generally starts in the Command window, but before you type any command, it is worth noting the current directory. The current directory address can be viewed directly from the main window (see Figure 1-8) or by typing this command in the Command window:

```
>> pwd
```

If required, you can change the current directory using the >> cd command. Here's an example:

>> cd C:\Users\David

Or you can click the path's directory (Figure 1-8) (C: ➤ Users ➤ David).

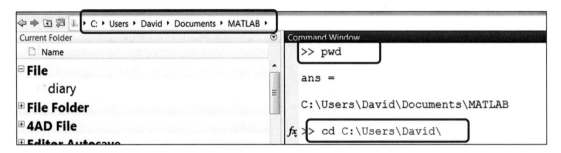

Figure 1-8. *Viewing and changing the current working directory*

In addition, you can add a few new paths to work within the working directory; you do this by using the next command, for instance, to add a path to an already existing folder:

>> addpath('C:\Users\David\Documents\MATLAB');

The command addpath() might be also helpful within scripts to read or load data from a specific folder or directory. For short commands and calculations or to view attributes of the available variables in the workspace and files in the current directory, you usually use the Command window. However, for a series of commands and longer scripts, it is much better and more efficient to use script editors, such as M-file and MLX-file editors.

The MATLAB application has a few files that can be recognized with their extensions. They are the M, MLX, MAT, BI, and FIG formats. M-files are used to write programs/scripts/function files, and files (Live M-files) are used to write programs/scripts/function files and see the computation results within the MLX file Editor window. MAT files are used to save all types of variables available in the workspace and can be accessed easily from M/MLX-files and the Command window. Among these files, BI files are used for the built-in files of MATLAB, and FIG files are used to save figure windows in MATLAB. In addition, the Simulink application has three types of files: MDL, SLX, and SLXC.

They are used to build and simulate Simulink models and can also be recalled/simulated from MATLAB without opening them. I will discuss all the essential features of these files and how to use them in later chapters.

While using the Command window for simple calculations and data generation or processing, you can press the up-arrow key to avoid retyping the previously entered commands and entries. For example, if you entered the following in the Command window:

```
>> A1 = [1, 2, 4, -5, 6]; B = A+2
```

and then needed to make changes to these entries, you can use the up-arrow (↑) key after typing >>A and MATLAB will automatically recall the previous entry.

Command Window and Variables

MATLAB is case-sensitive, and all its built-in commands are lowercase. When you perform computations or evaluations, you call, assign, or declare a name to the result of computation. The assigned name is the variable name. The result of your computations are saved in the MATLAB workspace under your given variable names. For example, >> A =13; B = A*2 means that the variable called A is equal to 13 and the variable called B is equal to 2 multiplied by A.

Using Variables

Variable names must start with a letter and can be combined with any integer numbers, such as 0, 1, 2, ... 9, and the underscore (_) sign. No other symbols can be used for variable names. The maximum length of the variable name can be 63 characters in total. For example, if there are two variables with the same variable names in the first 63 characters, MATLAB cannot differentiate them any variable. MATLAB treats the variables a and A as two different variables because of their ANSI/ASCII symbol conversions. Now let's get started working in the Command window by entering and assigning variable names, performing simple some basic arithmetic operations, and making changes in the output data formats.

```
>> A=3; B=-2; C=1/2; D = -1.5; ABCD=A^2+B/C+D;
>> ABCD              % ";" is missed and the content of the variable ABCD
                     is displayed
```

11

```
ABCD =
   3.5000
>> sqrt(ABCD)  % if a variable name is not assigned, "ans" is a default
name by MATLAB
ans =
     1.8708
>> ans+1  % if a variable name is not assigned, "ans" will substitute the
previous "ans".
ans =
     2.8708
```

As mentioned, the variable names cannot contain any symbols except for the underscore and cannot start with numbers or symbols. Here are a few examples of incorrect variable names starting with a number or a symbol and containing forbidden symbols:

```
>> % Wrong variable names
Did you mean:
>> 3A=2.8708
 3A=2.8708
   ↑
Invalid expression. Check for missing multiplication operator, missing or
unbalanced delimiters, or other syntax error. To construct matrices, use
brackets instead of parentheses.
Did you mean:
>> @A=2.8708
 @A=2.8708
    ↑
Incorrect use of '=' operator. Assign a value to a variable using '=' and
compare values for equality using '=='.
>> A$=2.8708
 A$=2.8708
   ↑
Error: Invalid text character. Check for unsupported symbol, invisible
character, or pasting of non-ASCII characters.
>> A#B=2.8708
```

```
A#B=2.8708
 ↑
```
Error: Invalid text character. Check for unsupported symbol, invisible
character, or pasting of non-ASCII characters.

The results of the user entries are displayed in the Command window and can be altered using these display format options: format long, format short, format long g, format bank, format hexadecimal, format rational, etc. Here are some examples for format options. When you change the display format types, the actual variable values do not change.

```
>> A=3; B=-2; C=1/2; D = -1.5;
>> ABCD=A^2+B/C+D;
>> format bank
>> ABCD
ABCD =
          3.50
>> format long eng
>> ABCD
ABCD =
     3.50000000000000e+000
>> sqrt(ABCD)
ans =
     1.87082869338697e+000
>> format short
>> sqrt(ABCD)
ans =
     1.8708
>> format bank
>> sqrt(ABCD)
ans =
          1.87
>> format long eng
>> sqrt(ABCD)
ans =
     1.87082869338697e+000
```

```
>> format compact
>> sqrt(ABCD)
ans =
    1.87082869338697e+000
>> format rat
>> sqrt(ABCD)
ans =
    1738/929
>> format hex
>> sqrt(ABCD)
ans =
    3ffdeeea11683f49
```

Note   MATLAB is case sensitive, and thus, it recognizes the variables called a and A as two different variables.

After entering one or two or more starting letters in the variable names or built-in commands/function names in the Command window, you can use the Tab key from the keyboard and then all available commands/functions including your developed function files. For example, if you typed >> AB and then pressed the Tab key, the rest of the ABCD variable calculation expression would appear as an option.

Another useful feature of the Command window is using the keyboard's up-arrow (↑) key to recall previously typed variables or commands. You simply type a few starting letters of any previously typed commands or function names and then press the up-arrow, as in (↑) >> f↑, that recalls the previously typed command: >> format long. Moreover, the up-arrow (↑) key can be associated with the Tab key to recall previously entered commands in the Command window.

The values and attributes of all entered variables in the Command window will be saved in the workspace until a user cleans up the workspace by deleting the variables with the command clear or clearvars or clear all or by using the right and left mouse button options by selecting the variables and deleting them. In addition, all of the variables and their attributes are saved in the workspace until the MATLAB package is closed.

There are three common commands for housekeeping in MATLAB.

- clc: For cleaning up the Command window and starting a blank Command window

- clear and clearvars: For removing all variables saved in the workspace

- clear all: For removing all variables and temporarily compiled and saved machine codes of M-files, breakpoints, and debug settings

All of these commands can be also used with M-files and MLX-files. It must be noted that the command clear all is not recommended to use within M-files and MLX-files unless it is necessary, because it will decrease the efficiency of code/scripts and unwanted behaviors of our created M-files when you declare to the command clear global. Let's look at some ways to employ these commands efficiently.

```
>> A=3; B=-2; C=1/2; D = -1.5; ABCD=A^2+B/C+D;
>> clear              % Removes all of the entries in the workspace and
                      workspace becomes all blank
>> clearvars          % The same as "clear"
>> clear variables    % The same as "clearvars"
>> clear all          % The same as "clear" and also removes already
                      compiled codes as well
>> clear  A B C D   % Removes variables: A B C D and leaves ABCD untouched
>> A=3; B=-2; C=1/2; D = -1.5; ABCD=A^2+B/C+D;
>> clearvars  -except A B C   % Removes all variables except for: A B C
```

MATLAB can use a wildcard asterisk (*) for variables and filenames. For example, to remove all variables whose names start with the letter *A*, then you would use the following command:

```
>> A=3; B=-2; C=1/2; D = -1.5; ABC=A^2+B/C; ABC=A^2+B*C+D;
>> clear  A*         % Removes all variables starting with A, i.e. A, ABC,
                     ABCD are removed
>> clearvars A*  % The same as: clear A*
```

All entered and saved variable names can be viewed from the Workspace window by typing the who or whos command. Here's an example:

```
 >> A=3; B=-2; C=1/2; D = -1.5; ABCD=A^2+B/C+D; ABC=A^2+B*C;
>> who                           % Variable names
```

```
  A   B   C   D   ABCD   ABC
>> whos                              % Variable names including their attributes
Name        Size               Bytes  Class      Attributes
  A         1x1                   8   double
  ABC       1x1                   8   double
  ABCD      1x1                   8   double
  B         1x1                   8   double
  C         1x1                   8   double
  D         1x1                   8   double
```

From these examples, it is clear that MATLAB reads every entry as an array/matrix. For example, a scalar is read by MATLAB as an array of size 1-by-1. This attribute of MATLAB is logically linked with its name MATrix LABoratory. MATLAB's default storage (memory allocation) is double precision, which is the maximum available space allocated. However, for memory efficiency and faster calculation purposes, other storage formats can be also used. MATLAB supports single precision or integer type int8...64 or uint8...64 formats. Table 1-1 shows how much data can be saved in every storage class type and what conversion function is used in MATLAB for each type.

Note   MATLAB's default storage type is a double. However, that can be changed into single precision or integer types such as int8…64 or uint8…uint64 by specifying or converting the values of variables/data.

Table 1-1. *Data Storage Format Types in MATLAB*

Class	Range of Values	Conversion Function
Signed 8-bit integer	-2^7 to 2^7-1	int8
Signed 16-bit integer	-2^{15} to $2^{15}-1$	int16
Signed 32-bit integer	-2^{31} to $2^{31}-1$	int32
Signed 64-bit integer	-2^{63} to $2^{63}-1$	int64
Unsigned 8-bit integer	0 to 2^8-1	uint8
Unsigned 16-bit integer	0 to $2^{16}-1$	uint16
Unsigned 32-bit integer	0 to $2^{32}-1$	uint32
Unsigned 64-bit integer	0 to $2^{64}-1$	uint64

When you have numerical data in a floating-point format, the double precision storage gives you the largest storage space for higher accuracy. Double precision can save up to 16 decimal digits. The double precision is the default storage format in MATLAB.

The single precision storage for floating-point data is more memory efficient and less accurate than the double precision. If only integers are used in your calculations or data processing, then it is more appropriate to use int8...int64 or uint8...uint64 depending on the largest value of your data. Here are some examples of how to specify the storage type while saving the values of the declared variables:

```
>> F01=127;Fint_08=int8(F01), Fnew1 = Fint_08+1
Fint_08 =
 int8
 127
Fnew =
 int8
 127
>> F16=65535; Fint_16=uint16(F16), Fnew2 = Fint_16+1
Fint_16 =
 uint16
 65535
Fnew2 =
 uint16
 65535
```

In calculations of the variables Fnew1 and Fnew2 from the int8- and uint16-formatted variable values, the allocated storage space (i.e., for the int8 maximum allocated storage space is $127 = 2^7 - 1$ and for uint16 it is $65535 = 2^{16} - 1$) cannot accommodate any more values. Therefore, these errors in calculations have taken place. Note that in such cases MATLAB does not show an error. Figure 1-9 shows all data storage types and data formats supported in MATLAB.

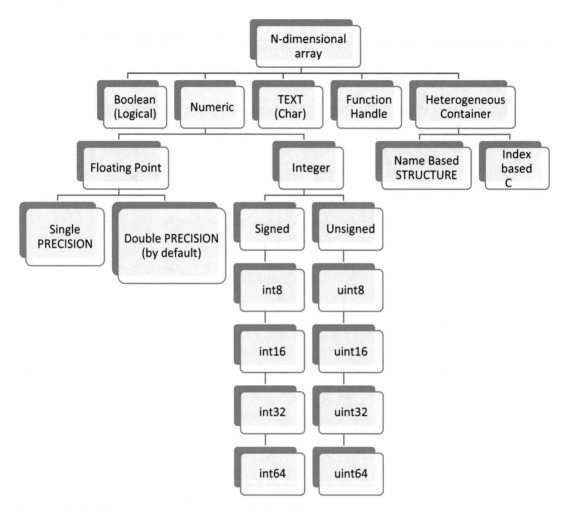

Figure 1-9. *Data storage options in MATLAB*

The Command History window is a good way to review all of the entries that will be kept unless you delete them. You can also change settings in the Preferences window to clear the history of entries after ending the session.

Finally, you can exit from MATLAB or quit the work session via one of these commands in the Command window:

```
>> exit
>> quit
```

Or press Ctrl+Q to quit.

An alternative is to click the X in the upper-right corner of the main window. This will close the whole package.

When to Use the Command Window

Use the Command window in these instances:

- To perform short calculations

- To view error and warning messages from the typed-in commands or after executing M-file and MLX-file and SLX/MDL Simulink models

- To view attributes of variables saved in the workspace and files in the current directory

- To view contents of the MATLAB-compatible files

- To execute MATLAB files, such as M-files, MLX-files, SLX/MDL-files, and MAT-files

- To get hands-on and quick help on the syntaxes of MATLAB commands/functions and user-created function files

- To adjust display formats of numerical data

- To add/remove a path/directory

- To create/delete or save variables and files

Let's look at several examples to show other operations you can perform in the Command window.

- To view and analyze some common errors and interpret the error messages, use this:

```
>> F16=65535; Fint_16=uint16(F16); Fnew2 = Fint_16+1;
>> Fnew+2 % The variable Fnew does not exist in the workspace
Undefined function or variable 'Fnew'.
>> clar F16 % Typo error: "clar" instead of "clear"
Undefined function or variable 'clar'.
>> CLear % Typo error: "CLear" instead of "clear". Note: MATLAB case-
sensitive
```

Undefined function or variable 'CLear'.
Did you mean:
>> clear % MATLAB automatically suggests closest command's correct syntax
>> B=-2; C=1/2; BC=B/.C; % Illegal operation: B/.C instead of B/C;
 B=-2; C=1/2; BC=B/.C;
                    ↑
Error: Unexpected MATLAB operator.
>> B=-2; C=1/2; BC=B /*C; % Illegal operation: B/*C instead of B/C;
 B=-2; C=1/2; BC=B /*C;
                      ↑
Error: Invalid use of operator.
>> % Let's create a two-row matrix containing two elements, viz. B, C in
the   >> % first row and F16 in the second row.
>> BCF = [B, C; F16] % Number of elements in row 1 does not match with the
ones in row 2
Error using vertcat
Dimensions of arrays being concatenated are not consistent.
>> % Let's try to create a row matrix with elements separated with ","
and   >> % space and "."
>> BCF = [B, C. F16] % Error is a misused "." instead of "," but not dot
indexing as shown
Dot indexing is not supported for variables of this type.
>> BCF = [B, C, F16] % This is the anticipated correct command.

- To save the variables saved in the Workspace window in a *.mat file,
  use this:

>> save MYdata.mat % Saves all variables residing in the workspace in
MYdata.mat file
>> save('MYdata.mat') % The same as above
>> save MYdata.mat F16 Fnew2 % Saves the variables F16, Fnew2 in MYdata.
mat file
>> save('MYdata.mat', 'F16 ', 'Fnew2 ') % The same as above
>> save MYdata.mat F* % Saves all variables whose name starts with F (in
the workspace)

- To obtain quick help, use this:

```
>> help format
 format Set output format.
 format with no inputs sets the output format to the default appropriate
 for the class of the variable. For float variables, the default is
 format SHORT. ...
>> help dir
 dir List directory.
 dir directory_name lists the files in a directory. Pathnames and
 asterisk wildcards may be used. A single asterisk in the path touching
>> help what
 what List MATLAB-specific files in directory.
 The command what, by itself, lists the MATLAB specific files found ...
>> help which
--- help for which ---
 which Locate functions and files.
 which ITEM displays the full path for ITEM. ITEM can include a partial
 path, complete path, relative path, or no path. If ITEM includes a
 partial path or no path, ...
```

- To view MATLAB-compatible files, use this:

```
>> type QQQ.txt % Note: the file QQQ.txt was available in the current
directory

CY  bBb   88
AH  AAAA+ 98
CWW AAAA+ 98

...
>> type MYfile.mlx % Note: the file MYfile.mlx was available in the current
directory

N=13;
M=randi(N,9);
stairs(M, 'bd-')
```

```
>> type myfun.m % Note: the file myfun.m was available in the current
directory
```

```
function f=myfun(x)
f=[2*x(1)-x(2)-exp(-x(1));
 -x(1)+2*x(2)-exp(-x(2))];
end
```

- To create, open, and execute the MATLAB files, such as M-files, MLX-files, MDL/MLX-files, and MAT-files, use this:

```
>> edit TRY1.m % To create a new M-file called TRY1.mlx
>> edit MYfile.mlx % To create a new MLX-file called MYfile.mlx
>> open('TRY1.m') % To open the file if it is residing in the current
directory
>> run('\...\TRY1.m') % Directory and a file name is needed, if it is
outside of the current dir.
>> TRY1 % To execute the file if it is residing in the current directory
>> open('MYfile.mlx') % To open the file if it is residing in the current
directory
>> MYfile % To execute the file if it is residing in the current directory
>> load MYdata.mat % Load contents of MYdata.mat (existing in the current
directory)
>> load('MYdata.mat') % The same as above
```

- To delete any files in the current directory or variables residing in the workspace, use the following:

Warning   Be careful when using the delete command because it deletes files that cannot be recovered.

```
>> delete TRY1.m % Deletes the file TRY1.m residing in the current
directory
>> delete MYfile.mlx % Deletes the file MYfile.mlx residing in the current
directory
```

```
>> delete QQQ.txt % Deletes the file QQQ.txt residing in the current
directory
>> delete *.txt % Deletes all *.txt files in the current directory
>> delete *.mlx % Deletes all *.mlx files in the current directory
>> delete DA*.txt % Deletes all *.txt files whose name starts with DA...
>> delete *.asv % Deletes all *.asv files (autosave) of MATLAB in the
current directory
```

- To view the current directory, change a directory, create a new directory, and remove a directory from the MATLAB path, use this:

```
>> MD = pwd % Shows the current directory and assigns to a character type
of variable: MD
>> cd C:\Users\sulaymon.eshkabilov\Documents\MATLAB % Change to this
directory
>> cd('C:\Users\sulaymon.eshkabilov\Documents\MATLAB') % The same as above
>> mkdir MYBook % Creates a new folder (directory) inside the current
directory
>> mkdir('MYBook') % The same as above
>> mkdir c:\Users\sulaymon.eshkabilov\BOOK % The same as above with a
full path
>> addpath C:\Documents % Adds this path (C:\Documents) to the
MATLAB's search
>> addpath('C:\Documents') % The same as above
>> rmdir('MYtask') % Removes the directory (folder: MYtask) including its
contents from the hard disk
>> rmdir c:\Users\sulaymon.eshkabilov\TASK % Removes the directory: TASK
```

Note   MATLAB supports wildcards (via the asterisk, *) when deleting and saving files and variables in the current directory and workspace. For example, >> delete M*.mat deletes all *.mat files whose name starts with M. >> save MYdata.mat B* saves all variables whose name starts with B. >> clearvars A* clears all variables whose name starts with A.

Many of the operations performed in the Command window, such as performing calculations and analyses and viewing variables or file contents, can be also done by other ways. For example, most of the previously listed operations carried out in the Command window can be also done via GUI tools, such as creating new variables [New Variable] or deleting them [Clear Workspace ▾] or opening them [Open Variable ▾].

Similarly, creating any MATLAB files with [New ▾] or M-files with [New Script] or opening existing MATLAB files with [Open ▾] GUI tools can be attained; and deleting the files can be done via right and left mouse button options, which is standard for Windows file manipulation operations. Viewing the current directory or changing it can be done also with. [⇐ ⇒ ⊡ ⊠ ▯ ▸ C: ▸ Users ▸]

One of the most essential functions of the Command window that cannot be done easily with GUI tools or other options is to view error and warning messages obtained while and after executing M-files, MLX-files, and MDL/SLX-files. This is essential for good programming. Another good use of the Command window is to obtain quick help on the syntax of MATLAB commands/functions.

Different Variables and Data Sets in MATLAB

MATLAB supports a few different data types, which can be numeric, character, logical, table, cell, structure, and function handle. The flowchart in Figure 1-10 shows the hierarchy of all data types that are supported in MATLAB and can be used for data storage. In the flow, there is one point worth mentioning, which is that function handles can also take vectors (row or column vectors) as well as scalar numbers.

As stated, MATLAB reads every entry (numerical and character types) as an array, and in the case of a storage type that is not specified, the default storage type is always double. Let's look at several examples of how to generate various data types supported and recognized in MATLAB. We'll work in the following order:

- Numerical data

- Logical arrays

- Character arrays/variables

- Table arrays

- Cell arrays

- Structure arrays

- Function handles

- Classes and graphic handles

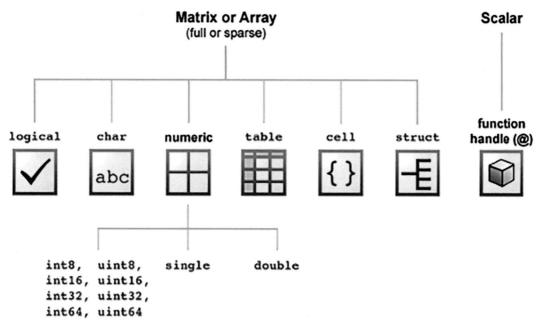

Figure 1-10. *Types of data (array) sets supported in MATLAB*

While demonstrating how to generate these arrays, all of the created variable/arrays types will be preserved up until the end of this section. Therefore, all variables/arrays are created once and preserved from all examples. Note that in some of the examples to generate random matrices, we employ random number generators of MATLAB, which will create different random numbers every time they are called. However, to have consistent random values for reproducibility purposes for variables and arrays, we set up the seed value of the random number generator: rng(). With the fixed seed value, the random number generators (rand(), randi(), randn(), and so forth) will generate permanent/fixed random values every time they are called.

Numerical Data/Arrays

There are many different ways to create and generate numerical arrays. For example, you can use direct entries from the Command window by typing all numerical values of an array and generating numerical entries using built-in functions and commands. Also, data can be imported from another file (.mat, .txt, .dat, .xls, .xlsx, .csv, .jpeg, .tiff, .png, .pdf, .au, etc.) or generated using existing data or imported data in MATLAB. Data can also be computed using specific mathematical expressions and arithmetic/matrix operations.

```
>> format short        % To display numerical elements of arrays in a
                         short format
>> A = 2; B = -3.25; C=(A+B)^2; % Entries and arithmetic operations.
>> ABC=[A, B, C] % Use of Existing Data: 1-by-3 array
ABC =
    2.0000    -3.2500     1.5625
>> Drow=[1, 2, 3, -4] % Just entry: row array. Comma "," element
separator in it.
Drow =
     1     2     3    -4
>> Erow=[-2 -1 0 Drow] % Entry and Use of existing data: Row Array. Space
is a separator.
Erow =
    -2    -1     0     1     2     3    -4
>> Fcol=[1; 2; 3; -4] % Entry: Column Array. Elements are separated
with ";".
Fcol =
     1
     2
     3
    -4
>> DE = [2*Drow; (3/4)*Drow] % Use of Existing Data and arithmetic
operations: 2-by-4 array
DE =
    2.0000    4.0000    6.0000    -8.0000
    0.7500    1.5000    2.2500    -3.0000
```

An alternative way of generating numerical arrays/data is to employ the built-in array generator functions of MATLAB.

```
>> Aone=ones(2, 6) % 2-by-6 array generated with elements of 1
Aone =
    1     1     1     1     1     1
    1     1     1     1     1     1
>> Bzero=zeros(5, 6) % 5-by-6 array generated with elements of 0
Bzero =
 0 0 0 0 0 0
 0 0 0 0 0 0
 0 0 0 0 0 0
 0 0 0 0 0 0
 0 0 0 0 0 0
>> Ceye=eye(6) % 6-by-6 eye matrix array generated
Ceye =
 1.00 0 0 0 0 0
 0 1.00 0 0 0 0
 0 0 1.00 0 0 0
 0 0 0 1.00 0 0
 0 0 0 0 1.00 0
 0 0 0 0 0 1.00
>> Seed = 1; rng(Seed);    % Seed value of the random number generator
rng() is set up in order to            % generate fixed random numbers
>> Dr1 =rand(9, 5)    % 9-by-5 array of uniform distributed random numbers
generated
Dr1 =
    0.4170    0.5388    0.1404    0.0391    0.6865
    0.7203    0.4192    0.1981    0.1698    0.8346
    0.0001    0.6852    0.8007    0.8781    0.0183
    0.3023    0.2045    0.9683    0.0983    0.7501
    0.1468    0.8781    0.3134    0.4211    0.9889
    0.0923    0.0274    0.6923    0.9579    0.7482
    0.1863    0.6705    0.8764    0.5332    0.2804
    0.3456    0.4173    0.8946    0.6919    0.7893
    0.3968    0.5587    0.0850    0.3155    0.1032
```

```
>> Dr2 =randn(4, 6) % 4-by-6 array of normally distributed random numbers
generated
Dr2 =
    -0.1551    -1.1714    -0.5581    -2.0187    -0.4852     0.0407
     0.6121    -0.6856    -0.0285     0.1997     0.5943     0.2830
    -1.0443     0.9262    -1.4763     0.4259    -0.2765     0.0636
    -0.3456    -1.4817     0.2589    -1.2700    -1.8576     0.4334
>> Dr3 =randi([-5, 5], 5,5) % 5-by-5 array of integer numbers ranging
between -5 ... 5
>> Dr3 =randi([-5, 5], 5,5)
Dr3 =
     1    -1     2    -3     1
     4    -4     4     4    -4
    -4     5     1    -1     5
    -4    -2     3     5    -1
     3     3    -2     2     1
```

Note that I have set up the seed value of the random number generator to generate the consistent values from the random number generators, such as rand(), randn(), and randi().

Moreover, there are many other standard matrix/array generators built into MATLAB, e.g., pascal(), krylov(), leslie(), cauchy(), clement(), lesp(), poisson(), neumann(), etc. To get some help on syntaxes of these array generators, type in the Command window:

```
>> help gallery
>> doc gallery
```

It must be noted that all these generated numerical entries/arrays are saved in double precision format. Changing the storage format type of any these created variables' data is quite simple and straightforward.

```
>> Dr3new=int8(Dr3) % Dr3new is Saved in int8
Dr3new =
  5×5 int8 matrix
     1    -1     2    -3     1
     4    -4     4     4    -4
```

```
 -4    5    1   -1    5
 -4   -2    3    5   -1
  3    3   -2    2    1
>> Dr2new=single(Dr3) % Dr2new is Saved in a single precision
Dr2new =
  5×5 single matrix
     1   -1    2   -3    1
     4   -4    4    4   -4
    -4    5    1   -1    5
    -4   -2    3    5   -1
     3    3   -2    2    1
>> A1New=uint8(Aone) % A1new is Saved in uint8
A1New =
  2×6 uint8 matrix
     1    1    1    1    1    1
     1    1    1    1    1    1
```

Now it is time to check the attributes/properties of the numerical entries created in the Command window and saved in the workspace. You can view them by typing the command whos command or viewing directly from the Workspace window, as shown in Figure 1-11.

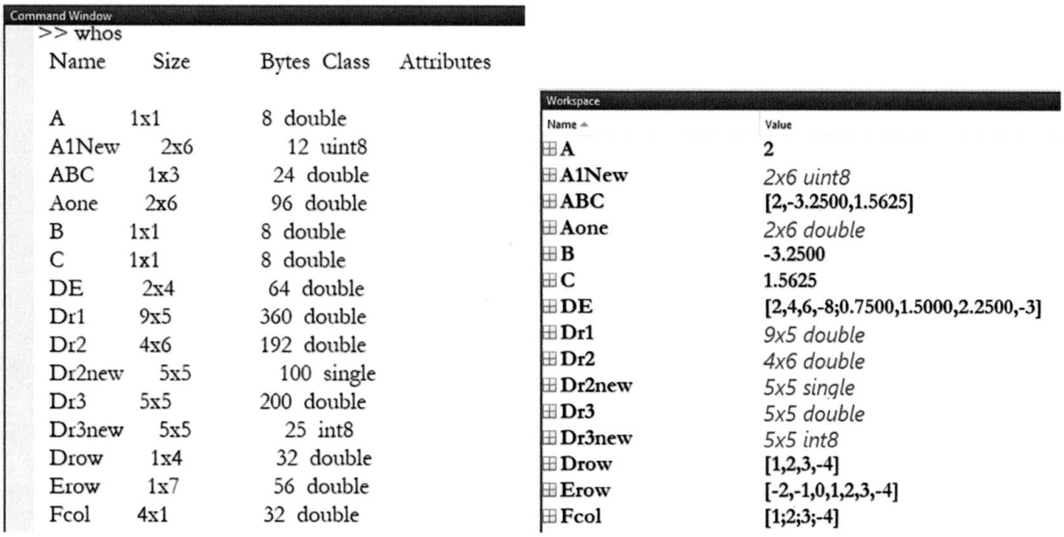

Figure 1-11. *Created numerical data types and variables residing in the workspace*

The most used data type in MATLAB is the numerical array. Therefore, it is essential to learn how to work with arrays of different sizes (many rows and many columns). To work with arrays, you should understand how to properly locate addresses of array elements, rows, and columns. Arrays are read as rows by columns. For example, >> DE(2, 1) means we are taking the element of DE residing on a second row and first column, i.e., 0.7500. As another example, >> DE(1, 4) means we have selected the element of DE residing on a first row and fourth columns, i.e., -8.

```
>> DE
DE =
      2.0000      4.0000      6.0000     -8.0000
      0.7500      1.5000      2.2500     -3.0000
>> DE(2,1)
ans =
      0.7500
>> DE(1,4)
ans =
     -8
```

Thus, an element in any array can be located with respect to the row and column. The colon operator (:) is very helpful to select all elements along rows or columns or both. For example, >> DE(1, :) selects all elements residing on the first row of DE, i.e., [2, 4, 6 -8], and DE(:, 3) selects all elements residing on the third column of DE, i.e., [6; 2.25].

```
>> DE(1, :)
ans =
     2     4     6    -8
>> DE(:, 3)
ans =
      6.0000
      2.2500
```

You can use the colon (:) operator to select all elements of matrices or arrays. For example, >> E(:,:) is equivalent to DE. The end keyword enables you to select elements up to the last one. For example, >> DE(1, 1:end) is equivalent to >> DE(1, :).

```
>> DE(1, 1:end)
ans =
     2     4     6    -8
>> DE(1, :)
ans =
     2     4     6    -8
```

Another example of how to select elements of matrices is Dr3new; the element residing in a second row and first column is 0. An alternative way of locating any element in any given array is the order count. Elements in arrays are counted on a column basis. For example, the order of the element number 2 in Dr3new will be 0, or element number 6 in Dr3new is 2. Note that I have used the random number generator to create Dr3new, and thus, your created Dr3new will differ. Again, let's look at several examples to manipulate arrays based on the previously created arrays, namely, Dr3new and Dr1.

```
>> Dr3new
Dr3new =
  5×5 int8 matrix
     3    -5     0     1    -3
     3    -5     1     5     3
     5    -5     4     1    -1
     2    -3    -4    -5     4
    -4     4    -2     3     3
>> Dr3new(2,1)
ans =
  int8
   3
>>Dr3new(2)
  int8
   3
>> Dr3new(6)
  int8
  -5
>>Dr3new(2, :) = 0 % This makes a second row of elements equal to 0
  5×5 int8 matrix
     3    -5     0     1    -3
     0     0     0     0     0
```

```
    5   -5    4    1   -1
    2   -3   -4   -5    4
   -4    4   -2    3    3
```

```
>> Dr3new(:,5)=1 % This makes fifth column of elements equal to 1
Dr3new =
  5×5 int8 matrix
    3   -5    0    1    1
    0    0    0    0    1
    5   -5    4    1    1
    2   -3   -4   -5    1
   -4    4   -2    3    1
```

```
>> Dr3new(end,:)=-5 % This makes the last row of elements equal to -5
Dr3new =
  5×5 int8 matrix
    3   -5    0    1    1
    0    0    0    0    1
    5   -5    4    1    1
    2   -3   -4   -5    1
   -5   -5   -5   -5   -5
```

```
>> Dr2new(1:5, 4:5)=2+2 % This makes the last two columns of elements
equal to 4
Dr2new =
  5×5 single matrix
     3    -5     0     4     4
     3    -5     1     4     4
     5    -5     4     4     4
     2    -3    -4     4     4
    -4     4    -2     4     4
```

```
>> Dr1(9, :)=[] % The last row elements are removed
>> Dr1(9,:) =[]
Dr1 =
    0.4479    0.4916    0.4142    0.1393    0.6237
    0.9086    0.0534    0.0500    0.8074    0.7509
```

```
    0.2936    0.5741    0.5359    0.3977    0.3489
    0.2878    0.1467    0.6638    0.1654    0.2699
    0.1300    0.5893    0.5149    0.9275    0.8959
    0.0194    0.6998    0.9446    0.3478    0.4281
    0.6788    0.1023    0.5866    0.7508    0.9648
    0.2116    0.4141    0.9034    0.7260    0.6634

>> Dr1(:, 4)=[ ] % The fourth column elements are removed
Dr1 =
    0.4479    0.4916    0.4142    0.6237
    0.9086    0.0534    0.0500    0.7509
    0.2936    0.5741    0.5359    0.3489
    0.2878    0.1467    0.6638    0.2699
    0.1300    0.5893    0.5149    0.8959
    0.0194    0.6998    0.9446    0.4281
    0.6788    0.1023    0.5866    0.9648
    0.2116    0.4141    0.9034    0.6634
```

In addition, you can create a new array from the elements of the existing arrays. Here's an example:

```
>> NewDr = [Dr1(1:3), Dr2new(1:3, 2:4)] % Some elements of Dr1 and Dr2new
are taken
NewDr =
  3×7 single matrix
  Columns 1 through 6
    0.4170    0.5388    0.1404    0.6865   -1.0000    2.0000
    0.7203    0.4192    0.1981    0.8346   -4.0000    4.0000
    0.0001    0.6852    0.8007    0.0183    5.0000    1.0000
  Column 7
    4.0000
    4.0000
    4.0000

>> NewDr2 = [Dr3new(:,:); Dr2new(2:end, :)] % All elements of Dr3new and
some from Dr2new
```

```
NewDr2 =
  9×5 int8 matrix
     1   -1    2   -3    1
     0    0    0    0    1
    -4    5    1   -1    1
    -4   -2    3    5    1
    -5   -5   -5   -5   -5
     4   -4    4    4    4
    -4    5    1    4    4
    -4   -2    3    4    4
     3    3   -2    4    4
```

Note in these examples, the colon (:) is one of the essential operators in managing and manipulating matrix and array elements. For example, NewDr(2, :) is equivalent to NewDr(2, 1:end). Both select all the elements along row 2. Likewise, NewDr(:, :) is equal to NewDr(1:end, 1:end). They both select all elements starting from the first one up to the last one.

These are a few examples of how to create arrays in the Command window. As stated, numerical arrays can be imported from other formatted data files, such as *.dat, *.txt, *.xls, *.xlsx, and *.csv, as well as image, audio and video files, such as *.jpg, *.tif, *.eps, *.png, *.bmp, *.wav, *.au, *.aif, *.mp3, *.mp4, *.ogg, etc.

Not a Number

While working and processing different data sets and analyzing experimental data, it is quite common to work with the not-a-number (NaN) values. NaN is the result of 0/0. There are also many other cases when NaN can be generated by MATLAB. The NaN is also present when some data is missing in the imported data set. So, how do you handle and work with NaN values in numerical arrays? There are a few ready-to-use functions/tools of MATLAB to handle the NaN. Let's take the following examples:

```
% Given:
A_var = [-8 10 NaN 9 4 -4 -7; 9 NaN 9 4 -10 9 0; -8 10 NaN 5 -10 -1 NaN]
A_var =
```

```
-8 10 NaN 9 4 -4 -7
9 NaN 9 4 -10 9 0
-8 10 NaN 5 -10 -1 NaN
```

How do you compute the summation of the given numerical array, A_var? Note that sum() computes the sum of columns of a matrix if the matrix has more than one row and column. If the matrix is the row matrix, then it computes the sum of all row elements.

```
>> sum(A_var)
ans =
 -7 NaN NaN 18 -16 4 NaN
```

Sometimes, you may need to remove NaN from our data. For instance, let's say you are analyzing measured data with some missing points (NaN). You would need to remove the NaN from our data. How do you address this problem?

One solution is to remove all NaN components of the array A_var and substitute them with 0. Otherwise, the summation will not give numerical results. You can substitute all NaN components using element-wise substitution, one by one. Or you can do it with a single command by recalling the indexes of all NaN components.

```
>> A_var(5)=0 % Element by element change
A_var =
 -8 10 NaN 9 4 -4 -7
 9 0 9 4 -10 9 0
 -8 10 NaN 5 -10 -1 NaN
```

```
>> A_var([7 9 21])=[0, 0, 0] % All at once or % A_var([7 9 21])=0
A_var =
 -8 10 0 9 4 -4 -7
 9 0 9 4 -10 9 0
 -8 10 0 5 -10 -1 0
```

Now the summation can be performed:

```
>> sum(A_var)
ans =
 -7 20 9 18 -16 4 -7
```

35

This answer is correct. This approach is quite straightforward, but for very large data sets it will become very tedious and too time-consuming or might be impossible.

Note   You can assign new values to some selected elements/components of arrays element by element or all at once by specifying the indexes (e.g., A_var ([7 9 21])=[0 0 0]) of the elements/components or using MATLAB's built-in function isnan() (e.g., A_var(isnan(A_var))=0).

Here is a second solution. nansum() is the MATLAB built-in function that handles the summation of numerical arrays with NaN components (elements).

```
>> A_var = [-8 10 NaN 9 4 -4 -7; 9 NaN 9 4 -10 9 0; -8 10 NaN 5 -10 -1 NaN]
A_var =
 -8 10 NaN 9 4 -4 -7
 9 NaN 9 4 -10 9 0
 -8 10 NaN 5 -10 -1 NaN
>> nansum(A_var)
ans =
 -7 20 9 18 -16 4 -7
```

An alternative solution to nansum() is using sum() with an option of omitnan.

```
>> A_var = [-8 10 NaN 9 4 -4 -7; 9 NaN 9 4 -10 9 0; -8 10 NaN 5 -10
-1 NaN];
>> sum(A_var, 'omitnan')
-7 20 9 18 -16 4 -7
```

Note that here using the omitnan option in sum() and nansum(), all NaN values are substituted with 0.

Moreover, there are several other MATLAB functions to compute mean values, standard deviations, covariance values, etc., of numerical data arrays with NaN elements.

Note   You can use the following MATLAB functions to compute maximum, mean, median, minimum, standard deviations, and variance values of a numerical array containing NaN elements: `nanmax()`, `nanmean()`, `nanmedian()`, `nanmin()`, `nanstd()`, and `nanvar()`. They work by ignoring all of the NaN elements in the given array.

There is also a third way, by using logical indexing with MATLAB's built-in function `isnan()`. It is widely employed when working with large data sets to sum the data, compute mean or average values, plot the values, and perform some other arithmetical matrix/array operations.

```
>> Index1=isnan(A_var) % Finds which elements of A_var are NaN and sets
them equal to 1
Index1 =
 3×7 logical array
 0 0 1 0 0 0 0
 0 1 0 0 0 0 0
 0 0 1 0 0 0 1
>> A_var(Index1)=0 % Assigning all NaN elements to "0"
A_var =
 -8 10 0 9 4 -4 -7
 9 0 9 4 -10 9 0
 -8 10 0 5 -10 -1 0
>> sum(A_var)
ans =
 -7 20 9 18 -16 4 -7
```

Here is another shorter way of using `isnan()` to find all NaN components of arrays and assign them equal to 0.

```
>> A_var = [-8 10 NaN 9 4 -4 -7; 9 NaN 9 4 -10 9 0; -8 10 NaN 5 -10
-1 NaN];
>> A_var(isnan(A_var)) = 0
```

```
A_var =
 -8 10 0 9 4 -4 -7
 9 0 9 4 -10 9 0
 -8 10 0 5 -10 -1 0
```

Here the function `isnan()` identifies which elements of A_var are NaN and which ones are not. The new logical array called Index contains 1s and 0s. The 1s represent NaN elements, and the 0s represent all other numerical elements.

There is an alternative logical indexing function introduced in recent versions of MATLAB: `ismissing()`. It is used to identify any missing data elements in numerical arrays.

```
Index2=ismissing(A_var) %Identifies all missing elements hidden behind NaN
Index2 =
 3×7 logical array
 0 0 1 0 0 0 0
 0 1 0 0 0 0 0
 0 0 1 0 0 0 1
>> A_var(Index2)     % Viewing the missing element values
ans =
 NaN
 NaN
 NaN
 NaN
>> A_var(Index2)=0 % Assigning the missing elements equal to 0
A_var =
 -8 10 0 9 4 -4 -7
 9 0 9 4 -10 9 0
 -8 10 0 5 -10 -1 0
```

Note   There are some differences in detecting NaN values using `isnan()` and `ismissing()`. If the variable containing NaN (some missing data) is a timetable-type array, `ismissing()` cannot detect NaN (also NaT) or ignores missing time data in the row vector of times.

The logical indexing approach is very efficient and flexible and can be applied for many other cases as well. For instance, you can easily identify all negative elements of a given data set or all elements within certain value ranges. Let's take the following numerical array (of size 7 – by -7) generated by randi() to separate out all elements that are greater than 3 but smaller than 9 and equate all them to 5.

```
>> B_var = randi([0, 25], 7) % Create uniform distributed integers within
[0, 25] of 7-by-7 size
B_var =
 25 1 17 21 1 11 12
 18 13 13 2 10 0 8
 13 2 25 3 13 25 24
 12 21 16 4 10 4 23
 1 21 20 10 17 2 1
 17 18 11 21 16 9 19
 1 3 11 20 7 5 6
>> IndexB = find(B_var>3 & B_var<9)
IndexB =
 25
 35
 39
 42
 44
 49
>> B_var(IndexB)=5
B_var =
 25 1 17 21 1 11 12
 18 13 13 2 10 0 5
 13 2 25 3 13 25 24
 12 21 16 5 10 5 23
 1 21 20 10 17 2 1
 17 18 11 21 16 9 19
 1 3 11 20 5 5 5
```

Note   Logical indexing is a very powerful and efficient tool in identifying the certain elements of numerical data sets/arrays/matrices according to their values and then assigning them new values.

Character Type of Variables

MATLAB can recognize characters based on ASCII/ANSI character symbols in the form of numerical arrays. Here's an example:

```
>> Ach1 = 'matlab'          % Character type of variable
Ach1 =
 ' matlab '
>> Bch2 = ' mathworks ' % Character type of variable
Bch2 =
 ' mathworks '
>> Cch3= ' matlab belongs to mathworks ' % Character type of variable
Cch3 =
 'matlab belongs to mathworks'
>> Dch4 = 'www.mathworks.com' % Character type of variable
Dch4 =
 ' www.mathworks.com '
```

These are character types of variables, but when the arithmetical operations are performed on these variables, they will become numerical arrays.

```
>> format short
>> Aa1 =Ach1+0
Aa1 =
 109    97   116   108    97    98
>>Ba2 = Bch2+0
Ba2 =

 109    97   116   104   119   111   114   107   115    32
>> Ca3=Cch3+0
Ca3 =
```

```
Columns 1 through 12
   109     97    116    108     97     98     32     98    101    108    111    110
 Columns 13 through 24
   103    115     32    116    111     32    109     97    116    104    119    111
 Columns 25 through 28
   114    107    115     32
Da4 = Dch4+0
 Columns 1 through 12
   119    119    119     46    109     97    116    104    119    111    114    107
 Columns 13 through 17
   115     46     99    111    109
```

So, these numbers represent the characters according to ASCII/ANSI standards. It is possible to get the character representation of the new variables with the MATLAB function of char().

```
>> char(Aa1)
ans =
 'matlab'
>> char(Ba2)
ans =
 'mathworks'
>> char(Ca3)
ans =
 'matlab belongs to mathworks'
>> char(Da4)
ans =
 'www.mathworks.com'
```

The following variables reside in the workspace:

```
>> whos
A 1x1 8 double
 A1New 2x6 12 uint8
 ABC 1x3 24 double
 Aa1 1x6 48 double
 Ach1 1x6 12 char
```

```
Aone 2x6 96 double
B 1x1 8 double
Ba2 1x9 72 double
Bch2 1x9 18 char
Bzero 5x6 240 double
Ca3 1x27 216 double
Cch3 1x27 54 char
Ceye 6x6 288 double
DE 2x4 64 double
Da4 1x17 136 double
Dch4 1x17 34 char
Dr1 8x4 256 double
Dr2 4x6 192 double
Dr2new 5x5 100 single
Dr3 5x5 200 double
Dr3new 5x5 25 int8
Drow 1x4 32 double
Erow 1x7 56 double
Fcol 4x1 32 double
NewDr 3x7 84 single
NewDr1 3x7 84 single
NewDr2 9x5 45 int8
ans 1x17 34 char
```

Note that these variables are kept and used in the coming sections to generate logical, table, cell, structure type of array variables.

Function Handle

The function handle is a special MATLAB data type and used to store a link to an expression or a function. By calling a function handle, we invoke the expression or function stored under that specific function handle. The function handle is one of the most useful features of MATLAB for various computations and programming aspects. For instance, they are used in various simulations for calculating functions and mathematical expressions, solving various equations and problems, and developing user interfaces. They are also widely employed in solving differential equations. Its syntax is rather intuitive and can be in two different forms.

(1) `Function_handle_name =@MYfunction;`

(2) `Function_handle_name=@(variable1, variable2, ...)`
`([expression1, expression2, ...]).`

Let's look at several examples of generating function handles.

```
F1 = @MY_function;
```

where `MY_function` is a function file or function expression or another function handle. Note that how to create function files and characteristics of function files is explained in detail in Chapter 2.

Here is an example:

```
function x = MY_function(a, b, c)
% MY_function.m is a function file that solves the quadratic equation
w.r.t. % a  user entries for a, b, c and outputs the found solutions.

D= b^2-4*a*c;
x1 = (-b+sqrt(D))/(2*a);
x2 = (-b-sqrt(D))/(2*a);
x = [x1; x2];
end
```

We can test the function handle F1 with the following command:

```
>> x = F1( 1, 2, 3)
x =
  -1.0000 + 1.4142i
  -1.0000 - 1.4142i
```

Here the function handle F1 calls the function file called `MY_function.m` and executes it with the user-specified input data for the `a`, `b`, and `c` variables.

Note that more detailed explanations on features of the function files (e.g., `MY_function.m`) and how to create them are given in Chapter 2.

Let's create a function handle for the following quadratic polynomial with for input arguments:

$$- f(x, a_1, a_2, a_3) = a_1 x^2 + a_2 x + a_3;$$

```
>> f =@(x,a1,a2,a3)(a1*x^2+a2*x+a3)
f =
 function_handle with value:
 @(x,a1,a2,a3)(a1*x^2+a2*x+a3)
```

Moreover, the function handles can be used to define a function of functions. For example, $H = 2e^{sin(x)}$ can be expressed in three different ways with the following function of functions:

```
>> ff1 = @(x) sin(x); ff2 = @(ff1)exp(ff1); ff3 = @(ff2)2*ff2;    % 1 - Way
>> gg1 = @(x)sin(x); gg2 = @(x)exp(gg1(x)); ff3 = @(x) 2*gg2(x);  % 2 - Way
>> hh3=@(x)2*exp(sin(x));                                          % 3 -Way
```

It is quite straightforward to perform computations from the function handles.

```
>> ff3(ff2(ff1(pi)))
ans =
    2.0000
>> gg3(pi)
ans =
    2.0000e+00
>> hh3(pi)
ans =
    2.0000e+00
>> x=1.3; a1=2; a2=-3; a3=13; f (x, a1, a2, a3)
ans =
 12.48
>> x=1.3; a1=[2, 3]; a2=[-3, 4]; a3=[11,13]; f2(x, a1, a2, a3)
ans =
    10.4800   23.2700
>> x=1:3; a1=2; a2=-3; a3=11; f2(x, a1, a2, a3)
Error using  ^  (line 51)
Incorrect dimensions for raising a matrix to a power.
Check that the matrix is square and the power is a
```

scalar. To perform elementwise matrix powers, use
'.^'.
Error in @(x,a1,a2,a3)(a1*x^2+a2*x+a3)

In the last part, for the vector or row array of entries for x, the expression of $f(x, a_1, a_2, a_3)$ needs to be fixed for elementwise matrix operations.

```
>> f =@(x,a1,a2,a3)(a1*x.^2+a2*x+a3)
f =
  function_handle with value:
    @(x,a1,a2,a3)(a1*x.^2+a2*x+a3)>> x=1:3; a1=2; a2=-3; a3=11; f(x,
a1, a2, a3)
ans =
    10    13    20
```

An alternative version of the function handle is the inline function that is similar to the function handle. Note that the inline function will be removed in future releases of MATLAB, and it is recommended to use the anonymous function (function handle) instead.

```
>> F1=inline('a1*x^2+a2*x+a3', 'a1', 'a2', 'a3', 'x')
F =
 Inline function:
 F (a1,a2,a3,x) = a1*x^2+a2*x+a3
```

Now, the previously created inline function can be evaluated with specific values of variables, x, a1, a2, and a3.

```
>> x=1.3; a1=2; a2=-3; a3=13; F(a1, a2, a3, x)
ans =
    12.4800
```

It is important while employing the function handles to follow the order of the variables. In other words, while calling them, you need to follow the sequence of the input variable values. Here's an example:

```
>> x=1.3; a1=2; a2=-3; a3=13;
>> f =@(x,a1,a2,a3)(a1*x.^2+a2*x+a3);
>> f(x, a1, a2, a3)
```

45

```
ans =
    12.4800
>> f(a1, a2, a3, x)
ans =
    15.3000
```

The function handles can also take different predefined input variables.

```
>> f =@(x,a1,a2,a3)(a1*x.^2+a2*x+a3);
>> y=1.3; b1=2; c2=-3; d3=13;
>> f(y,b1,c2,d3)
ans =
    12.4800
```

Note that in this example, the function handle f is taking the predefined variables y, b1, c2, and d3 instead of x, a1, a2, and a3.

Figure 1-12 shows the list of variables and function handles created in this section residing in the Workspace window.

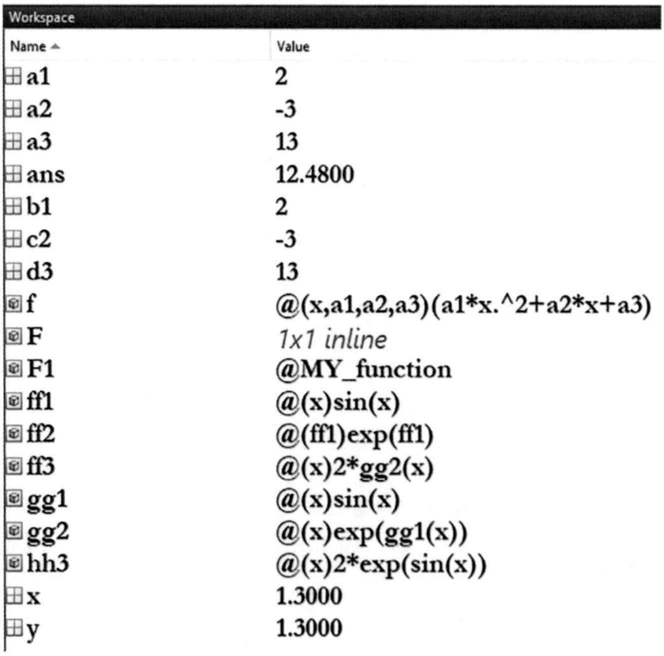

Workspace	
Name ▲	Value
⊞ a1	2
⊞ a2	-3
⊞ a3	13
⊞ ans	12.4800
⊞ b1	2
⊞ c2	-3
⊞ d3	13
▣ f	@(x,a1,a2,a3)(a1*x.^2+a2*x+a3)
▣ F	1x1 inline
▣ F1	@MY_function
▣ ff1	@(x)sin(x)
▣ ff2	@(ff1)exp(ff1)
▣ ff3	@(x)2*gg2(x)
▣ gg1	@(x)sin(x)
▣ gg2	@(x)exp(gg1(x))
▣ hh3	@(x)2*exp(sin(x))
⊞ x	1.3000
⊞ y	1.3000

Figure 1-12. *Created variables residing in the workspace*

Broader applications and essential uses of the function handles will be discussed in Chapter 2 and Chapter 8.

Logical Arrays

Logic and logical arrays are important for programming and Boolean operations. The local answers are 0s and 1s. 0 means that a statement or condition is not true, and 1 means that a statement is true. There are a few ways logic arrays can be generated in MATLAB. For example, you can apply the comparative analysis within numerical arrays or verify them or define their types. The following Boolean logical arrays are based on the existing variables created in the previous sections:

```
>>A = 2; B = -3.25; C=(A+B)^2; ABC=[A, B, C];
>> isnumeric(A)
ans =
 logical
 1
>> Ach1 = 'matlab';
>> isstr(Ach1)
ans =
 logical
 1
>> ABC_logic=ABC>=2
ABC_logic =
 1×3 logical array
 1 0 0
>> Seed = 1; rng(Seed);   % Setting up the random number generator to
generate permanent numbers.
>> Dr1 = rand(4, 5)
>> Dr1 = rand(4, 5)
Dr1 =
    0.0500    0.9446    0.1393    0.9275    0.8833
    0.5359    0.5866    0.8074    0.3478    0.6237
    0.6638    0.9034    0.3977    0.7508    0.7509
    0.5149    0.1375    0.1654    0.7260    0.3489
>> Dr1Logic=Dr1>0.5 & Dr1<=0.71
```

```
>> Dr1Logic=Dr1>0.5 & Dr1<=0.71
Dr1Logic =
  4×5 logical array
   0   0   0   0   0
   1   1   0   0   1
   1   0   0   0   0
   1   0   0   0   0
```

Note that we have set the seed value of the random number generator rng() in order to generate permanent element values with the uniform random number generator rand().

You can use the evaluated logical arrays indexing of the arrays and find out which values meet the set conditions and which ones do not. For example, in the previous examples, ABC_logic represents that the first element of ABC meets the set condition and is greater or equal to 2. Similarly, the array Dr1Logic means three elements in row 2, one element in row 5 are greater than 0.5 and at the same time, they are less than or equal to 0.71. Now we can find out which elements with their specific element order meet the set conditions by using indexing operations. Note that a more detailed explanation of logical operators is given in Chapter 2.

```
>> Index=find(Dr1>0.5 & Dr1<=0.71)
>> Index=find(Dr1>0.5 & Dr1<=0.71)    % Index numbers indicate which element
                                        has met the condition.
Index =
     2
     3
     4
     6
    18>> Dr1(Index)                     % The element meeting the set conditions
ans =
    0.5359
    0.6638
    0.5149
    0.5866
    0.6237
```

An alternative way of finding the true values is as follows:

```
>> Dr1Logic.*Dr1
ans =
         0         0         0         0         0
    0.5359    0.5866         0         0    0.6237
    0.6638         0         0         0         0
    0.5149         0         0         0         0
```

As another example, let's find out where 0 elements in an array generated by the random integer number generator that is used to generate integer numbers within [-2 2] to fill out the five-by-five square matrix:

```
>> Seed = 1; rng(Seed);    % Setting up the random number generator to
generate permanent numbers.
>> Hr=randi([-2, 2], 5)
Hr =
     0    -2     0     1     2
     1    -2     1     0     2
    -2    -1    -1     0    -1
    -1    -1     2    -2     1
    -2     0    -2    -2     2>> Ind=(~Hr)    % Locates which elements of
                                            Hr are equal to "0"
Ind =
  5×5 logical array
   1   0   1   0   0
   0   0   0   1   0
   0   0   0   1   0
   0   0   0   0   0
   0   1   0   0   0
```

In this section, we have generated the variables, namely, numerical and logic matrices and character strings, as shown in Figure 1-13.

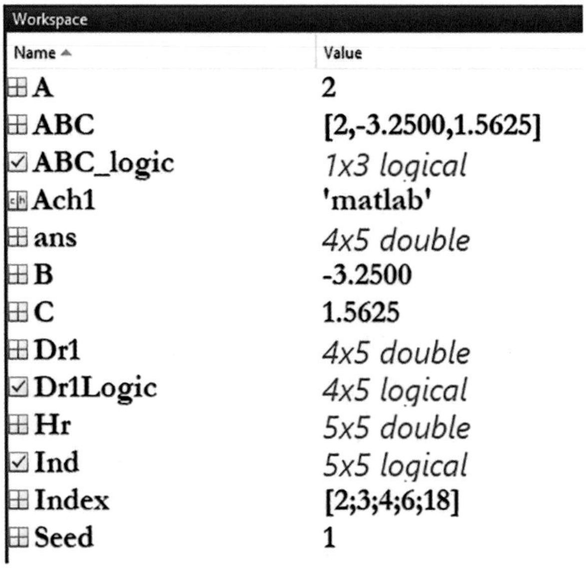

Figure 1-13. *Created variables residing in the workspace*

Other essential uses of logical arrays and indexing in examples are discussed in other sections of the book.

Table Arrays

The table arrays are a newly introduced tool of MATLAB, and thus, in older versions of MATLAB, table arrays cannot be generated. The table arrays are particularly useful for preparing reports and displaying/presenting the simulation/analysis results obtained in arrays of several columns and rows, each of which represents certain variables. They are used to collect heterogeneous data and metadata into a single container in a tabular data format. The table arrays can accommodate variables of different types, sizes, units, etc. They are used most frequently in machine learning and deep learning.

The table arrays are often used to store experimental data, with rows representing different observations and columns representing different measured variables. It displays the arrays in a more informative and tabulated format by indicating names of the columns and rows by respective assigned names. Let's look at several examples of how to create table arrays with the command table() by creating a new numerical array A1new and from the existing variables created in the previous sections.

```
>> A1new = [1 2 3; 3 4 5; 7 8 9];
>> A1Tab=table(A1new)
A1Tab =
  3×1 table
      A1new

    _____

    1    2    3
    3    4    5
    7    8    9
>> B1new = A1new/5;
>> B1Tab=table(B1new)
B1Tab =
  3×1 table
        B1new

    _____

    0.2    0.4    0.6
    0.6    0.8      1
    1.4    1.6    1.8
```

In the previous cases, the table arrays A1Tab and B1tab have been created as column tables from A1New and B1New, respectively. It is also possible to obtain/create tables from the existing arrays (arrays, cells, and structures) by using the array2table(), cell2table(), and struct2table() commands.

```
>> A1Tab2=array2table(A1New) % Column names are not specified
>> A1Tab2=array2table(A1new)
A1Tab2 =
  3×3 table
    A1new1     A1new2     A1new3

    _____     _____     _____

      1          2          3
      3          4          5
      7          8          9
```

```
>> Ach1Tab=array2table(A1new, 'variablenames',{'a','b','c'})
Ach1Tab =
  3×3 table
    a    b    c

    _    _    _
    1    2    3
    3    4    5
    7    8    9
>> C1new = [A1new, B1new]
C1new =
    1.0000    2.0000    3.0000    0.2000    0.4000    0.6000
    3.0000    4.0000    5.0000    0.6000    0.8000    1.0000
    7.0000    8.0000    9.0000    1.4000    1.6000    1.8000
>> Dr1Tab=array2table(C1new, 'variablenames', {'v1', 'v2', 'v3', 'v4',
'v5', 'v6'})
Dr1Tab =
  3×6 table
    v1    v2    v3    v4     v5     v6

    __    __    __    ___    ___    ___
    1     2     3     0.2    0.4    0.6
    3     4     5     0.6    0.8     1
    7     8     9     1.4    1.6    1.8
```

It is also possible to rename variables saved in the table arrays using renamevars() following a pattern of TableArrayName = renamevars(TableArrayName, 'OldVarName', 'NewVarName').

```
>> A1Tab=array2table(A1new)
A1Tab =
  3×3 table
    A1new1    A1new2    A1new3

    _____    _____    _____
      1         2         3
      3         4         5
      7         8         9
>> A1Tab = renamevars(A1Tab, 'A1new1', 'a')
```

52

```
A1Tab =
  3×3 table
    a     A1new2     A1new3

    _     _____     _____

    1       2          3
    3       4          5
    7       8          9
>> A1Tab = renamevars(A1Tab, 'A1new2', 'b');
>> A1Tab = renamevars(A1Tab, 'A1new3', 'c')
A1Tab =
  3×3 table
    a    b    c

    _    _    _
    1    2    3
    3    4    5
    7    8    9
```

You can also remove any column or row of a created table array in a few different ways.

```
R1Tab = array2table(A1new)
R1Tab =
  3×3 table
    A1new1     A1new2     A1new3

    _____     _____     _____

      1          2          3
      3          4          5
      7          8          9
>> R1Tab.A1new3 = [ ]    % Removes the variable A1new3
R1Tab =
  3×2 table
    A1new1     A1new2

    _____     _____

      1          2
      3          4
      7          8
```

```
>> >> R1Tab(2,:)=[ ]    % Removes row 2
R1Tab =
  2×2 table
    A1new1    A1new2

    _____    _____

      1         2
      7         8
>> R1Tab=removevars(R1Tab, 'A1new1')   % Removes the variable A1new1
R1Tab =
  2×1 table
    A1new2

    _____

      2
      8
>> R1Tab=removevars(R1Tab, {'A1new2'})   % Removes the variable A1new2
R1Tab =
  2×0 empty table
>> clearvars R1Tab    % Deletes the table array R1Tab
```

Note that the command syntax `removevars()` is available starting from the MATLAB2018a version.

It is also possible to convert table arrays into arrays and cell arrays by using the `tabel2array()` and `table2cell()` commands, respectively. Understanding and working with table arrays will be of great help not only when you are preparing reports but also when you are importing with the data import wizard 📥 Import Data or recommended data import function `readtable()` and manipulating various data sets from the external files (e.g., `.txt`, `.xls`, `.xlsx`, `.csv`, `.dat`, etc.) into the MATLAB workspace.

Cell Arrays

Cell arrays are useful to accommodate various types (numerical, character, logical, table, and function handle) of arrays in different cells of one cell type variable by preserving all attributes of each variable unchanged. They might be handy to carry or pass various data sets inside one variable. Cell arrays contain indexed data containers such as cells accommodating lists of text, character strings, combinations of text and numerical data, and numerical arrays, function handles, structure arrays, and tables. One of the most essential features of the cell arrays is that they require curly brackets to be used in specifying cell addresses. Another important feature of the cell-type arrays is that in many cases imported/read data by MATLAB will be in cell array mode.

Let's look at several examples of creating cell arrays with different types of arrays discussed earlier and generate the new ones.

```
>> Acell = cell(5,5) % Creates an empty cell of size 5-by-5
>> Acell = cell(3,4) % Each cell of Acell will be filled with Arrays
Acell =
  3×4 cell array
    {0×0 double}    {0×0 double}    {0×0 double}    {0×0 double}
    {0×0 double}    {0×0 double}    {0×0 double}    {0×0 double}
    {0×0 double}    {0×0 double}    {0×0 double}    {0×0 double}
>> A0 = 13; A1new = [1 2 3; 3 4 5; 7 8 9];
>> Acell{1,1}=A0  % Cell (1, 1) is filled with A0
Acell =
  3×4 cell array
    {[      13]}    {0×0 double}    {0×0 double}    {0×0 double}
    {0×0 double}    {0×0 double}    {0×0 double}    {0×0 double}
    {0×0 double}    {0×0 double}    {0×0 double}    {0×0 double}
>> Acell{1,2}=A1new  % Cell (1, 2) is filled with A1new
Acell =
  3×4 cell array
    {[      13]}    {3×3 double}    {0×0 double}    {0×0 double}
    {0×0 double}    {0×0 double}    {0×0 double}    {0×0 double}
    {0×0 double}    {0×0 double}    {0×0 double}    {0×0 double}
>> Ach1Tab=array2table(A1new, 'variablenames',{'a','b','c'});
>> Acell{1,3}=Ach1Tab  % Cell (1, 3) is filled with Ach1Tab
Acell =
  3×4 cell array
    {[      13]}    {3×3 double}    {3×3 table }    {0×0 double}
    {0×0 double}    {0×0 double}    {0×0 double}    {0×0 double}
    {0×0 double}    {0×0 double}    {0×0 double}    {0×0 double}
```

```
>> C1 = 'matlab';
>> Acell{1,4}=C1  % Cell (1, 4) is filled with C1
Acell =
  3×4 cell array
    {[        13]}    {3×3 double}    {3×3 table }    {'matlab'  }
    {0×0 double}    {0×0 double}    {0×0 double}    {0×0 double}
    {0×0 double}    {0×0 double}    {0×0 double}    {0×0 double}
>> D1 = A1new>3 & A1new<9
D1 =
  3×3 logical array
   0   0   0
   0   1   1
   1   1   0
>> Acell{2,2}=D1  % Cell (2, 2) is filled with logical array D1
Acell =
  3×4 cell array
    {[        13]}    {3×3 double }    {3×3 table }    {'matlab'  }
    {0×0 double}    {3×3 logical}    {0×0 double}    {0×0 double}
    {0×0 double}    {0×0 double }    {0×0 double}    {0×0 double}
>> f=@(x, a1, a2, a3)(a1*x^2+a2*x+a3);
>> Acell{2,3}=f  % Cell (2, 3) is filled with function handle f
Acell =
  3×4 cell array
    {[        13]}    {3×3 double }    {3×3 table       }    {'matlab'  }
    {0×0 double}    {3×3 logical}    {function_handle}    {0×0 double}
    {0×0 double}    {0×0 double }    {0×0 double      }    {0×0 double}
>> Acell{3,1}= 'This is a cell array' % Cell (3, 1) is filled with
characters
Acell =
 3×4 cell array
 {[              13]} {3×3 double }    {3×3 table       }    {'matlab'  }
 {0×0 double       } {3×3 logical}    {function_handle}    {0×0 double}
 {'This is a cell ...'} {0×0 double }    {0×0 double      }    {0×0 double}
```

```
>> Acell{3,2}=rand(5,6) % Cell(3,2) is filled with a numerical array
Acell =
  3×4 cell array
  {[                  13]}   {3×3 double }   {3×3 table       }   {'matlab' }
  {0×0 double         }     {3×3 logical}   {function_handle}   {0×0 double}
  {'This is a cell ...'}    {5×6 double }   {0×0 double     }    {0×0 double}
```

To access and view the contents of the created cell arrays, use the following commands:

```
>> Acell{1,1}
ans =
    13
>> Acell{1,4}
ans =
    'matlab'
>> Acell{2,3}
ans =
  function_handle with value:
    @(x,a1,a2,a3)(a1*x^2+a2*x+a3)>> Acell{7}
ans =
  3×3 table
    a    b    c

    _    _    _
    1    2    3
    3    4    5
    7    8    9
>> Acell{8}
ans =
  function_handle with value:
    @(x,a1,a2,a3)(a1*x^2+a2*x+a3)
```

By double-clicking the cell variable name in the workspace, the contents of the cell can be viewed, as shown in Figure 1-14.

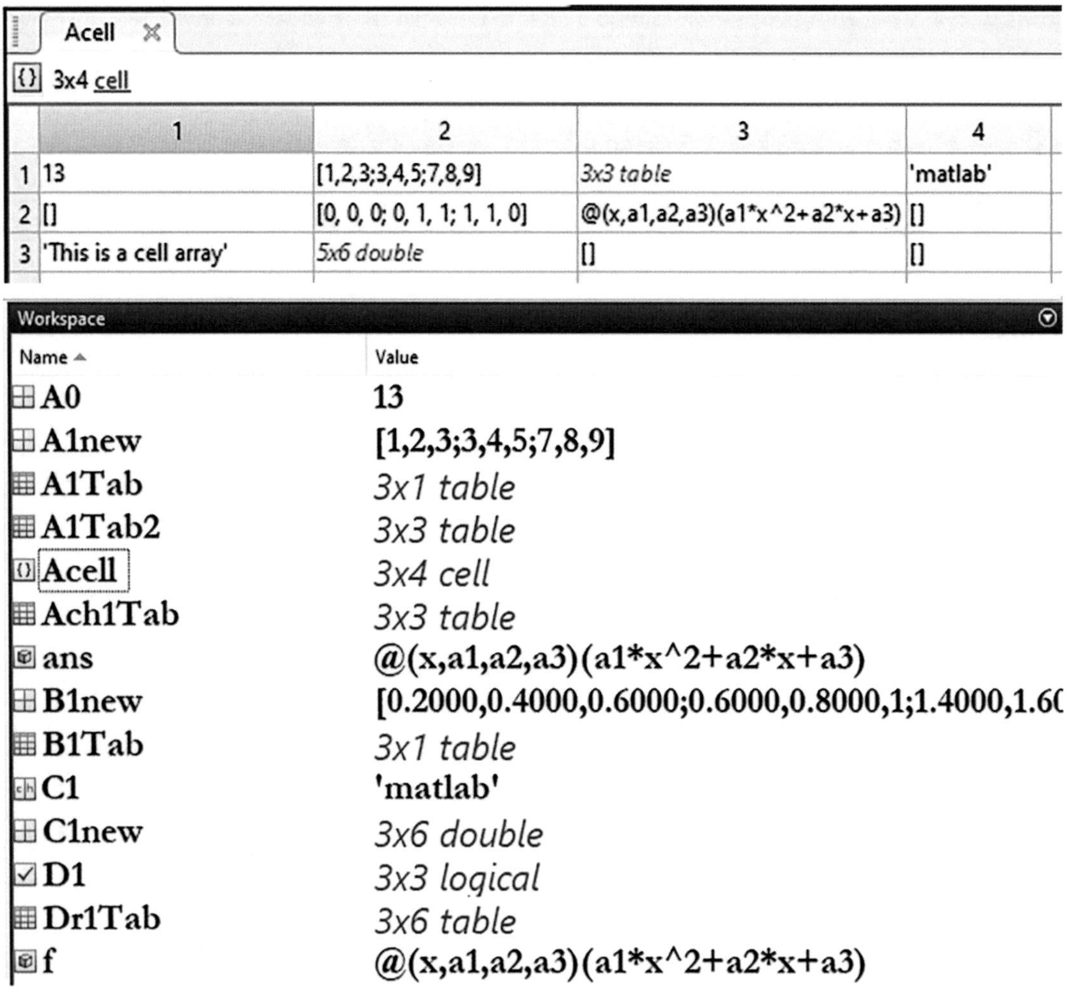

	Acell ✕			
{} 3x4 cell				

	1	2	3	4
1	13	[1,2,3;3,4,5;7,8,9]	*3x3 table*	'matlab'
2	[]	[0, 0, 0; 0, 1, 1; 1, 1, 0]	@(x,a1,a2,a3)(a1*x^2+a2*x+a3)	[]
3	'This is a cell array'	*5x6 double*	[]	[]

Workspace	
Name ▲	Value
⊞ A0	13
⊞ A1new	[1,2,3;3,4,5;7,8,9]
⊞ A1Tab	*3x1 table*
⊞ A1Tab2	*3x3 table*
{} Acell	*3x4 cell*
⊞ Ach1Tab	*3x3 table*
▣ ans	@(x,a1,a2,a3)(a1*x^2+a2*x+a3)
⊞ B1new	[0.2000,0.4000,0.6000;0.6000,0.8000,1;1.4000,1.6(
⊞ B1Tab	*3x1 table*
▣ C1	'matlab'
⊞ C1new	*3x6 double*
☑ D1	*3x3 logical*
⊞ Dr1Tab	*3x6 table*
▣ f	@(x,a1,a2,a3)(a1*x^2+a2*x+a3)

Figure 1-14. *Variables in the workspace and contents of the Acell cell array*

You can change the contents of the cell array via double-clicking in each cell and entering the new values or contents. You can also change the contents (elements) of the cell array by recalling the cell address and assigning new values. You can also empty some cells of the cell array by assigning an empty matrix to them.

```
>> BCell{1}=[1, 2, 3; 5 6 7];   % Cell 1 of BCell
>> BCell{2}=randi(5, 5, 3);   % Cell 2 of BCell
>> BCell{1}   % View the content of BCell's cell 1
ans =
     1     2     3
     5     6     7
```

```
>> BCell{2}   % View the content of BCell's cell 2
ans =
     5     1     1
     5     2     5
     1     3     5
     5     5     3
     4     5     5
>> BCell{1}(2,3)=13;   % Change the element residing in row 2 and column 3
of cell 1 in BCell
>> BCell{1}   % View cell 1 of BCell
ans =
     1     2     3
     5     6    13
>> BCell{2}(4:5,:)=0;      % Change the elements of row 4 and 5 of cell
2 in BCell
>> BCell{2}   % View cell 2 of BCell
 ans =
     5     1     1
     5     2     5
     1     3     5
     0     0     0
     0     0     0
>> BCell{1}(1,:)=[ ];   % Remove row 1 in cell 1 of BCell
>> BCell{1}        % View cell 1 of BCell
ans =
     5     6    13
>> BCell{2} = [ ]; % Empty cell 2
>> BCell{2}           % View cell 2 of BCell
ans =
    []
>> clearvars BCell  % Delete BCell variable
```

Structure Arrays

Structure arrays can accommodate all of the previously created arrays and entry (variable) types, namely, all types of numeric, logical, character, table, cell, and function handles. They can store data not only of different types but also of different sizes. One of the important aspects of the structure arrays is that they are suitable for code generation. Moreover, they are useful in programming, data processing, data acquisition, and reading the outputs of Simulink models. In addition, many MATLAB toolboxes and their functions produce a various structure array type of outputs after their simulations. Therefore, it is necessary to understand how to handle the structure arrays efficiently.

Note the cell array can also accommodate structure arrays.

Structure arrays store data in different fields or field names that we can access by their names. Here's an example:

```
% Person 1: Name - SE; DOB - June 6, 1982; Profession - professor
WHO(1).Name = 'SE';
WHO(1).DOB = '06.06.1982';
WHO(1).Profession = 'professor';
% Person 2: Name - NE; DOB - Dec 12, 1992; Profession - designer
WHO(2).Name = 'NE';
WHO(2).DOB = '12.12.1982';
WHO(2).Profession = 'designer';
```

The structure variable WHO contains information about two people. The data contains their names, dates of birth, and professions.

```
>>WHO =
  1×2 struct array with fields:
    Name
    DOB
    Profession
>> WHO(1)
ans =
  struct with fields:
        Name: 'SE'
         DOB: '06.06.1982'
    Profession: 'professor'
```

```
>> WHO(2)
ans =
  struct with fields:
        Name: 'NE'
         DOB: '12.12.1982'
   Profession: 'designer'
```

Note that we have saved in the workspace only one variable that is in a structure array form. If you want to have access to a specific field of the created structure, then you call that field name along with the structure name.

```
>> WHO(1).Name    % To access to the name field of a person 1
ans =
    'SE'
>> WHO(1).Profession    % To access to the profession field of a person 1
ans =
    'professor'
>> WHO(2).DOB    % To access to the DOB field of a person 2
ans =
    '12.12.1982'
```

Besides this approach of creating structure arrays, there are several other ways to set up or create them. Let's look at some examples. Another way is to create an empty structure with the command struct(). The empty structure will be filled with variables and their values with the command syntax of struct('FieldName', VALUE). Note that the field names (variable names) must not contain empty spaces or symbols except for the underscore sign (_).

```
>>  Astr1 = struct()
Astr1 =
  struct with no fields.
>> % Now we can assign/set up values and data fields inside the created
empty structure: Astr1.
>> a = 2.1; A = [1, 2; 3, 4]; B = A<a; f = @(x)(a*x^2+x-a); C{1} =rand(3);
C{2} = 'matlab';
>> D = 'mathworks.com';
>> F = table(magic(3));
```

```
>>  Astr1 = struct('Number', a, 'Matrix', A, 'Logic', B, 'F_Handle', f,
'Cell', C, 'Char', D, 'Table', F)
Astr1 =
  1×2 struct array with fields:
    Number
    Matrix
    Logic
    F_Handle
    Cell
    Char
    Table
```

Another way to create structure arrays is to assign their fields individually, as shown in the beginning with the example of creating the structure variable WHO.

```
>>  Bstr.Number = a; Bstr.Matrix = A; Bstr.Logic = B; Bstr.F_Handle=f; Bstr.
Cell=C; Bstr.Char = D;
>>  Bstr.Table = F;
>>  Bstr
Bstr =
  struct with fields:
      Number: 2.1000e+00
      Matrix: [2×2 double]
       Logic: [2×2 logical]
    F_Handle: @(x)(a*x^2+x-a)
        Cell: {[3×3 double]  'matlab'}
        Char: 'mathworks.com'
       Table: [3×1 table]
```

Now we can compare the two different ways we created the structure variables Astr and Bstr. Another way to create structure arrays is conversions, in other words, to convert existing cell-type variables or table-type variables into structure-type variables with the cell2struct() and table2struct() commands.

```
>>  % Cell variable C contains: Matrix C and String character 'matlab'
>>  C{1} =rand(3); C{2} = 'matlab';
>>  F_Names = {'Matrix', 'Char'};     % Field names/Headers
```

```
>> Cstr = cell2struct(C, Headers, 2) % 2 means two cells embedded
Cstr =
  struct with fields:
    Matrix: [3×3 double]
      Char: 'matlab'
```

Now let's see how to convert a table array into a structure array with the table2struct() command.

```
>> F = table(magic(3));
>> Dstr = table2struct(F)
Dstr =
  3×1 struct array with fields:
    Var1
```

Figure 1-15 shows the list of variables created and saved in the workspace and the contents of the created structure arrays, namely, Astr and Bstr.

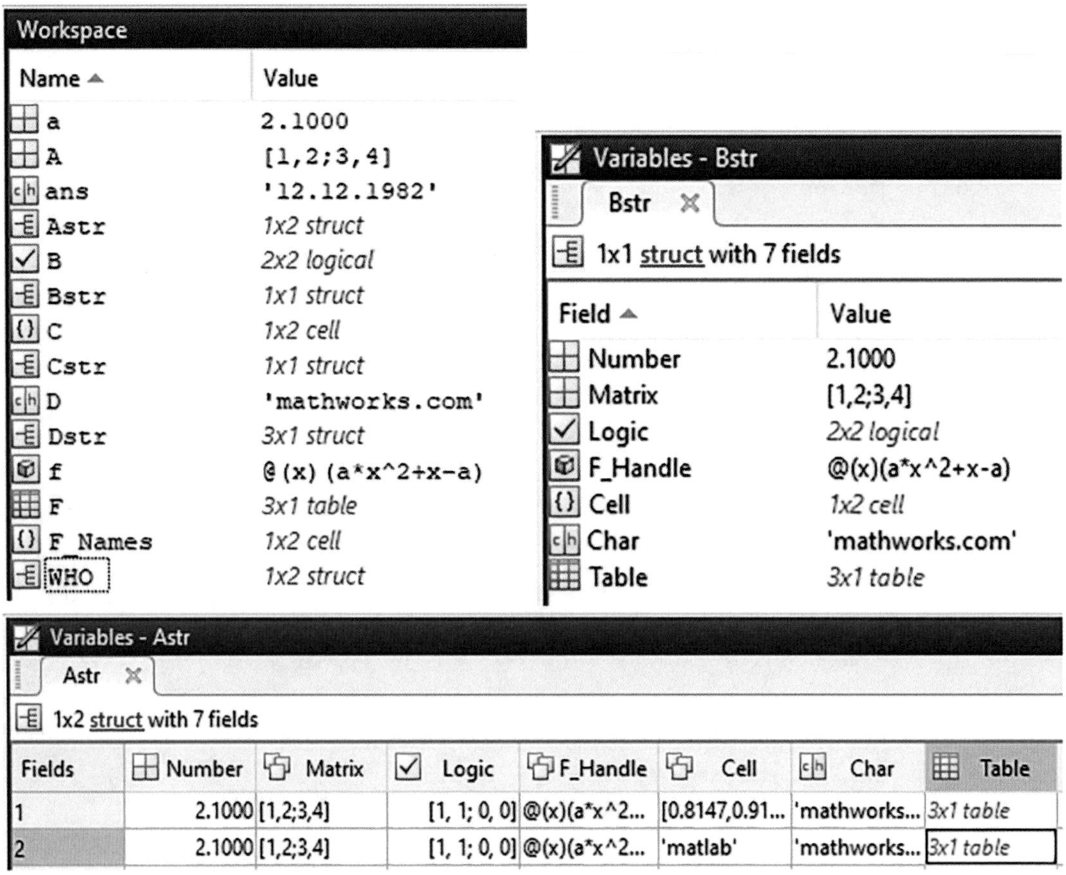

Figure 1-15. *Created variables and Astr and Bstr structure arrays*

This section has demonstrated via examples how to create seven different types of variables (arrays): numerical arrays (scalar and array type variables), character strings, logical arrays, table arrays, cell and structure arrays, and function handles. You can remove from the workspace any of the created variables by using the clear or clearvars command or using the right-mouse button options to delete. From the Command window, we clear the variables from the workspace with the following commands:

```
>> clearvars a A  B  % Removes variables a, A, B
>> clear ans C D      % Removes variables ans, C, D
```

Note   When you delete the variables from the workspace using the `clearvars` or `clear` command, the comma is not used between the variable names.

From the attributes of the created and saved variables in the workspace (Figures 1-12, 1-13, 1-14, 1-15), you can read the variable type (scalar, array, logical, table, cell, structure, character, function handle), its storage type (double, single, uint8, int8), and its size (how many rows and columns or cells, etc.). Moreover, the symbols representing each variable type shown in Figures 1-12, 1-13, 1-14, and 1-15 demonstrate the MATLAB supported data (array) types shown in Figure 1-9 and 1-10.

It must be noted that many of these arrays can be converted from one type into another as you have seen in some of the examples. For example, a cell array can be transferred into table array via a `cell2table()` function, or similarly, a structure array can be converted into a table array via `struct2table()`; vice versa, a table array can be converted into a cell and table array via `table2cell()` and `table2struct()`, respectively.

Complex Numbers

Two letters, *i* and *j* or 1*i* and 1*j*, are reserved for notating imaginary numbers. Therefore, it is advised not to use these letters for assigning variable names. An alternative safe approach to assigning a complex number is to multiply it by `sqrt(-1)`. For example, to obtain 3.76+2.35i, use one of the following commands:

```
>> A = 3.76+2.35i;                % Way 1
>> A = 3.76+2.35*sqrt(-1);        % Way 2
```

Precision

MATLAB's precision is not absolute.

For instance:

```
>> sin(pi)
ans =
 1.2246e-016
```

In MATLAB, *sinsin* (π) *is not equal to* 0. That is because the number π is represented by the double precision number in MATLAB. We can demonstrate the precision issue by performing the calculations of the Pythagorean theorem: $1 = (t) + (t)$.

```
t=0:pi/50:2*pi; F=1-(sin(t).^2+cos(t).^2); plot(t, F)
```

In the previous expressions, *t* is a time vector containing a row of elements, such as $[0, \pi/50, \ldots 2\pi]$. Some values of F are zero, and others are nonzero even though they are very small numbers. The reason for this is that all of the trigonometric functions including exponential and logarithmic functions are approximated by a polynomial of degree 13 with only odd powers of the argument variable (in this example *t*). For instance, $sin(t) \approx t - c_1t^3 + c_2t^5 + \ldots + c_6t^{13} = p(t)$. The computation algorithm for all of these functions is implemented based on fdlibm, a "Freely Distributable Math Library" developed at Sun Microsystems by K. C. Ng and others (see for more information www.netlib.org/fdlibm).

It must be noted that MATLAB's accuracy (precision) level depends on which data storage type is chosen to save data. For instance:

```
>> int8(128)*-5
ans =
 -128
```

The allocated data storage int8 can hold up to $2^8 - 1$ integer numbers. All MATLAB supported data storage types are shown in Figure 1-15.

M-file and MLX-file Editors

In the context of the book, the terms *code*, *script*, and *program* are used interchangeably to refer to the M-files with the extension of *.m and the MLX-files with the extension of *.mlx, including function and executable files. Because of their extensions, these files are called *M-files* and *MLX-files*. In the previous examples, all of the operations are done in the Command window. However, for programming and writing, editing, and debugging, M-file and MLX-file editors will be of great help due to their many helpful tools and hints in writing fast and more efficient code, scripts, and programs.

The overall functionality of M-files and MLX-files is similar except for one important feature. The MLX file editor window can display the outputs of calculations/simulations within the MLX editor window and indicate most common command syntax-related errors in its left output window. The M-file editor shows all errors in the Command window after the M-file's execution. Moreover, the MLX-file editor can show interactively all inserted equations via the equation editor, inserted images and hyperlinked texts right in the same window, and others. The outputs from both files will be shown in the workspace. Both files can be used interchangeably. Let's start reviewing M-file and MLX-file editor windows and tools.

M-file Editor

M-file editor window menu and GUI tools are grouped into three tabs: Editor, Publish, and View (as shown in Figure 1-16, 1-17, 1-18, respectively). Note that there are three main menu subgroupings, HOME, PLOTS, and APPS, which belong to the main MATLAB window that has been shown in the initial sections.

In the M-file editor's main tools menu (see Figure 1-16), there are five subsections: File, Navigate, Edit, Breakpoints, and Run. All of the tools in each subsection are quite intuitive. For example, the FILE subsection has GUI tools used to open a new file or existing files, save the current file, find M-files, compare different versions of the M-files with the same names, and print out the current M-file. Similarly, the NAVIGATE subsection tools help a user to move the cursor within the current file and find keywords and if necessary to substitute them with other words. The EDIT subsection tools insert a new section into the current file, add or remove comment lines, or wrap comments and put indents to make the file more readable. The BREAKPOINTS subsection has tools to choose from the drop-down options for debugging/editing the current M-file code contents, not comments. Finally, the RUN subsection has GUI tools to run different cell sections of the current M-file step by step and run the current M-file and measure the evaluation time in different sections of the file. It should be noted that in writing M-files, the EDITOR window tools are mainly used.

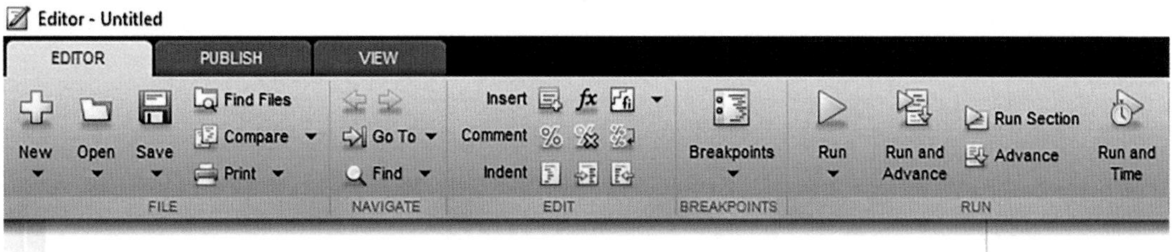

Figure 1-16. *M-file editor's main tools menu*

The PUBLISH tools, shown in Figure 1-17, are used to generate report files in different file formats such as HTML, DOC, PPT, PDF, etc. The PUBLISH window's GUI tools are very intuitive and similar to many document-editing software applications. FILE contains the Save (Save, Save As, Save All, Save Copy As...) drop-down options; the INESRT section contains Section and Section with Title; the INSERT INLINE MARKUP section contains B (bold), I (Italic), M (Monospaced), Hyperlinked, and Inline LaTeX; Insert Block Markup contains Bullet List, Numbered List, Image, Preformatted Text, Code, Display LaTeX; and PUBLISH contains Publish (options).

Figure 1-17. *M-file editor's Publish tools menu*

The VIEW tools, as shown in Figure 1-18, are used to display several windows of M-files and MLX-files (documents) within one window area. You can split the view window side by side or top to bottom by using the Tiles, Document Tabs, and Split Document tools. The check marks in the Display subsection are handy to display data tips, show line numbers, and highlight the current line while editing M-files.

Figure 1-18. *M-file editor's View tools menu*

MLX-file Editor

The MLX-file (live editor) editor tab, shown in Figures 1-19, 1-20, and 1-21, contains many of the M-file editor tools along with several other different GUI tools. The LIVE EDITOR (see Figure 1-19) has one main different subsection from the M-file editor (see Figure 1-16), which is called TEXT. This contains most of the functions of the PUBLISH tools of the M-file editor (see Figure 1-17).

Figure 1-19. *MLX-file editor's main tools menu*

Moreover, the MLX-editor's INSERT subwindow (see Figure 1-20) has a few handy tools to write/edit equations directly in the MLX-file contexts and to insert subsections with comments. You can insert control GUI tools, such as slider and drop-down boxes within the current MLX-file content.

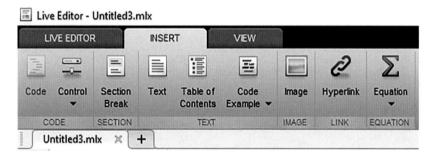

Figure 1-20. *MLX-file editor's Publish tools menu*

The VIEW window (Figure 1-21) of the MLX-editor has some similar tools of M-file editor's VIEW (Figure 1-21) and some other tools, such as Document Tabs, Display, Output, and Layout, by which a user can display script (code) line numbers (excluding the comment lines), Datatips, Full Screen View, Clear all Outputs, show or hide outputs, or show outputs inline or on the right, respectively.

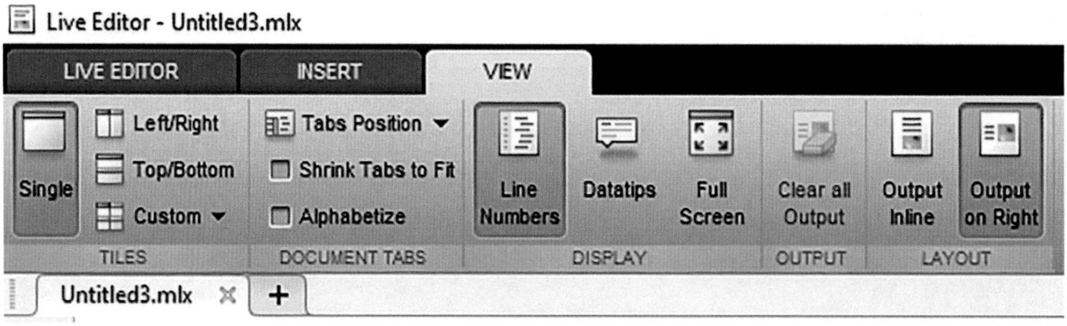

Live Editor - Untitled3.mlx

LIVE EDITOR	INSERT	VIEW	

Single | Left/Right Top/Bottom Custom ▾ | Tabs Position ▾ ☐ Shrink Tabs to Fit ☐ Alphabetize | Line Numbers | Datatips | Full Screen | Clear all Output | Output Inline | Output on Right

TILES | DOCUMENT TABS | DISPLAY | OUTPUT | LAYOUT

Untitled3.mlx ✕   +

Figure 1-21. *MLX-file editor's view tools menu*

The M-file or MLX-file can be created in several different ways, by using the GUI

buttons Script or Live Script , respectively, or by typing >> edit in the Command window and pressing Enter.

To demonstrate some of the previously mentioned tools and options of the M-file and MLX-file editors, let's look at the following example to demonstrate that MATLAB's precision is not absolute via the Pythagorean theorem:

$$\alpha = -\frac{\pi}{2}...\frac{\pi}{2}, \Delta\alpha = \frac{\pi}{100}; F(\alpha) = 1 - (\alpha + \alpha\ );$$

First, we write the solution script of this simple example in the M-file editor and then publish the results. Subsequently, we perform the same simulations in the MLX-editor again to demonstrate similar and different features of both editors.

The algorithm of solving this exercise is composed of the following six steps:

1. Insert some comments describing the given problem statement.

2. Define the input variable: $\alpha = -\frac{\pi}{2}...\frac{\pi}{2}, \Delta\alpha = \frac{\pi}{100};$.

3. Perform the computation: $F(\alpha) = 1 - (\sin^2(\alpha) + \cos^2(\alpha)\ )$.

4. Define for which values of the input variable α the function values of $F(\alpha) = 0$.

5. Plot the simulation results: α versus $F(\alpha)$.

6. Publish all the obtained results including the whole script.

This is the script written in the M-file editor (Figure 1-16.a) directly:

```
% Step 1. Problem statement
% MATLAB's precision is not absolute.
% Pythagorean Theorem: F = 1 - (sin^2 (alpha) + cos^2(alpha));
% Input variable: alpha = -0.5*pi ... 0.5*pi

% Step 2. Define the input variable
alpha = -pi/2:pi/100:pi/2;

% Step 3. Perform the computation
F = 1-(sin(alpha).^2+cos(alpha).^2);

% Step 4. Define for which values of alpha, F(alpha) = 0.
Findex=find(F==0);
F0=F(Findex);

% Step 5. Plot the simulation results: alpha vs. F(alpha)
plot(alpha, F, 'b-', alpha(Findex), F0, 'ro'),
legend('\alpha vs F(\alpha)', 'F(\alpha) = 0')
title('Simulation of MATLAB"s precision via the Pythagorean Theorem
'), grid on
xlabel('\alpha'), ylabel('F(\alpha) = 1-(sin(\alpha)^2+cos(\alpha)^2)')

% Step 6. Publish all of the obtained results including the whole script
```

Note that in this script, we have used % to insert comments (nonexecutable information) and \ to insert and display the Greek letters.

% Comments

Comments are not executable and contain additional information for the users. The sign % is used to place comments and remarks or any additional information within M-files and MLX files or callback functions or in the Command window. The comments can be added on a separate line or behind command syntaxes and have to start with %. If the sign % is placed double (%%) followed with a space, that automatically makes the following comments bold. Moreover, inserting %% at the beginning of a line leaves a blank space and creates a cell mode in the script. That also creates an option of the subsection feature in the M-file editor. We discuss in detail the cell mode options and their advantages in Chapter 2. Note that there are several other functionalities of the % sign. It is used for format specifications for write, display, and read purposes that we discuss in Chapter 2.

This is slightly edited with the PUBLISH tools (Figure 1-17), such as Bold **B** Bold for steps 1, 2, and 6 under Insert Inline Markup and Publish (PUBLISH); see Figure 1-17. Note that to make any selected lines of comments bold, you first select the line and then hit the **B** Bold button. Note that in this script, we used the LaTeX commands to insert the Greek letters (α, π) and the equation $F = 1 - (\alpha + cos^2 \alpha)$.

Note that M/MLX editors are compatible with most common LaTeX mathematical mode commands. The LaTeX-compatible mathematical commands and symbols can be inserted for plot titles, axis labels, graphic notes, and so forth that we discuss in examples in the following chapters (programming, plots, ODEs). For example, to insert the expression $x^2 + y^2 = R^2$, we enter the expression x^2+y^2=R^2. To insert α, β, Ω, Ψ, we type \alpha, \beta, \Omega, and \Psi. Moreover, to insert the equations with Greek letters, the notations need to start with $$ signs and end with $ (see steps 1 and 5). For more information and help on various mathematical expressions and symbols to write in LaTeX math, type >> doc latex in the Command window.

See [6] also for more information on how to handle the LaTeX. Note that %% at the beginning initializes automatic recognition of the LaTeX-compatible mathematical commands within step 1.

```
%%
%% *% Step 1. Problem statement*
% MATLAB's precision is not absolute.
% Pythagorean Theorem: F = 1 - (sin^2 $$\alpha$ + cos^2 $$\alpha$ );
% Input variable: $$\alpha$ = -0.5* $$\pi$ ...0.5* $$\pi$
%% *% Step 2. Define the input variable*
alpha = -pi/2:pi/100:pi/2;
%% *% Step 3. Perform the computation*
F = 1-(sin(alpha).^2+cos(alpha).^2);
%% *% Step 4. Define for which values of $$\alpha$ $$F(\alpha) =0$*
Findex=find(F==0); F0=F(Findex);
%% *% Step 5. Plot the simulation results: $$\alpha$ vs. $$F(\alpha)$*
plot(alpha, F, 'b-', alpha(Findex), F0, 'ro'),
legend('\alpha vs F(\alpha)', 'F(\alpha) = 0')
```

```
title('The Pythagorean Theorem '), grid on
xlabel('\alpha'), ylabel('F(\alpha) = 1-(sin^2(\alpha)+cos^2(\alpha))')
%% *%Step 6. Publish all of the obtained results including the
whole script*
```

After completing the script writing in the editor, save the M-file with the valid file

Publish

name (e.g., P1.m). Now, to publish it in an HTML format, click the Publish

button on the PUBLISH tab (see Figure 1-7). After clicking the button, the script is executed automatically, and its HTML formatted report will be generated, as shown in Figures 1-22 and 1-23.

Web Browser - P1

P1 ✕ +

Location: file:///C:/Users/sulaymon.eshkabilov/Downloads/html/P1.html

Contents

- % Step 1. Problem statement
- % Step 2. Define the input variable
- % Step 3. Perform the computation
- % Step 4. Define for which values of α $F(\alpha) = 0$
- % Step 5. Plot the simulation results: α vs. $F(\alpha)$
- % Step 6. Publish all of the obtained results including the whole script

% Step 1. Problem statement

MATLAB's precision is not absolute. Pythagorean Theorem: F = 1 - (sin^2 α + cos^2 α); Input variable: α = -0.5* π ...0.5* π

% Step 2. Define the input variable

```
alpha = -pi/2:pi/100:pi/2;
```

% Step 3. Perform the computation

```
F = 1-(sin(alpha).^2+cos(alpha).^2);
```

Figure 1-22. *MATLAB generated an HTML-formatted report of the M-file called P1.m*

% Step 4. Define for which values of α $F(\alpha) = 0$

```
Findex=find(F==0); F0=F(Findex);
```

% Step 5. Plot the simulation results: α vs. $F(\alpha)$

```
plot(alpha, F, 'b-', alpha(Findex), F0, 'ro'),
legend('\alpha vs F(\alpha)', 'F(\alpha) = 0')
title('The Pythagorean Theorem '), grid on
xlabel('\alpha'), ylabel('F(\alpha) = 1-(sin^2(\alpha)+cos^2(\alpha))')
```

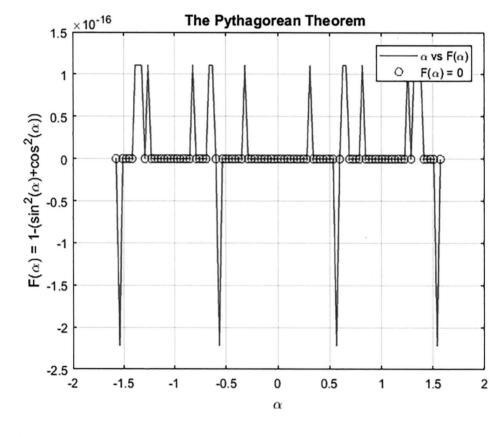

% Step 6. Publish all of the obtained results including the whole script

Figure 1-23. *MATLAB generated an HTML-formatted report of the M-file called P1.m*

Note that the formatted bold lines (Figure 1-18 a, b), which are starting lines of cell modes preceded with %%, have been recognized by the M-file editor automatically and put in contents and hyperlinked, such as `% Step 1 ... Step 2 ... Step 6.`

Now let's try the same procedures with the MLX-file editor. Note that the comments are edited using the Text tab tools of the MLX-editor, as shown in Figure 1-19. The parts

of the script are entered as text (comments) by using `Text` . Step 1 ... Step 2 ... Step 3... Step 6 lines are bolded with **B** .

Step 1. Problem statement.

MATLAB's precision is not absolute.

Pythagorean theorem: F = 1 - (sin^2 (alpha) + cos^2(alpha));

Input variable: alpha = -0.5*pi ... 0.5*pi

Step 2. Define the input variable.

Step 3. Perform the computation.

Step 4. Define for which values of alpha, F(alpha) = 0.

Step 5. Plot the simulation results: alpha versus F(alpha).

Step 6. Publish all of the obtained results including the whole script.

Note that there is an alternative way to make the chosen lines bold, which is to use %% followed with a blank space, as in the M-file editor. In this case, such editing is automatically detected as a header of the following section of the script. If you enter the following in the Code section of the MLX-editor:

```
%% Step 2. Define the input variable.
```

and press Enter, MATLAB automatically creates this bolded text header:

Step 2. Define the input variable.

Now, in between step 2 and step 3, and step 5 and step 6, the following executable

commands are inserted by putting the cursor on the desired line and clicking the `Code` button on the Code subtab (see Figure 1-19). Finally, the complete code is obtained.

Step 1. Problem statement.

MATLAB's precision is not absolute.

Pythagorean theorem: F = 1 - (sin^2 (alpha) + cos^2(alpha));

Input variable: alpha = -0.5*pi ... 0.5*pi

Step 2. Define the input variable.

```
alpha = -pi/2:pi/100:pi/2;
```

Step 3. Perform the computation.

```
F = 1-(sin(alpha).^2+cos(alpha).^2);
```

Step 4. Define for which values of alpha, F(alpha) = 0.

```
Findex=find(F==0);F0=F(Findex);
```

Step 5. Plot the simulation results: alpha versus F(alpha).

```
plot(alpha, F, 'b-', alpha(Findex), F0, 'ro'),
legend('\alpha vs F(\alpha)', 'F(\alpha) = 0')
title('The Pythagorean Theorem'), grid on
xlabel('\alpha'), ylabel('F(\alpha) = 1-(sin(alpha)^2+cos(alpha)^2)')
```

Step 6. Publish all of the obtained results including the whole script.

Save the file as an *.mlx file (e.g., call P2.mlx). Insert the mathematical expressions with the Equation Tools on the Insert subtab (shown in Figure 1-20) and by using $\boxed{\Sigma}$ Equation.

When you press the button $\boxed{\Sigma}$ Equation, the menu of symbols, structures, and matrices will be opened, as shown in the following image:

Now you put the cursor where you want to insert the mathematical expressions and insert the symbols by selecting the necessary symbols or by using LaTeX expressions. For example, \alpha gives α. Enter the expressions of the Pythagorean formulation, input the variable range, and finalize the script.

Step 1. Problem statement.

```
MATLAB's precision is not absolute.
```

The Pythagorean theorem: $F = 1 - (\sin^2\alpha + \cos^2\alpha)$

Input variable: $\alpha = -0.5\alpha \ldots 0.5\alpha$

Step 2. Define the input variable.

```
alpha = -pi/2:pi/100:pi/2;
```

Step 3. Perform the computation.

```
F = 1-(sin(alpha).^2+cos(alpha).^2);
```

Step 4. Define for which values of α, $F(\alpha) = 0$.

```
Findex=find(F==0);
F0=F(Findex);
```

Step 5. Plot the simulation results: α vs. $F(\alpha)$.

```
plot(alpha, F, 'b-', alpha(Findex), F0, 'ro'),
legend('\alpha vs F(\alpha)', 'F(\alpha) = 0')
title('The Pythagorean Theorem '), grid on
xlabel('\alpha'), ylabel('F(\alpha) = 1-(sin^2(alpha) +cos^2(alpha))')
```

Step 6. Publish all of the obtained results including the whole script.

After executing (by pressing the Run button shown in Figure 1-19), you will get the script outputs via Output Inline in LAYOUT. See Figure 1-24 and Figure 1-25.

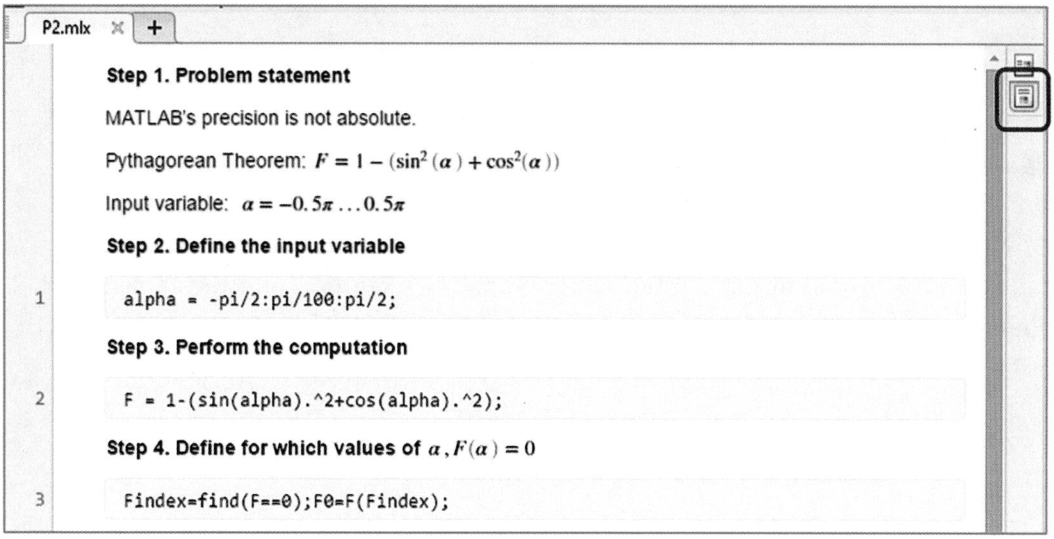

Figure 1-24. *MLX-editor output of P2.mlx*

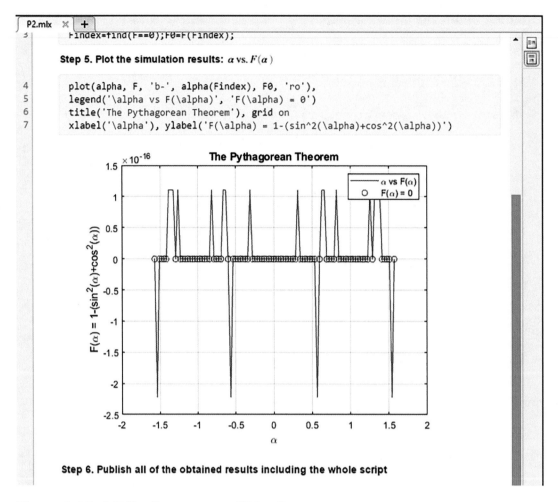

Figure 1-25. *MLX-editor output of P2.mlx*

Note that in the MLX- editor all of the executable and nonexecutable lines of the script are identified automatically and put in separate sections. There are some other salient issues on hints, warnings, and error messages of the M/MLX editors that we discuss in Chapter 2.

Closing the MATLAB Window

Quitting MATLAB is simple. There are several commands that can be used to complete your work in MATLAB and close all the windows. You can type `>> quit`, type `>> exit`, or press Ctrl+Q. You can also click the X in the upper-left corner of the main MATLAB window or call the `>> finishdlg` function from the Command window and click the yes button in the opened GUI window.

Note that all variables residing in the workspace will be cleared upon exiting/quitting MATLAB. They will be lost and not be recovered by default the next time MATLAB is started. However, they can be saved to a `*.mat` file and loaded back into the workspace later. The command history of entered commands is saved automatically, and all of the entered commands in the Command window can be accessed the next time you launch MATLAB. If you are interested in saving the number of commands that can be adjusted via MATLAB preferences, choose Preferences ➤ Command History.

Summary

This chapter introduced the MATLAB environment, including settings, variables, several most used commands, and M-file/MLX-file GUI tools. You also learned about assigning variables and values from the Command window and working in the M/MLX-file editor windows. In addition, the chapter explored data types, formats, and structures as well as ways to use built-in MATLAB commands and functions. In particular, it covered help search options and commonly used commands, including `help`, `helpwin`, `helpbrowser`, `doc`, `lookfor`, `clear`, `clear all`, `dir`, `pwd`, `cd`, `ls`, `save`, `load`, `clearvars`, `edit`, `format`, `char`, `size`, `who`, `whos`, `input`, `what`, and `exit/quit`.

References

[1] http://www.mathworks.com/matlabcentral/fileexchange/authors

[2] https://www.mathworks.com/matlabcentral/answers/

[3] http://ctms.engin.umich.edu/CTMS/index.php

[4] http://www.mit.edu/people/abbe/matlab/

[5] https://www.mathworks.com/company/newsletters/
articles/the-tetragamma-function-and-numerical-
craftsmanship.html

[6] https://www.mathworks.com/help/matlab/creating_plots/
greek-letters-and-special-characters-in-graph-text.html

Exercises for Self-Testing

Exercise 1

Perform the following steps:

1. Find the Vibrating Logo demo from the preinstalled MATLAB demos and run the demo. Hint: `membrane`.

2. Locate Product Overview from the help library of the package.

3. Change the font type and size of the Command window.

4. Change the font size and color of comments in the M-file editor.

5. Make the numerical data display format in long eng in the Command window using Preferences and commands in the Command window.

Exercise 2

Do the following steps:

1. Use the help library to find out how to add a new path for search. Add a new path for search: `C:\Users\Public`. Hint: `addpath`.

2. Use the MATLAB help browser to find how to create a new directory. Create a new directory called `my_new_dir` inside directory `C:\Users\Public`. Hint: `mkdir`.

3. Change the current directory to the newly created directory `C:\Users\Public` using the Command window. Hint `cd`.

Exercise 3

Get help on the exp (exponential) function using the Command window. Use the help, lookfor, doc, and help browser commands, and then compare the results of the four help options.

Exercise 4

Create and open an *.m file called learn.m in the M-file editor window using the Command window. (Hint: >> edit ...). Insert two commands in it that display the current date and time. Hint: date, clock.

Exercise 5

Create a shortcut (a set of favorites commands) that opens a new M-file named My_Shortcut and simultaneously closes all figure windows and clears the Command window and Workspace window from all previously entered data and commands.

Exercise 6

Change how the data formats display in the Command window from the Preferences. Make it a hexadecimal *format*. Hint: format.

Exercise 7

What are the commands used to clean up the Workspace, Command, and History windows?

Exercise 8

Given $x = 2.25$, $y = 3.1$, and $z = 13.20$, use MATLAB to evaluate the following expressions:

$$A = \sqrt[y]{xyz - \frac{z}{xy}} + x^2 y^{\frac{3}{2}} z^{\frac{5}{4}}; B = e^{-xyz} ln\left(\frac{xy}{z}\right).$$

Use the array editor to change the assigned values for x, y, and z and reexecute the expressions to compute A and B.

Exercise 9

Create a script and save it as my_first_program.m.

1. Your program should contain an input variable that is the length of a square side as a variable parameter in meters.

2. Your program should calculate the area of a square and the volume of a cube.

3. Your program should output the calculated results (area and volume) in metric (m^2, m^3) and British (in^2, in^3) systems by using conversion, e.g., 1 $inch = 25.4\ mm$.

4. Execute your created script (my_first_program.m) from the Command window.

5. Execute your created script (my_first_program.m) from the M-file Editor window.

Exercise 10

Do/answer the following:

- Which command displays what is stored in your MATLAB workspace?

- Create $P = \dfrac{\pi}{4}$ and save it in uint64 format.

- Explain why the value of P in uint64 is equal to 1. How do you fix this issue?

- If the memory space is your concern, what is the best format to use to save integer values ranging between -2^16 to 2^16?

Exercise 11

Save all computation results from Exercises 8, 9, and 10 in a MAT file called my_FIRSTwork.mat and clean up your MATLAB workspace from all the variables except for A and B from Exercise 8.

Exercise 12

Given:

C1 = Hello. I'm from NYC. What is your name?

D1= I'm Dan.

C2 = Nice to meet you

D2= Nice to meet u 2

create a conversation displayed in the Command window in the form of a dialogue:

'Hello.' 'I'm from NYC.'

'What is your name?' 'I'm Dan.'

'Nice to meet you.' 'Nice to meet u 2.'

Exercise 13

Do the following steps:

1. Change the display data formats for the Command window and make it *long e.*

2. Where are the Preferences settings of MATLAB saved?

3. What is the main function of the M-file finishdlg.m, and what commands does it contain?

Exercise 14

Perform the following steps:

1. Write a command in the Command window that creates and opens an M-file called Ex14.m.

2. Edit your M-file Ex14.m so that it contains a command that changes the display format type.

3. Edit your M-file Ex14.m so that it contains a command that changes the current directory.

4. Edit your M-file Ex14.m so that it contains a command that displays the current date and time.

5. Edit your M-file Ex14.m so that it contains a command that saves the current date and time data under two variable names, Day and T, respectively.

6. Edit your M-file Ex14.m so that it contains a command that stores the variables Day and T in the file called Ex14.mat.

7. Edit your M-file Ex14.m so that it contains commands that clean up the Command and Workspace windows and quits the MATLAB session.

Exercise 15

Given:

```
>> day_ONE='MatlabDAY'; day_DUE=day_ONE+1.0
day_DUE =
 78 98 117 109 98 99 69 66 90
```

1. Why is the answer day_DUE numeric data and equal to a 10-element row matrix?

2. What do these numbers represent?

3. How can you obtain the original characters given in day_ONE and display them in the Command window? Hint: char().

Exercise 16

Given:

```
>> A=[1,2; -12.0, 3]; mat2str(A); SNA=ans+0; char(SNA)
```

What is hidden behind the variable SNA?

Exercise 17

Create a structure type variable called E17 composed of cell and numeric array elements, such as MATLIB and classes.

1. Create a cell called `Matlab` that is composed of two subcells `{'Day#1', 'Start'}`.

2. Create a numeric array called `classes` containing the elements $\left[\pi; 2\pi; \sqrt{\pi} + 2i\right]$. Note that i represents an imaginary number.

Exercise 18

Create a function handle and inline function of the following mathematical expressions:

1. $h(\theta, t) = 1.3 * e^{-ti\theta}$. Note that i - represents an imaginary number.

2. $Z(x, y, a, b) = ax^2 + by^2$.

3. $U(t, \omega, A, \ B) = A sin(\omega t) + B cos(\omega t)$.

Exercise 19

It is analytically proven that $cos2\alpha = 2\alpha - 1$. Use MATLAB to compute the equality for the different values $[\, 0, \ldots \frac{\pi}{2}, \pi \ldots \frac{3\pi}{2}, 2\pi \ldots \frac{5\pi}{2}, 3\pi \ldots 5\pi\,]$ of α and define the values of α in which the accuracy of MATLAB calculations does not represent equality.

Exercise 20

Use MATLAB to compute the expression $\sqrt[5]{1.5 * 10^{24} - 10^{24} i}$ in the most accurate way. Note that it is in the fifth root.

Exercise 21

Why do the following outputs look "strange"?

```
>> A=[4/5, 'matlab'+0, sin(pi)]
A =
 Columns 1 through 3
```

```
3fe999999999999a 405b400000000000 4058400000000000
Columns 4 through 6
405d000000000000 405b000000000000 4058400000000000
Columns 7 through 8
4058800000000000 3ca1a62633145c07
```

How do you fix this problem and make the results look readable?

Exercise 22

Create a five-by-five matrix called [A] by using randi() within [1, 20] and divide it by 3. Display [A] as rational numbers as shown below. Note that your answer array numbers (in the numerator) of [A] will differ from the ones shown here. Why does your answer differ from the one shown here?

```
A =
    10/3      6     13/3    25/3    20/3
     2/3   16/3    17/3    13/3    11/3
       5      0    10/3     8/3     2/3
       4    1/3       7     2/3       2
       6    8/3       6       5       1
```

Exercise 23

Create the next array in the most efficient way (at least in two different ways). Note the display format of A2 elements.

```
A2 =
    1   2   3   4   5   6   7   8   9   10   11   12   13
    1   2   3   4   5   6   7   8   9   10   11   12   13
    1   2   3   4   5   6   7   8   9   10   11   12   13
    1   2   3   4   5   6   7   8   9   10   11   12   13
    1   2   3   4   5   6   7   8   9   10   11   12   13
    1   2   3   4   5   6   7   8   9   10   11   12   13
    1   2   3   4   5   6   7   8   9   10   11   12   13
    1   2   3   4   5   6   7   8   9   10   11   12   13
    1   2   3   4   5   6   7   8   9   10   11   12   13
    1   2   3   4   5   6   7   8   9   10   11   12   13
    1   2   3   4   5   6   7   8   9   10   11   12   13
    1   2   3   4   5   6   7   8   9   10   11   12   13
    1   2   3   4   5   6   7   8   9   10   11   12   13
```

Exercise 24

By using `randi()`, create a 15-by15 array (called A3) with elements ranging from -125 to 127 and save it in the most memory-efficient way with the name A3.

Exercise 25

Create the following HTML report using M-file editor tools:

Contents

- % Step 1. Problem statement
- % Step 2. Perform the computation

% Step 1. Problem statement

(1) Compute: G $\xi = e^{cos(A\xi)}$; (2) Compute: $a^2 = b^2 + c^2$; (2) Compute: $sin\ \beta = \dfrac{a}{c}$; $cos\ \beta = \dfrac{a}{b}$;

% Step 2. Perform the computation

```
b = 3; c = 4; a = sqrt(b^2+c^2);
F = 1-(sin(alpha).^2+cos(alpha).^2);
sinB=a/c; cosB = a/b;
```

Exercise 26

Given:

A5 = [1 2 3; 4 5 6; 7, 8, 9]

1. Obtain B5 from A5 by two arithmetic operations: B5 = [16, 9 4; 1 0 1; 4 9 16].

2. Obtain C5 from A5 and B5 by using relational logic (<, >) and arithmetic operations (+ 13): C5 = [13 13 13; 14 14 14; 14 13 13].

Exercise 27

Create three numerical (row matrix) arrays (variables called AJ, IS, LJ) so that when you subtract 3 from each of them and one conversion operation, you should obtain Al-Khwarizmi, Ibn Sina, and Lennart Johansson.

Exercise 28

Save all of your created variables (A, A2, A3, A4, A5, A6) from Exercises 22 to 27 in a
*.mat file named with your last name, e.g., Jones_HW2.mat.

Exercise 29

Create the matrix A in the most efficient way:

```
A = 7×7
        1    2    3    4    5    6    7
        1    2    3    4    5    6    7
        1    2    3    4    5    6    7
        1    2    3    4    5    6    7
        1    2    3    4    5    6    7
        1    2    3    4    5    6    7
        1    2    3    4    5    6    7
```

Obtain a new A matrix in a most efficient way:

```
A = 7×7
        1    2    3    0    5    6    7
        1    2    3    0    5    6    7
        1    2    3    0    5    6    7
        0    0    0    0    0    0    0
        1    2    3    0    5    6    7
        1    2    3    0    5    6    7
        1    2    3    0    5    6    7
```

Exercise 30

Create a cell array (called A) containing three variables: a=4/5, b='matlab'+0, c=sin(π).
Create a structure (called B) containing four variables: a, b, c, and A. Show how to get
access to the variables a, b, and c residing inside A and B.

Exercise 31

Create the following variables and entries in the MLX-file editor:

Function handle F: $F(\omega_1, \omega_2, \theta) = cos(\omega_1\theta) - sin(\omega_2\theta)$

Identity matrix: $I = [1\ 0\ 0\ 0\ 1\ 0\ 0\ 0\ 1]$

Magic numbers: $M = [8\ 1\ 6\ 3\ 5\ 7\ 4\ 9\ 2]$

Multiply *I* matrix by 2 and subtract from the *M* matrix and call the new matrix by *MI*:
MI = [6 1 6 3 3 7 4 9 0]

Logic array *L* by locating/comparing the elements of MI that are greater than 1 and less than 6: *L* = [0 0 0 1 1 0 1 0 0]

1. Create a table array TM from M.

2. Create a cell array CA containing F, I, M, MI, L, TM.

3. Create structure SA containing F, I, M, MI, L, TM, CA.

Exercise 32

Write down how to get to Layout (1), Preferences (2), and Quick Access (3), as shown in the screenshots. **Note** what the Fonts and Colors are used in (2).

Exercise 33

Create these shown files (MLX/M-files) and write down the steps for how to display the results, as shown in the figure in the Live Editor window here in this exercise.

Note that there are four windows displaying MLX and M-files, equations, Greek letters, plot figures, hyperlinks, data-tips, and how to insert an image.

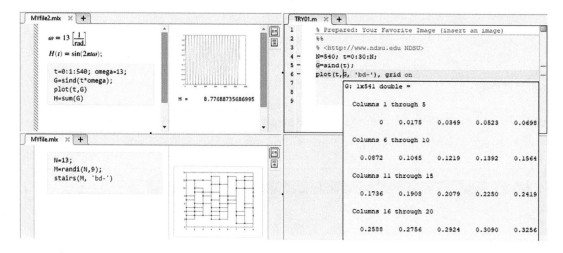

Exercise 34

Answer the following true/false questions:

- MATLAB's default numerical format (int8, uint8, int16, uint16, ... single, double) depends on the operating system of the computer in which MATLAB is installed. (True/False)

- Stored variables and their associated attributes change with the change of a display format type in the Command window. (True/False)

- The MATLAB user interface is customizable. (True/False)

- MATLAB supports cell arrays only if they are numerical data but not strings. (True/False)

- It is possible to put a table array variable into a cell array. (True/False)

We are changing the values of entries [C] by changing a display format. (True/False).
>>A = 1.1; B = [2 3; 1 2]; C = B/A; format hex; C)

We have changed the values of entries D, E by changing a display format. (True/False).
>>D=uint8(255); D=255; D+1 =255; E = [12/14, 3/5; 1/3, 4/9]; format rat; E)

- The commands `clearvars` and `clear all` are the same and don't have any difference. (True/False)

- The command `clear A*` deletes all `.m`, `.mlx`, and `.mat` files starting with a file name of A. (True/False)

Exercise 35

Answer the following array size and representation-related questions:

- Given a cell array ABBA containing 10 cells, which command will recall the elements residing in cell 3 of ABBA? (Give a command.)

- Given a 5-by-5 array (matrix) called A, `A(4:end, 3:4)` will produce a matrix of what size?

- How do you create a linearly spaced data array: a=(-13, -12, -11, ... 11, 12, 13) and b=(0, 1/13, 2/13, 3/13, ... 24/13, 25/13, 2) without typing all the elements? Note how to obtain the rational format type of the array b.

Exercise 36

1. Obtain the logical array C = [1 1 1 1 1 0 0 0 0 0 0] from the array A = [-5, -4, -3 ... 3, 4, 5] whose elements are linear (equally spaced).

2. Obtain the numerical array D = [3 5 0; 0 0 4] from the array E = [3 5 7; 1 1 4].

Exercise 37

Given in the Command window:

```
>> x=linspace(-13, 0), F(x)=2*x^2+2*x-1;
```

There are two potential errors. Find errors and fix them. What is the size of x and F now?

Exercise 38

Write the answer commands to the following questions:

- How do you change a working directory to `C:\Users\Public` and add this path for a search path?

- How do you create a new directory called `MYdir` inside the `C:\Users\Public` directory?

- How do you find out what (variables) are stored in your MATLAB workspace?

- How do you remove the created directory `MYdir` using the Command window?

Exercise 39

Answer the following array-related questions on equally spaced data points:

1. Change the last two columns (column 8 and 9) of D2 (D2=zeros(9); D2(6,:)=1:9;) given previously to have the elements: $[e^0, e^1, e^2, ..., e^8]$, $[tan\ (e^0), tan\ (e^1), tan\ (e^2), ..., tan\ (e^8)]$, respectively in the most efficient way.

2. Generate these vector spaces in two different ways: [-100, -90, -80, ... 100], [-100, -99, -98, ... 100].

3. Generate an equally spaced 500 data points within $[-\pi . . \pi]$.

Exercise 40

Answer the following logical indexing and logical array-related questions:

1. Given >> A=magic(3); C1=le(A,7), C2=A<=7, what do 0 and 1 mean in all logical arrays shown previously for each individual case?

2. Given >> A1=12.12; C='nan'; B=[1 2; 0, 3i]; D=B/0; AA1=isfinite(A1), CC1=isnan(C), DD1=isinf(D), DD2=isnan(D), what do 0 and 1 mean in all logical arrays shown previously for each individual case?

3. Given >> AG=randi([-13, 25], 3, 2); BAG=(AG>0 & AG<13), why do your answers differ when you run the previous commands to define AG and BAG?

4. Given >> GAB=find(AG>0 & AG<13), AG(GAB), what numbers are behind GAB, and how are they related to AG?

5. Given >> 13>10; -1.2<=7.8; -11+13>=3, why are we getting 0s and 1s?

6. Given >> AA=randi([-13, 13], 10, 2); AA(AA<=0); % OR >> IN=(AA<=0); AA(find(IN)), what does IN represent with regard to [AA]?

7. Given >>B = randi([0, 13], 5); k=find(B>=3 & B<=5), what numbers are in k with reference to B?

Exercise 41

Create an * .m file called ARRAY_1.m and write the following:

1. Write the command to clean up the Workspace and Command windows of MATLAB, and then display the current date and time in the Command window.

2. Create array A. Write in it the commands to generate the following arrays: A1 (1-by-10) with the operator :, A2 (10-by-1) with linspace(), A3 (2-by-10) with eye().

3. Create array B. Write in it the commands to create the following arrays: B1 (5-by-6) with `randi()` elements ranging between [-1....1], B2 (5-by-6) with `rand()`, and B3 (5-by-10) with `randn()`.

4. Create array C. Write in it the commands to generate the following arrays: C1 (5-by-10) with `magic()` and `repmat()`, C2 (6-by-10) with `eye()`, and C3 (10-by-10) with `ones()`.

5. Write in it the commands performing all possible (arithmetic array) operations (+, -, *, /, .*, ./, ^, .^) with A1, A2, and A3 (at least three operations) and call these new matrices: A1new1, A1new2, A1new3, A2new1, A2new2, A2new3, A3new1, A3new2, A3new3. Hints: use `transpose()` and `rot90()` while performing arithmetic array operations.

6. Write in it the commands performing all possible (arithmetic array) operations (+, -, *, /, .*, ./, ^, .^, sum, mean) with B1, B2 and B3 (at least three operations) and call new matrices: B1new1, B1new2, B1new3, B2new1, B2new2, B2new3, B3new1, B3new2, B3new3. Hint: use `fliplr()` and `transpose()` while performing arithmetic array operations.

7. Create AB1, AB2, and AB3 matrices from A1, A2, A3, and B1, B2, and B3. Also, use part of any A1, A2, A3 and B1, B2, B3 arrays. Note that every AB1, AB2, AB3, ABC4, ABC5 should contain some elements from arrays A and B. Hint: use `flipud()` and `repmat()` while creating the arrays AB1, AB2, and AB3.

8. Create ABC1, ABC2, and ABC3 matrices by combining/concatenating the previously created arrays: A1, A2, A3 and B1, B2, B3 and C1, C2, C3. You should also use part of any A1, A2, A3 and B1, B2, B3 and C1, C2, C3 arrays. Note that every ABC1, ABC2, ABC3 should contain some elements from the A, B and C matrices from Parts 1, 2, and 3.

Exercise 42

Create an *.mlx file called Ex42.mlx and write in it:

1. The command that creates a cell array called Q1 with six empty column cells.

2. The command gives binary representations of 12321 and 987654321, and a command writing these numbers including their binary representations in cell 1, 2, 3, 4 of Q1, respectively.

3. The command converts the binary representations of 12321 and 987654321 into a numerical type of array by using MATLAB's conversion commands, namely, str2num( ) or str2double( ), and the command writing the two converted numerical arrays in cell 5 and 6 of Q1, respectively.

4. The command generating the following array by using pascal() (MATLAB built-in matrix function):

   H =

1.00	0	0	0	0
1.00	-1.00	0	0	0
1.00	-2.00	1.00	0	0
1.00	-3.00	3.00	-1.00	0
1.00	-4.00	6.00	-4.00	1.00

   And by applying logical indexing (logical array) and elementwise matrix multiplication operations, the following array:

   HLG =

1.00	0	0	0	0
1.00	0	0	0	0
1.00	0	1.00	0	0
1.00	0	3.00	0	0
1.00	0	6.00	0	1.00

5. The command that creates a structure array called S5 and commands writing it: Q1, H, HLG.

6. Explain: why do the converted numbers (in step 3) from the binary representations (of 12321 and 987654321) differ from the original decimal numbers, i.e., 12321 and 987654321?

Exercise 43

Create an *.mlx file (called Ex43.mlx) and add all the necessary comments, such as questions, equations, explanations, and other relevant remarks, and also, write in it the following commands:

1. Generate two COLUMN arrays with 202 equally spaced data points in two different ways: $= -2\pi...2\pi$; $\beta = -360^0...360^0$.

2. Compute these three equations (take the values of α, β from step 1): $F(\alpha) = e^{sin(\alpha)}$; $H(\beta) = e^{cos(\beta)}$; $S = 1 - (sin^2\alpha + cos^2\beta)$. Note: β is given in degrees not in radians, and thus, do not forget to convert it into radians. Also, insert the equations by using the equation editor of *.mlx.

3. Create an array (called Solution) of five columns containing α, $F(\alpha)$, β, $H(\beta)$, S.

4. Create a table of arrays. The table of variables should be called TVall and has to be in the following format:

alpha	F	beta	H	S
-6.2832	1	-360	2.7183	0
-6.1563	1.135	-352.73	2.6965	-2.2204e-16
-6.0293	1.2855	-345.45	2.6325	0
-5.9024	1.4501	-338.18	2.5304	-4.4409e-16
-5.7755	1.6261	-330.91	2.3961	-4.4409e-16
-5.6485	1.8092	-323.64	2.2373	-2.2204e-16
-5.5216	1.9939	-316.36	2.0621	-2.2204e-16
-5.3947	2.1731	-309.09	1.8786	-8.8818e-16
-5.2677	2.339	-301.82	1.6942	0
-5.1408	2.4834	-294.55	1.515	-4.4409e-16
-5.0139	2.5984	-287.27	1.3457	2.2204e-16
-4.8869	2.6773	-280	1.1896	-2.2204e-16
4 74	2 7152	272 73	1 0487	0

5. Find all of the positive values of F, S, and H, and corresponding α, β values and save all of them in a cell array variable called FSH_pos.

6. Find all absolute zero values of S and corresponding α, β values. Save them in an array called ABS_0 with three COLUMNS of the found S, α, β values.

7. Create a structure of arrays called ABFSH_struct containing SOLUTION, TVall, FSH_pos, and ABS_0 from steps 3, 4, 5, 6.

8. Clear all variables in the workspace except for α, β, F, H S, SOLUTION, TVall, FSH_pos, and ABFSH_struct from the previous steps. Save these variables in an *.mat called Ex43.mat.

Exercise 44

Create the *.m file (called Ex44.m) that should perform the following operations:

1. Clear up the workspace and Command window from all entries.

2. Close all open figure windows.

3. Create a new directory: C:\Documents\Ex44.

4. Change the current directory of MATLAB to a newly created directory: C:\Documents\Ex44.

5. Compute $t = [0, 3\pi]$ $with \Delta t = \pi/200$,

$$f_1(t) = 3sin\left(\frac{3t}{2}\right), \quad f_2(t) = \left(\frac{2t}{5}\right).$$

6. Save the computation results $(t, f_1(t), f_2(t))$ in a .mat file called Ex44.mat.

Exercise 45

Create an *.m file (called Ex45.m) that performs the following operations:

1. Changes the current directory of MATLAB to the MATLAB's root directory (Hint: matlabroot)

2. Removes the directory created in Exercise 44: C:\Documents\Ex44

3. Displays MATLAB's root directory in the Command window

4. Creates two function handles: $r = \dfrac{1}{2}at^2 + v_o t + r_0$
    and $A = P\left(1 + \dfrac{r}{n}\right)^{nt}$

5. Saves the current path and the previous created function handles
    in a structure array called EX45

CHAPTER 2

Programming Essentials

This chapter covers the most essential and widely used programming tools, operators, and control statements in MATLAB. In addition, the chapter covers modeling essentials in Simulink, the development of a graphical user interface (GUI), and the development of MATLAB executable files and stand-alone applications. Also, the chapter shows a number of simple examples demonstrating efficient ways to program and model in MATLAB/Simulink to save computation time as well as how to create short and compact code/scripts. In the process, I will give a few essential hints and show different approaches for writing robust programs and scripts. Throughout the book, key terms such as *script, code, program, M-file, MLX-file*, and *function file* are used frequently to refer to the programs written for MATLAB. The script and M/MLX-files, including the function files, are meant to be source code readable by users, not the machine code understood by a computer.

Writing M/MLX-Files

The previous chapter discussed various entries, such as arrays, characters, cells, structures, tables, and logic arrays, via examples in the Command window. Also, I gave a general overview of M/MLX-file editors by writing several short pieces of code (scripts) that performed computations. Scripts (M/MLX-files) can also be written in Notepad or WordPad. They become M-files as soon as they are named with a file extension of *.m or *.mlx.

As discussed in the previous chapter, the M/MLX-file editors have many easy-to-use tools and some easy-to-understand options that can be employed while writing and debugging scripts. The M/MLX-file editors' most used tools are the ability to use cell modes, set up breakpoints, evaluate selected lines of cells or scripts, use automatic error detection options for misspelled command names and missing brackets in algebraic

© Sulaymon Eshkabilov 2022
S. Eshkabilov, *Beginning MATLAB and Simulink*, https://doi.org/10.1007/978-1-4842-8748-4_2

operations, detect all used and unused/unreferred to but introduced variables, and get useful hints to improve the script performance and display of warning hints on unnecessarily assigned variables, see the overloaded display of results, use variable values within loops, use a profile viewer, and so forth.

All of these tools of the M/MLX-file editors help you avoid syntax errors while writing scripts. In addition, there are many other tools that help you save time and effort on the mechanical parts such as writing reports and publishing reports in HTML or PDF formats, for example.

Before starting to write some code, let's consider the most essential steps in any programming language. The process of writing code starts with a pen and paper and is composed of the steps shown in Figure 2-1.

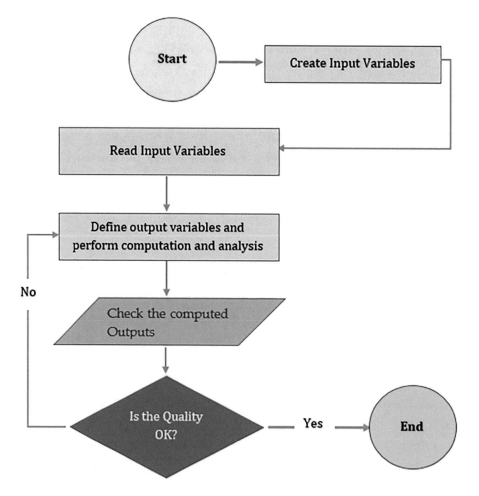

Figure 2-1. *General process of code/program writing*

This flowchart shows the following steps:

1) Clarify the problem statement.

2) Clarify/create/declare input variables: `var1`, `var2`, `var3` ... `varN`.

3) Read the values of `var1`, `var2`, `var3`, ... , `varN`.

4) Define the output variables: `out1`, `out2`, `out3`, ... `outN`. Perform computation, evaluation, and analysis operations.

5) Check the quality/correctness of the obtained results/output variables.

6) If the quality/correctness of the achieved results is not adequate, then go back to step 3. Repeat steps 3 through 5 until the expected quality/correctness is attained.

7) End and report on the results.

This algorithm is a generalized process and might also include considering objectives, specifics, and nuances, such as input/output data types, sources, evaluation /computation operations, etc. Once the (general) algorithm is well-defined, you can begin coding in the given programming language. When writing code, the most time- and effort-consuming part is to carry out the verification operations in steps 3 to 5. This is called *debugging*, and it helps you locate errors or flaws in the calculation/evaluation and analysis operations from step 4.

Specifically, *debugging* is the process of correcting the syntax of the code, script, or program with respect to the programming language, correcting the calculation/ computation operations with respect to the given problem statement, and, if required, adjusting the precision of the output. Debugging is not so straightforward in many instances, but MATLAB's M/MLX file editors have a few helpful tools that detect general errors made in your scripts. At the same time, many errors cannot be detected without executing and analyzing the obtained outputs. There is no single solution approach to finding all possible bugs (flaws and mistakes) made when writing code. One of the most common ways to see if the code is performing as anticipated is to use test examples and then verify the output.

Throughout the book, considerable attention is given to debugging. It must be noted that it is impossible to write code without any flaws. Therefore, debugging is a "must-have" step. Moreover, you will hone your programming skills when you write code that solves different problems and make errors that you can then find and fix.

Note   Always start writing your scripts with simple operations/commands to perform the most essential parts of the exercise. Once the essential parts of the exercise are complete, you can add more details. It's good programming practice to move from the simple to the complex.

How to Create an M/MLX-File

There are a few ways to create a new script (in this case, an M/MLX-file).

1) By typing >> edit in the Command window and pressing the Enter key on the keyboard.

2) By hitting Ctrl+N on the keyboard.

3) By clicking New Script for the M-file.

4) By clicking New Live Script for the MLX-file.

5) By clicking New and selecting from the drop-down options Script   Ctrl+N for the M-file or Live Script for the MLX-file or fx Function for a function file with an *.m file extension or fx Live Function for a function file with an *.mlx file extension

6) By collecting the commands typed in the Command window via these commands: diary on, diary NewFileName, diary off

Warnings in Scripts

While writing M/MLX-files, including the function files, the M-file and MLX (Live)-file, editors automatically generate some warning signs that are in many instances very helpful hints to improve the efficiency of scripts and locate some missing or overlooked arguments. They do not prevent the scripts from being executed, though. These warnings are indicated by underlined wavy lines and hyphens ▭ and rectangular boxes

▣ on the right side of the M-file Editor window. They are by default orange. Similarly, the MLX-editor (Live Editor) indicates warnings by underlined wavy lines, as well as hyphens ▭ and triangular warning signs ⚠ on the right-side bar of the Editor window. Note that their color type can be adjusted via the Preferences settings. There are a few common warnings that are detected automatically by the M/MLX-file editors. They are as follows:

- To suppress the display of outputs in the Command window detected by the M-file editor only.

```
1    %% Warnings
2    A = 2
3    B = [A+1, 3]
4    C = A + B
5    ⚠ Terminate statement with semicolon to suppress output (within a script). [Details ▼] [Fix]
```

- To suggest memory allocation (e.g., the variables A and B are underlined with an orange wave line) when a variable size changes/ increases in the loop iteration within [for ... end] and [while ... end].

```
1    %% Warnings with memory allocation
2    for ii = 1:100
3        for jj = 1:200
4            A(ii, jj) = ii^2+jj;
5    ⚠ The variable 'A' appears to change size on every loop iteration (within a script). Consider preallocating for speed. [Details ▼]
6        end
7    end
```

```
1    %% Warnings with memory allocation
2    ii=1; jj=1;
3    while ii ~= 100
4        for jj=1:2000
5            B(ii, jj) = ii^2+jj;
6    ⚠ The variable 'B' appears to change size on every loop iteration (within a script). Consider preallocating for speed. [Details ▼]
7        end
8        ii=ii+1;
9    end
```

- To cancel the premature ending of the command with comma. For example, the comma after the grid on the command (on line 4) needs to be replaced with a semicolon.

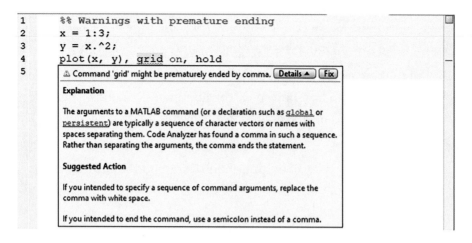

- To remove unnecessary semicolons. For example, at the end of line 1, the ; is unnecessary.

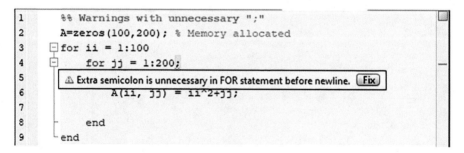

- To remove unnecessary semicolons in the [for ... end] loop's index declaration (on line 4).

```
1       %% Warnings with unnecessary ";"
2       A=zeros(100,200); % Memory allocated
3    □ for ii = 1:100
4    □     for jj = 1:200;
5          ⚠ Extra semicolon is unnecessary in FOR statement before newline.  Fix
6                A(ii, jj) = ii^2+jj;
7
8         end
9    └ end
```

- To indicate unused but assigned variable names (G on line 5) within M-files.

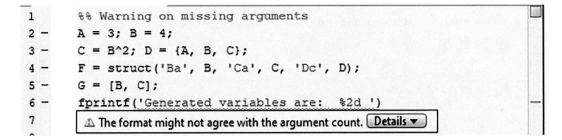

- To indicate missing arguments when formatting `fprintf()`.

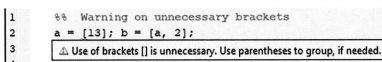

```
1   %% Warning on missing arguments
2 -  A = 3; B = 4;
3 -  C = B^2; D = {A, B, C};
4 -  F = struct('Ba', B, 'Ca', C, 'Dc', D);
5 -  G = [B, C];
6 -  fprintf('Generated variables are:   %2d ')
7     ⚠ The format might not agree with the argument count. [ Details ▼ ]
```

- To indicate unnecessary brackets (a = [13] on line 2).

```
1   %%  Warning on unnecessary brackets
2   a = [13]; b = [a, 2];
3     ⚠ Use of brackets [] is unnecessary. Use parentheses to group, if needed.
```

- To indicate an unrecommended function. For example, `xlswrite()` on line 2 is not recommended; instead, it is recommended to use `writematrix()` or `writecell()`. Note that the unrecommended function detecting warning option is available starting from the MATLAB R2022a version.

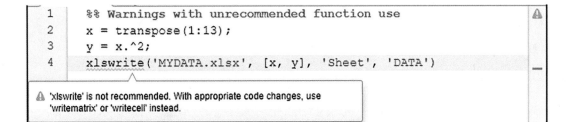

```
1   %% Warnings with unrecommended function use
2   x = transpose(1:13);
3   y = x.^2;
4   xlswrite('MYDATA.xlsx', [x, y], 'Sheet', 'DATA')

⚠ 'xlswrite' is not recommended. With appropriate code changes, use
  'writematrix' or 'writecell' instead.
```

It must be noted that some of the warning signs detected by the M-file editor will not be picked up by MLX-editor (Live Editor). For example, the missed ; used to suppress the display of the output in the Command window is not applicable to the MLX-editor (Live Editor).

On the other hand, other warnings such memory allocation warnings, unnecessary semicolons, missing arguments when formatting `fprintf( )d`, and unnecessary brackets are detected and highlighted very explicitly with the MLX-editor with an ⚠ icon on the right side of the Editor window.

- To advise the memory allocation.

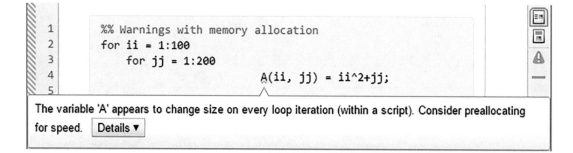

```
1    %% Warnings with memory allocation
2    for ii = 1:100
3        for jj = 1:200
4            A(ii, jj) = ii^2+jj;
5
```

The variable 'A' appears to change size on every loop iteration (within a script). Consider preallocating for speed. Details ▼

- To indicate the prematurely ended command with a comma.

```
1    %% Warnings with unrecommended function use
2    x = transpose(1:13);
3    y = x.^2;
4    plot(x,y), grid on, hold
```

⚠ Line 4: Command might be prematurely ended by comma. Details ▼   Fix

- To indicate unnecessary semicolons in the function statement.

```
1    function [A, B] = MY_fun(a, b);
2
3
4
5    A = a^2; B = b^2;
6    end
```

Extra semicolon is unnecessary in FUNCTION statement before newline.   Fix

- To indicate unnecessary semicolons in defining indexes for the [for ... end] loop.

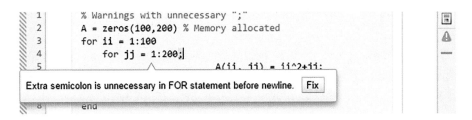

- To indicate a missing argument in fprintf().

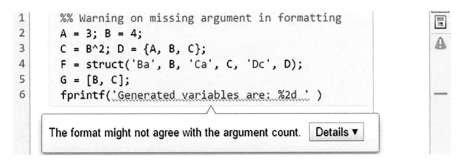

- To indicate unnecessary bracket.

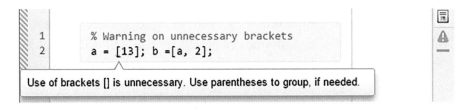

- To indicate an unrecommended function use. For example, xlswrite() on line 2 is not recommended. Instead, it is recommended to use writematrix() or writecell(). Note that the unrecommended function detecting warning option is available starting only from the MATLAB R2022a version.

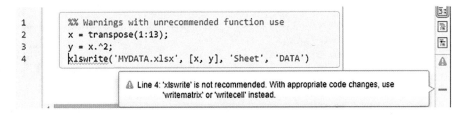

Errors in Scripts

In MATLAB, scripts can contain code to perform various computations and analyses and to define functions. Let's look at a few simple examples of how to write scripts in the M-file and MLX-file editors and see how to locate/fix common errors occurring while writing scripts.

Example 1

Let's solve a quadratic equation represented in a general form: $ax^2 + bx + c = 0$. First, open an M-file editor and type in the following commands:

```
a=input('Enter, a = ');
b=input('Enter, b = ');
c=input('Enter, c = ');
D=b^2-4*a*c;
disp(['Discriminant of the equation is: ' num2str(D)])
```

Once the file is saved with a file extension of *.m, then it can be executed. When it is executed, this code prompts the user for three input (input()) entries and then computes the discriminant of the quadratic equation with the user entries for a, b, and c and displays the result in the Command window. It should be noted that the command disp() on line 5 is optional and is used to display the computation result in the Command window with some comments. The command disp() does not make any changes in the output. There are two more computing steps left in this code, namely, computing the two roots of the quadratic equation. The remaining steps can be inserted after line 5. If there are some illegal operations/errors while writing the script, the M/MLX-file editor will automatically detect them and underline them with red waves.

```
1    a=input('Enter, a = ');
2    b=input('Enter, b = ');
3    c=input('Enter, c = ');
4    D=b^2-4*a*c;
5    disp(['Discriminant of the equation is: ' num2str(D)])
6    x1=(-b+sqrt(D))/2a;
7    x2=(-b-sqrt(D))/(2*a)
8
```

There is one error on line 6 (the red wavy line under a), where the multiplication sign is missing, and one warning shown with an orange wavy line under the = sign on line 7, where a semicolon (;) is recommended to suppress the display of the output from this line.

Error and warning messages like the ones on lines 6 and 7 with red and orange highlights are shown on the right edge of the editor's scroll bar. If there is a red wavy line showing errors, the script (M/MLX-file) cannot be executed. If there are any warning signs with orange wavy underlines, that script can be executed without a problem. A few different types of typos or illegal operations are detected automatically by the M/MLX editors, but they cannot be fixed automatically. Thus, you have to understand and work out such issues.

```
1    a=input('Enter, a = ');
2    b=input('Enter, b = ');
3    c=input('Enter, c = ');
4    D=b^2-4*a*c;
5    disp(['Discriminant of the equation is: ' num2str(D)])
6    x1=(-b+sqrt(D))/2a;
7    ❂ Parse error at a: usage might be invalid MATLAB syntax.
8
```

The warning message on line 7 can be fixed either by putting a semicolon where the cursor is or by clicking the Fix button.

```
1    a=input('Enter, a = ');
2    b=input('Enter, b = ');
3    c=input('Enter, c = ');
4    D=b^2-4*a*c;
5    disp(['Discriminant of the equation is: ' num2str(D)])
6    x1=(-b+sqrt(D))/2a;
7    x2=(-b-sqrt(D))/(2*a)|
8    ⚠ Terminate statement with semicolon to suppress output (within a script).  Details ▼   Fix
```

Finally, here is the fixed script. The green square in the right corner indicates that the syntax of all the typed-in commands are correct, and the script is ready for execution.

```
1    a=input('Enter, a = ');
2    b=input('Enter, b = ');
3    c=input('Enter, c = ');
4    D=b^2-4*a*c;
5    disp(['Discriminant of the equation is: ' num2str(D)])
6    x1=(-b+sqrt(D))/(2*a);
7    x2=(-b-sqrt(D))/(2*a);
8
```

C

Now after saving this script with a file name of Eqn.m, it can be executed by clicking

the ▷ **Run** button on the Editor's main menu, by pressing Ctrl+Enter on the keyboard, or by calling the script by its name (>> run('Eqn')) from the Command window directly. Another way of executing the code is pressing the F5 functional key on the keyboard. After executing it, it prompts for the input in the Command window. You would enter the values **1**, **2**, and **3** for a, b, and c, respectively.

```
Enter, a = 1
Enter, b = 2
Enter, c = 3
Discriminant of the equation is: -8
```

Then the whole computation is completed, and this is what is obtained in the Command window. There are also some other variables saved in the Workspace window that are shown.

Workspace	
Name ▲	Value
a	1
b	2
c	3
D	-8
x1	-1.0000 + 1.4142i
x2	-1.0000 - 1.4142i

All of the entries and processed/computed outputs from the scripts and M/MLX-files are saved automatically in the workspace. Except for when the function files are executed, not all results are saved in the workspace apart from the specified output variables in the function file. This issue will be addressed in the Function Files section of this chapter.

Example 2

Compute the following expression by writing an M-file (script): $ab\sqrt[3]{\dfrac{(ax-by)^2}{cd-f}}$, where $x=\left[\dfrac{1}{5},\dfrac{2}{5},\dfrac{3}{5},\dfrac{4}{5}\right]$, $y=\left[\dfrac{1}{3},\dfrac{2}{3},1,\dfrac{4}{3}\right]$, $c=2, d=2.5,$ and $f=2$ and the values a, b with the user entries (scalars). According to the given values of the variables, x and y are row arrays and the other variables (a, b, c, d, f) are scalars.

Like with the quadratic equation, the input prompts are included in the script.

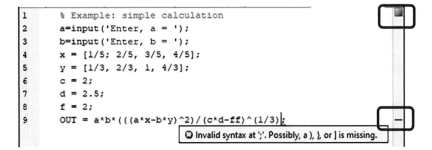

This is the created short script to compute the given assignment. By taking a quick look at this script, you can see that there is one error detected by the MATLAB editor. It is invalid syntax related to a missing parenthesis at the end of line 9 that is true, and in fact, the missing parenthesis is before the first power raise (^) sign. Here is the fixed code with the green square in the upper-right corner:

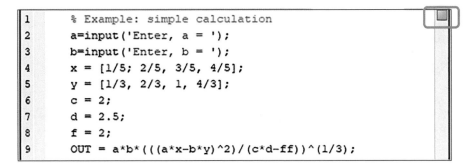

Now everything appears to be correct according to the editor syntax. However, there are still several errors.

- On line 4 while defining the elements of x, a semicolon is used as an element separator that must be a comma or just a space.

- The variable f is defined on line 8, but in the expression on line 9 ff is used instead.

- The computation expression on line 9 is performed with the variables a, b, c, d, and f, which are scalars, and the variables x and y, which are row arrays. That is not correct. This line has to contain element-wise operations over the row array variables x and y. After fixing these errors, the script will be in the following form:

```
1       % Example: simple calculation
2       a=input('Enter, a = ');
3       b=input('Enter, b = ');
4       x = [1/5, 2/5, 3/5, 4/5];
5       y = [1/3, 2/3, 1, 4/3];
6       c = 2;
7       d = 2.5;
8       f = 2;
9       OUT = a*b*(((a*x-b*y).^2)/(c*d-f)).^(1/3);
```

Now the script can be executed by pressing Ctrl+Enter or F5 on the keyboard without

saving it or by clicking ▷ Run after saving it. Here are the results (input entries from the Command window):

```
Enter, a = 1
Enter, b = 2R
```

Here are the variables in the workspace:

Workspace	
Name ▲	Value
a	1
b	2
c	2
d	2.5000
f	2
OUT	[0.8343,1.3244,1.7354,2.1023]
x	[0.2000,0.4000,0.6000,0.8000]
y	[0.3333,0.6667,1,1.3333]

Example 3

Let's compute the mathematical expression's values and plot them by using the MLX-file editor and creating the MLX-file. Given: $H(t) = sinsin\,(\omega t)$; $\omega = 3$; $t = [0^0, 450^0]$ with $\Delta t = 1^0$.

Note that the given argument values of t are in degrees, not in radians. In MATLAB there are two functions, namely, $sin()$ and $sind()$, to compute sine function values with input arguments in radians and degrees, respectively. Therefore, in such cases, there are two approaches: one should always use the right MATLAB function or one should employ the conversion function from degrees to radians (deg2rad()) or vice versa (rad2deg()).

Now the MLX-file (script) is created in a live editor by considering the input argument *t*:

```
t = 0:1:450
omega = 3
H = sind(omega)
 ⑦ sind(X)
```

Angle in degrees	
⊞ omega	
🔓 OPENmlx	
𝑓𝑥 orderfields	Order fields of structure array
𝑓𝑥 ones	Create array of all ones
𝑓𝑥 ode113	Solve nonstiff differential equations ...
𝑓𝑥 ode15i	Solve fully implicit differential equatio...
𝑓𝑥 ode15s	Solve stiff differential equations and ...

```
1        t = 0;1:450
2        omega = 3
3        H = sind(omega,*t)
4
```

Parse error at '*': usage might be invalid MATLAB syntax.

Note that the error is shown by the wavy underline, and ❗ and ▬ in red on the right side of the MLX-editor. After fixing the error (the comma before *), the corrected code will be in the following form:

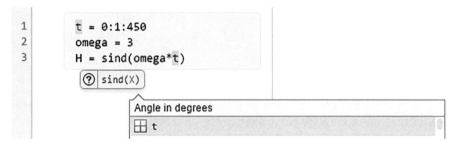

```
1        t = 0:1:450
2        omega = 3
3        H = sind(omega*t)
           ⑦ sind(X)
```

Angle in degrees
⊞ t

There are several essential differences between the M-file editor and MLX-file editor. One of them is that the MLX-editor can detect automatically the previously defined variables after typing the first letter of the variable name as an input argument with an additional hint showing the argument type to be inserted (e.g., *t* is the angle in degrees).

Another feature of the MLX-file editor is that it does not show a warning message if the semicolon is not placed at the end of commands assigning variable values. For example, $t = 0{:}1{:}450$, which we saw while working with the M-file editor. The MLX-file editor displays the outputs not in the Command window but in the MLX-editor's right-side window. At the same time, all of the variable values and computed expression values are saved in the workspace after executing the MLX-files just like with the M-files. The execution of the MLX-files is similar to M-files that can be done by pressing Ctrl+Enter on the keyboard or by clicking the Run button. Here is the final script with the plot() command and its computed results:

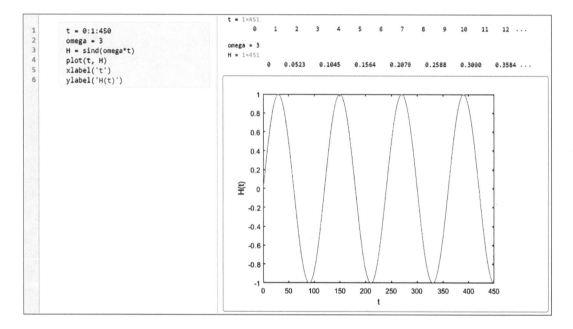

Example 4

Write an MLX-file that analyzes and computes the following acceleration equation $a(v)$ (acceleration as a function of velocity) of a skydiver:

$$a(v) = g\left(1 - \frac{v^2}{3600}\right)$$

$$g = 9.81\,\frac{m}{s^2}$$

1. Plot *a* versus *v*.
2. Compute the terminal speed of a skydiver when $a(v) \cong 0 \dfrac{m}{s^2}$.
3. Take the vertical velocity *v* to be independent variable with the

   step size of $\Delta v = 0.1 \dfrac{m}{s}$ within $\left[0, \ 100 \dfrac{m}{s} \right]$.
4. Display the terminal speed value on the plot.

Here is the initial code with one error on line 8 because of a mistyped variable name (*Ttt* instead of *Tt*), indicated by the exclamation point ⬤ icon. This type of error (a mistyped variable name) can be detected in the live editor only after executing the script, not while writing it.

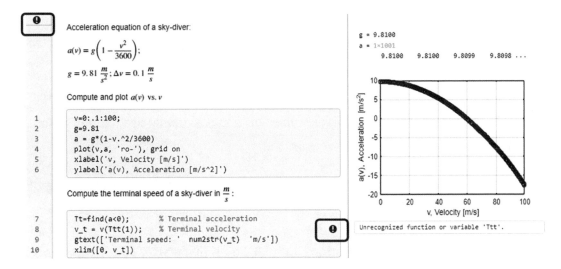

Acceleration equation of a sky-diver:

$$a(v) = g\left(1 - \frac{v^2}{3600}\right);$$

$$g = 9.81 \frac{m}{s^2}; \Delta v = 0.1 \frac{m}{s}$$

Compute and plot $a(v)$ vs. v

```
1   v=0:.1:100;
2   g=9.81
3   a = g*(1-v.^2/3600)
4   plot(v,a, 'ro-'), grid on
5   xlabel('v, Velocity [m/s]')
6   ylabel('a(v), Acceleration [m/s^2]')
```

Compute the terminal speed of a sky-diver in $\frac{m}{s}$:

```
7   Tt=find(a<0);          % Terminal acceleration
8   v_t = v(Ttt(1));       % Terminal velocity
9   gtext(['Terminal speed: '  num2str(v_t)  'm/s'])
10  xlim([0, v_t])
```

Unrecognized function or variable 'Ttt'.

Note that the error with the mistyped variable name (*Ttt* instead of *Tt*) showed up after executing the MLX-file editor. The results of calculations from the MLX-editor, unlike the M-file editor, do not show up in the Command window. Therefore, it is a good idea to execute the live script frequently to catch any overlooked rules and errors while typing and editing the scripts in the MLX-editor.

Here are the corrected and final solution scripts for this exercise in the MLX-editor:

Acceleration equation of a sky-diver:

$$a(v) = g\left(1 - \frac{v^2}{3600}\right);$$

$$g = 9.81 \ \frac{m}{s^2}; \Delta v = 0.1 \ \frac{m}{s}$$

Compute and plot $a(v)$ vs. v

```
1    v=0:.1:100;
2    g=9.81
3    a = g*(1-v.^2/3600)
4    plot(v,a, 'ro-'), grid on
5    xlabel('v, Velocity [m/s]')
6    ylabel('a(v), Acceleration [m/s^2]')
```

g = 9.8100

a = 1×1001

   9.8100   9.8100   9.8099   9.8098 ...

Compute the terminal speed of a sky-diver in $\frac{m}{s}$:

```
7    Tt=find(a<0);        % Terminal acceleration
8    v_t = v(Tt(1));      % Terminal velocity
9    gtext(['Terminal speed: '  num2str(v_t)  'm/s'])
10   xlim([0, v_t])
```

The outputs are displayed within the same window on the right side, including the plot figures. In this exercise, there are a few lines of (nonexecutable) text; comments and

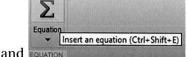

equations are added between executable commands by using the

and EQUATION equation editor GUI tools of the MLX-editor, which are not available in M-file editor.

Example 5

It is possible to call within an M-file another M/MLX file or files. Let's look at an example of writing M- and MLX files to compute the following expressions. You'll also obtain their computation results by calling/executing another M/MLX-file. This will demonstrate how you link/connect several scripts (M/MLX files) or, in other words, how you execute several scripts, obtain their simulation results, and use them within a single script.

1) Compute $F(\alpha) = e^{\sin(\alpha)}$; $H(\beta) = e^{\cos(\beta)}$; $S = 1 - (\sin^2\alpha + \cos^2\beta)$; for $\alpha = -2\pi...2\pi$; $\beta = -360^0...360^0$ by writing an MLX-file.

2) Compute $(at) - \sin \sin (bt)$, for $a = 3$, $b = 2$, $-13 \leq t \leq -3$ with $\Delta t = \pi/50$ by writing an M-file.

3) Compute $f(t) = \cos \cos (20t) - \sin (10t)$, for $t \in [-\pi, \pi]$ with $\Delta t = \pi/50$, and plot the computation results t versus $f(t)$ by writing an M-file.

4) Compute $$T(s) = \frac{13e^{-2s}}{s^2 + \delta \omega_n s + \omega_n^2} \text{ for } \delta = [0, \; 0.5, \; 1, \; 1.5, 2];$$

$\omega_n = 3\left[\dfrac{rad}{\sec}\right]; s = [0,25], \Delta s = 5*10^{-2}$ by writing MLX-file.

5) Write an M-file that executes all of the M and MLX-files from the previous tasks and saves all the computation results in a single array called A1 (for step 1), a table array called Btab2 (for step 2), a cell array called Ccell3 (for step 3), and a structure array called Dstruct4 (for step 4).

Here are the solution scripts in the M/MLX-files of these five tasks. This MLX-file, called ET1.mlx, is the solution of task 1:

(1) Compute (ET1.mlx):
$F(\alpha) = e^{\sin(\alpha)}; H(\beta) = e^{\cos(\beta)}; S = 1 - (\sin^2 \alpha + \cos^2 \beta); \alpha = -2\pi \ldots 2\pi ; \beta = -360^0 \ldots 360^0$

```
1   alpha=linspace(-2*pi, 2*pi, 360); beta = linspace(-360, 360, 360);
2   F=exp(sin(alpha)); H=exp(cosd(beta)); S = 1-(sin(alpha)^2+cosd(beta)^2);
```

At initial glance, this script (ET1.mlx) looks error-free, and the MLX editor does not show any problematic issues. However, there are two errors (in array power operations) on line 2 in computing S, which will be detected by the editor only after you execute the script.

(1) Compute (ET1.mlx):
$F(\alpha) = e^{\sin(\alpha)}; H(\beta) = e^{\cos(\beta)}; S = 1 - (\sin^2 \alpha + \cos^2 \beta); \alpha = -2\pi \ldots 2\pi ; \beta = -360^0 \ldots 360^0$

```
1   alpha=linspace(-2*pi, 2*pi, 360); beta = linspace(-360, 360, 360);
2   F=exp(sin(alpha)); H=exp(cosd(beta)); S = 1-(sin(alpha)^2+cosd(beta)^2);
```

```
Error using  ^
Incorrect dimensions for raising a matrix to a power. Check that the matrix is square and the power is a scalar. To perform
powers, use '.^'.
```

Here is the corrected script:

(1) Compute (ET1.mlx):

$$F(\alpha) = e^{\sin(\alpha)}; H(\beta) = e^{\cos(\beta)}; S = 1 - (\sin^2\alpha + \cos^2\beta); \alpha = -2\pi \ldots 2\pi; \beta = -360^0 \ldots 360^0$$

```
alpha=linspace(-2*pi, 2*pi, 360); beta = linspace(-360, 360, 360);
F=exp(sin(alpha)); H=exp(cosd(beta)); S = 1-(sin(alpha).^2+cosd(beta).^2);|
```

Here is the initial version of the script that solves task 2:

```
1        %% ET2.m
2        % Task 2. Compute the values of g(a, b, t)
3 -      t=-3:pi/50;13;
4 -      a = input(' Enter the value of a  = ');
5 -      b = input(' Enter the value of b  = ');
6 -      g=cos(a*t)-sin(b*t);
```

A quick glance at this code shows that it is correctly typed and ready to execute. That can be verified with the green box ◻ on the right side of the editor's window. Even with the execution of the script, no error will show up in the Command window. However, checking the obtained results of *g* shows that many anticipated data points are missing in the results. The error is on line 3. The semicolon is typed in instead of the colon operator to create an array of *t*. This is an implicit error or bug in the code that cannot be detected by the M-file editor. Here is the corrected code:

```
1        %% ET2.m
2        % Task 2. Compute the values of g(a, b, t)
3 -      t=-3:pi/50:13;
4 -      a = input(' Enter the value of a  = ');
5 -      b = input(' Enter the value of b  = ');
6 -      g=cos(a*t)-sin(b*t);
```

Here is the initial version of the answer script for task 3:

```
1       %% ET3.m
2       % Task 3. Compute the values of f(t)
3 -     t=-pi:pi/200:pi;
4 -     f=cos(20*t)-sin(10*t);
5       plot(t; f, 'b-'), title('Results of Task (3)'), grid on
6       ⊗ Parse error at ')': usage might be invalid MATLAB syntax.
```

In this script, the editor detects two errors and one warning sign on line 5. The script contents can be executed up to line 5, and the following error message is displayed in the Command window:

```
Error: File: ET3.m Line: 5 Column: 7
Invalid expression. When calling a function or indexing a variable, use
parentheses. Otherwise, check for mismatched delimiters.
```

Actually, there was one error: a misplaced semicolon instead of a comma within the `plot()` command on line 5: `plot( t; f, 'b-')`. There are two common functionalities of the semicolon for MATLAB, one of which is the termination of display results in the Command window and the end of row elements of an array. Therefore, in this example, the semicolon inside `plot()` is misplaced. Here is the corrected script:

```
1       %% ET3.m
2       % Task 3. Compute the values of f(t)
3 -     t=-pi:pi/50:pi;
4 -     f=cos(20*t)-sin(10*t);
5 -     plot(t,f, 'bo-'), title('Results of Task (3)'), grid on
```

Here is the initial solution script of task 4:

(4) Compute (ET4.mlx):

$$T(s) = 13\frac{e^{-2s}}{s^2 + \delta\,\omega_n s + \omega_n^2}; \delta = [0, 0.5, 1, 1.5, 2]; \omega_n = 3; s = [0, 25]; \triangle s = 5 * 10^{-2}$$

```
1       delta = [0, 0.5, 1, 1.5; 2]; s=0:5e-2:25; omegaN=3;
2       T(:,1)=13*exp(-2*s)./(s.^+delta(1)*s*omegaN+omegaN^2);
3       T(:    All matrix rows must be the same length.   Details ▼    egaN^2);
4       T(:                                                            egaN^2);
5       T(:,4)=13*exp(-2*s)./(s.^2+delta(4)*s*omegaN+omegaN^2);
6       T(:,5)=13*exp(-2*s)./(s.^2+delta(5)*s*omegaN+omegaN^2);
```

It automatically detected an error (the mistyped semicolon instead of a comma) as an element separator of the array. With this error, the script cannot be executed.

Here is the corrected version of the script:

(4) Compute (ET4.mlx):

$$T(s) = 13 \frac{e^{-2s}}{s^2 + \delta\omega_n s + \omega_n^2}; \delta = [0, 0.5, 1, 1.5, 2]; \omega_n = 3; s = [0, 25]; \Delta s = 5 * 10^{-2}$$

```
delta = [0, 0.5, 1, 1.5, 2]; s=0:5e-2:25; omegaN=3;
T(:,1)=13*exp(-2*s)./(s.^2+delta(1)*s*omegaN+omegaN^2);
T(:,2)=13*exp(-2*s)./(s.^2+delta(2)*s*omegaN+omegaN^2);
T(:,3)=13*exp(-2*s)./(s.^2+delta(3)*s*omegaN+omegaN^2);
T(:,4)=13*exp(-2*s)./(s.^2+delta(4)*s*omegaN+omegaN^2);
T(:,5)=13*exp(-2*s)./(s.^2+delta(5)*s*omegaN+omegaN^2);|
```

Here is the script of task 5. It calls the scripts in tasks 1 to 4.

```
1     %% ET5.m
2     % Task 5. Execute all M/MLX files (ET1.mlx, ET2.m, ET3.m, ET4.mlx)
3     % and save the results
4  -  clearvars; close all
5     % Step 1. Call and execute the MLX file called: ET1.mlx
6  -  run('ET1.mlx');
7     % Collect all computation results
8  -  A1 = [alpha'; F'; beta'; H'; S'];
9
10    % Step 2. Call and execute the M file called: ET2.m
11 -  run('ET2.mlx');
12 -  Btab2 = table(t',g','variablenames', {'t', 'g'});
13
14    % Step 3. Call and execute the M-file called: ET3.m
15 -  run('ET3.m');
16 -  Ccell3{1,1} = t; Ccell3{1,2} = f;
17
18    % Step 4. Call and execute the MLX-file called: ET4.m
19 -  run('ET4.mlx');
20 -  Dstruct4.In = s; Dstruct4.Fun = T; Dstruct4.delta = delta;
21 -  Dstruct4.omega=omegaN;
22 -  whos   % Show all of the computed outcomes in the command window|
```

The script editor does not show any problems or issues. However, when it is executed, the following error message pops up in the Command window:

```
Error using run (line 87)
RUN cannot execute the file 'ET2.mlx'. RUN requires a valid MATLAB script
Error in ET5 (line 11)
run('ET2.mlx');
```

This error message indicates that there is a problem recognizing the file. That means the file name ET2.mlx is not the correct file name (the file extension is wrong); it has to be ET2.m instead.

Here is the corrected script of task 5:

```
1    %% ET5.m
2    % Task 5. Execute all M/MLX files (ET1.mlx, ET2.m, ET3.m, ET4.mlx)
3    % and save the results
4  - clearvars; close all
5    % Step 1. Call and execute the MLX file called: ET1.mlx
6  - run('ET1.mlx');
7    % Collect all computation results
8  - A1 = [alpha'; F'; beta'; H'; S'];
9
10   % Step 2. Call and execute the M file called: ET2.m
11 - run('ET2.m');
12 - Btab2 = table(t',g','variablenames', {'t', 'g'});
13
14   % Step 3. Call and execute the M-file called: ET3.m
15 - run('ET3.m');
16 - Ccell3{1,1} = t; Ccell3{1,2} = f;
17
18   % Step 4. Call and execute the MLX-file called: ET4.m
19 - run('ET4.mlx');
20 - Dstruct4.In = s; Dstruct4.Fun = T; Dstruct4.delta = delta;
21 - Dstruct4.omega=omegaN;
22 - whos   % Show all of the computed outcomes in the command window
```

By executing the last file, ET5.m, all other four scripts are also called and executed consecutively. Here are the computed outcomes from all scripts in the Command window:

```
Enter the value of a  =  2
 Enter the value of b  =  3
```

Name	Size	Bytes	Class	Attributes
A1	1800x1	14400	double	
Btab2	255x2	5162	table	
Ccell3	1x2	1840	cell	
Dstruct4	1x1	24800	struct	
F	1x360	2880	double	
H	1x360	2880	double	
S	1x360	2880	double	
T	501x5	20040	double	

a	1x1	8	double
alpha	1x360	2880	double
b	1x1	8	double
beta	1x360	2880	double
delta	1x5	40	double
f	1x101	808	double
g	1x255	2040	double
omegaN	1x1	8	double
s	1x501	4008	double
t	1x101	808	double

Also, there is a plot figure (not shown here) from task 3 (ET3.m).

Note that there are many other common errors made while writing scripts that will be highlighted throughout the book. Moreover, there are a few common mistakes made while creating the function files that will be highlighted in the section dedicated to the function files.

Via these simple examples, you have seen how scripts, such as M-files and MLX-files, and the tools of M and MLX (Live) editors can be employed while writing scripts to detect common errors automatically. There are some other errors that are not detected by the M/MLX editors automatically that can be found only after executing the scripts. These include various operations (arithmetic, matrix, and array) that are performed with mismatched sizes of variables or improperly spelled MATLAB commands.

Note   Finding and fixing the errors you make while writing code is a good exercise for learning how to write great programs.

Cell Mode

In the M/MLX editors, the cell mode option is a handy tool to write well-structured code/scripts. Working in cell mode is simple and can be accomplished by typing %% and leaving one cursor space. This creates a new cell by default. Writing long scripts separated in cells for every separate operation helps you execute your code cell by cell and detect where the bugs/flaws are. It also helps you visually separate the code into distinct blocks that can be highlighted one at a time. This helps with the readability of the code while editing it. Let's consider the following example.

The equation for an ellipse centered at the origin of the Cartesian coordinates (x, y) is $\dfrac{x^2}{a^2}+\dfrac{y^2}{b^2}=1\,(A)$, where a and b are constants that determine the shape of the ellipse. The variables x, y are defined by $x(u)=\dfrac{a\left(1-u^2\right)}{u^2+1}, y(u)=\dfrac{2bu}{u^2+1}$, and $u=tg\left(\dfrac{t}{2}\right)$ where $0 \le t \le 2\pi$.

1) Compute the ellipse given in equation in (A) for $a = 1.5, b=3.5$ and plot it.

2) Plot and display the points of intersection of the two ellipses described by $x^2+\dfrac{y^2}{4}=1$ and $\dfrac{x^2}{25}+y^2=1$.

3) Compute three ellipses defined in (A) as a three-column array for these cases: $a = [1, 2, 3], b=[3.5, 1, 2]$.

Here is the solution script of this exercise created in three cell modes representing answers for each subsection of the exercise. Each cell contains an answer script for each part of the given exercise. Cell 1 (part 1), composed of lines 2–9, computes the ellipse in (A) with respect to the values of a and b. Cell 2 (part 2), composed of lines 10–27, computes two ellipses and plots their values and, subsequently, displays intersecting points of the two ellipses. Cell 3 (part 3) computes three ellipses with respect to the values of a and b and plots the computed ellipses.

```
1       % Cell Mode based code writing
2       %% Part 1. Compute the Ellipse with a = 1.5 and b =3.5, and plot it
3 -     clearvars, close all
4 -     t = 0:pi/100:2*pi;
5 -     a=1.5; b=3.5;
6 -     u = tan(t/2);
7 -     x = a*(1 - u.^2)./(u.^2+1);
8 -     y = 2*b*u./(u.^2+1);
9 -     plot(x, y, 'bo-'), grid on
10      %% Part 2. Compute the Ellipses with a = [1, 5], b =[2, 1], and plot them
11 -    clearvars, close all
12 -    a = [1, 5]; b = [2, 1];
13 -    t = 0:pi/100:2*pi;
14 -    u = tan(t/2);
15      % Ellipse 1
16 -    x1 = a(1)*(1 - u.^2)./(u.^2+1);
17 -    y1 = 2*b(1)*u./(u.^2+1);
18      % Ellipse 2
19 -    x2 = a(2)*(1 - u.^2)./(u.^2+1);
20 -    y2 = 2*b(2)*u./(u.^2+1);
21 -    plot(x1,y1, 'b-', 'linewidth', 2), hold on
22 -    plot(x2, y2, 'r--', 'linewidth', 2.5), grid on
23 -    legend('Ellipse 1', 'Ellipse 2')
24 -    [px, py] = ginput(4);   % Picking the intersection points
25 -    gtext(['Point of intersection: (' num2str(px(1)) ',' num2str(py(1)) ')'])
26 -    gtext(['Point of intersection: (' num2str(px(2)) ',' num2str(py(2)) ')'])
27 -    gtext(['Point of intersection: (' num2str(px(3)) ',' num2str(py(3)) ')'])
28 -    gtext(['Point of intersection: (' num2str(px(4)) ',' num2str(py(4)) ')'])
29      %% Part 3. Compute the Ellipses with a = [1, 2, 3], b =[3.5, 1, 2], and plot them
30 -    clearvars, close all
31 -    a = [1, 2, 3]; b = [3.5, 1, 2];
32 -    t = 0:pi/100:2*pi;
33 -    u = tan(t/2);
34      % Ellipse 1, 2, 3
35 -    x = [a(1)*(1 - u.^2)./(u.^2+1); a(2)*(1 - u.^2)./(u.^2+1); ...
36          a(3)*(1 - u.^2)./(u.^2+1)]';
37 -    y = [2*b(1)*u./(u.^2+1); 2*b(2)*u./(u.^2+1); 2*b(3)*u./(u.^2+1)]';
38 -    plot(x(:,1),y(:,1), 'r-', x(:,2),y(:,2), 'g--',x(:,3),y(:,3), ...
39 -        'b-.', 'linewidth', 2), grid on
40 -    legend('Ellipse 1', 'Ellipse 2',  'Ellipse 3')
```

After writing the code in cell mode, you can execute each cell separately by using the Ctrl+Enter keys on the keyboard when the cursor is within the cell that is meant to be

executed or using the  Run Section and Advance and Run and Advance tools of the M-file editor. Note that an alternative way of executing the selected part of a code is to hit the F9 key on the keyboard. The advantages of using cell mode is that you can execute and test each cell separately. If there are any errors, you can fix them without any interference with other cells. You can save considerable time and effort when you debug and correct long scripts.

Note   When you are writing long, extended scripts and debugging them, it is more efficient to use cell mode.

As noted, the M/MLX-file editors have many useful tools. For instance, they automatically detect common typographical errors. They detect and warn a user about the following:

- Mistyped MATLAB built-in function/command names

- Unclosed mathematical expressions with brackets, namely, (), {}, [ ]

- Misused/improperly used or missed mathematical operators, namely, +-*/\, ..., ( ), {}, [ ]

- Invalid syntaxes

- Unused variable names

- Suppress display of (lengthy) outputs

- Which entries are numeric data, which entries are characters or text messages, and which entries are comments

- Warning messages that let you know how to improve computation efficiency of a script

In addition, the M/MLX-file editors have the following tools:

- Debugging modes, such as set/clear breakpoints, set/modify conditional breakpoints, enable/disable breakpoints, stop if errors/ warnings

- Working in a cell mode

- Text editing options under text

- Report writing by save and publish options

Debugging Mode

The dbstop function temporarily halts the execution of a script and provides the user with an opportunity to examine the local workspace. While debugging a script, you can set the breakpoints at specific lines in the Editor window by clicking that line in the left gutter. There are more than a dozen forms of dbstop function uses, as shown here:

```
(1)  dbstop in FILE at LINENO
(2)  dbstop in FILE at LINENO@
(3)  dbstop in FILE at LINENO@N
(4)  dbstop in FILE at SUBFUN
(5)  dbstop in FILE
(6)  dbstop in FILE at LINENO if EXPRESSION
(7)  dbstop in FILE at LINENO@ if EXPRESSION
(8)  dbstop in FILE at LINENO@N if EXPRESSION
(9)  dbstop in FILE at SUBFUN if EXPRESSION
(10) dbstop in FILE if EXPRESSION
(11) dbstop if error
(12) dbstop if caught error
(13) dbstop if warning
(14) dbstop if naninf  or  dbstop if infnan
(15) dbstop if error IDENTIFIER
(16) dbstop if caught error IDENTIFIER
(17) dbstop if warning IDENTIFIER
```

The name FILE is the file name in which you want to perform the debugging operations. It has to be specified as a character vector of the string scalar. FILE can also include a full or partial path to the file directory. LINENO is a line number within FILE (the script in which the debugging operation is being performed), and N is an integer specifying the Nth anonymous function on the line. SUBFUN is the name of a subfunction within FILE. EXPRESSION is an executable conditional expression, specified as a character vector or string scalar. IDENTIFIER is a MATLAB message identifier (see the help for ERROR for a description of message identifiers). The AT and IN keywords are optional.

M-Lint Code Check

The command `mlin()` is used to check MATLAB code files for possible problems. When you run an M-Lint code check from the M-file Editor window, by choosing Tools ➤ M-Lint ➤ Show M-Lint Report, it will prompt you with a full report (on your M-file) of warning messages. For instance, it will check which variable is unused, which variable's size changes on every loop iteration, what the opportunities are for improvement, which one of the employed built-in functions of MATLAB is deprecated, and so forth. The general command syntax of `mlint()` is as follows: >>mlint('filename'). Starting from MATLAB 2019a, `mlint()` is not recommended, and the `checkcode()` command is recommended instead.

Let's consider the following example to demonstrate how to use the M-Lint code check tool:

Given $\{2x_1 - 3x_2 + x_3 = 5\ 4x_1 + 2x_2 - 2x_3 = 3\ 6x_1 + 3x_2 + 2x_3 = 11$, write the code to solve the given system of linear equations, and import the computed solutions along with the given data into an external Microsoft Excel file.

Here is the solution script:

```
1    % Code_Check.m. Code checking with mlint() and checkcode()
2    % Given: [A]*{x} = [b] system
3    % Solve for {x}
4    A = [2 -3  1; 4 2 -2; 6 3 2];
5    b = [5; 3; 11];
6    x = inv(A)*b;
7    xlswrite('Solution_Data.xlsx', [A, x, b])
```

The code (`Code_Check.m`) contains two warnings automatically detected by the M-file editor. They are related to the `inv()` and `xlswrite()` functions.

When you call this created code (`Code_Check.m`) using `mlint()` and `checkcode()`, you will get the following analysis report:

```
>> mlint('Code_Check.m')
L 6 (C 5-7): INV(A)*b can be slower and less accurate than A\b. Consider
using A\b for INV(A)*b or b/A for b*INV(A).
L 7 (C 1-8): 'xlswrite' is not recommended. With appropriate code changes,
use 'writematrix' or 'writecell' instead.

>> checkcode('Code_Check.m')
```

L 6 (C 5-7): INV(A)*b can be slower and less accurate than A\b. Consider using A\b for INV(A)*b or b/A for b*INV(A).
L 7 (C 1-8): 'xlswrite' is not recommended. With appropriate code changes, use 'writematrix' or 'writecell' instead.

The analysis report also has recommendations. If you follow the recommendations and make the respective changes in the code (Code_Check.m), it looks like this:

```
1      % Code_Check.m. Code checking with mlint() and checkcode()
2      % Given: [A]*{x} = [b] system
3      % Solve for {x}
4      A = [2 -3  1; 4 2 -2; 6 3 2];
5      b = [5; 3; 11];
6      x = A\b;
7      writecell('Solution_Data.xlsx', [A, x, b])
```

Now, the M-file editor shows no more warnings. Let's run code checking again with the updated code (Code_Check.m).

```
>> mlint('Code_Check.m')
>> checkcode('Code_Check.m')
```

The code checking produces no recommendations, which means the updated code is perfected.

Code Profiling

Code profiling measures where a program spends time and where the problems/errors are. By identifying the performance of your program, you can improve it. Code profiling can be started from the M-file Editor window by clicking the icon. The Code Profiling (Profiler) window pops up. The profile report contains the whole benchmarking report of your script (M-file) including extensive information on each command and operation, including how much the CPU spends obtaining the results. By studying a profile report of your script, you can learn how to improve simulation time efficiency. The profile mode is off by default and can be also switched on by using the >> profile on command. The profile summary report of any M-file or Simulink model can be generated and viewed with general syntaxes, for example:

```
1     % Test.m. Code Profile testing                        ✓
2     profile on
3     % Script - M-file starts here:
4     clc; close all; clearvars
5     M = magic(25);
6     Msq = zeros(size(M));
7 ⊟   for ii = 1:size(M, 1)
8 ⊟       for jj = 1:size(M,2)
9                Msq(ii,jj)=M(ii,jj)^2;
10 ⊦       end
11 ⊦   end
12    % Script - M-file ends here:
13    profile viewer
```

Here's another example:

```
>> profile on; Test; profile viewer
```

Here's an example for Simulink models:

```
>> profile on; sim('My_Model'); profile viewer
```

Note   You can insert any M-file or MLX-file name (e.g., My_function.m, My_Code.mlx) or function file name (e.g., My_function.m) or Simulink model name (e.g., My_Model.mdl, Model_Sim.slx) between the two commands profile on; and profile viewer. Here's an example: >> profile on; My_Code; profile viewer.

After running the code (Test.m), the comprehensive profile report of the code (Test.m) will pop up, as shown in Figure 2-2.

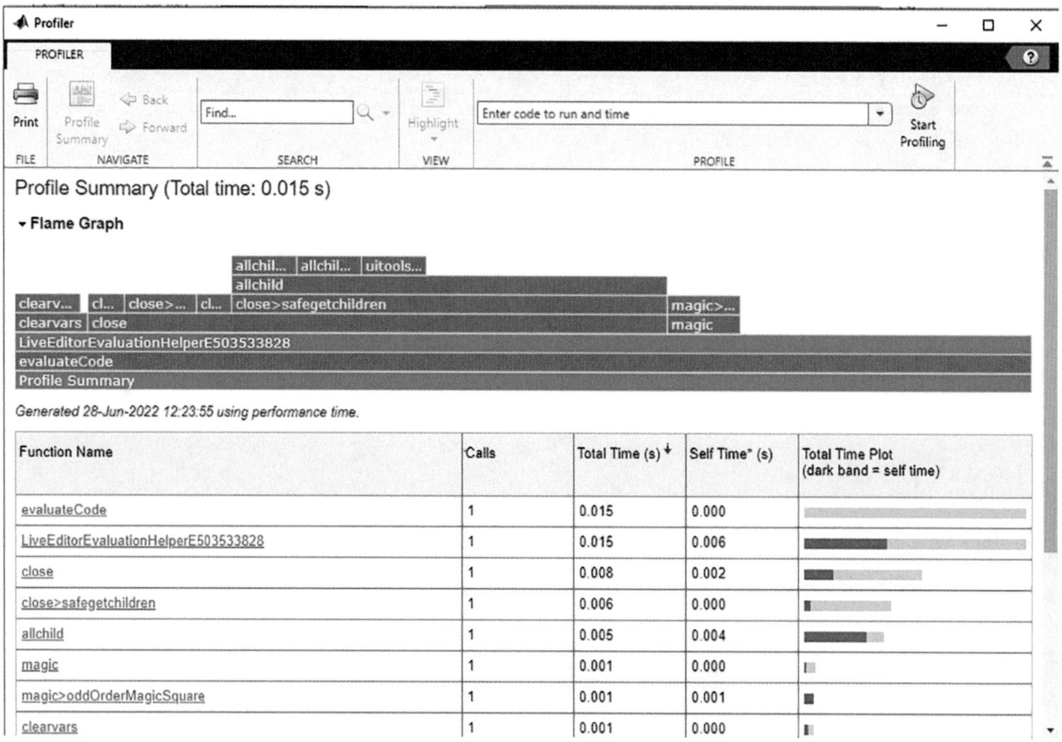

Figure 2-2. *Profiler view results*

The total processing time is 0.015 seconds. From the report, you can see for each function that the evaluation function took a significant amount of processing time. You can also view how much processing time is spent in each subfunction by clicking the function name in the profile report.

Dependency Report

The dependency report is used to identify all functions, scripts, and external programs (applications) that are called/used within our program. This report finds information about dependent files and tools of the current file that we have employed within our script built-in functions from MATLAB toolboxes and other M-files and function files from our current folder. This will be helpful before sharing our work with other users who may not have all the M-files, function files, and toolboxes that we have in our computer. The dependency report shows dependencies among MATLAB files in a directory and can be called with the following steps:

1. Select Desktop ➤ Current Directory and navigate to the directory containing MATLAB files for which we want to see the dependency report.

2. Click the Current Folder tab, and from the right-mouse button options, select Reports, and then select Dependency Report.

3. The dependency report of all MATLAB files will be shown automatically.

P-Codes

In MATLAB and Simulink, except for M-files (function files), with a file extension of `*.m`, `*.mlx`, `*.mat`, `*.mdl`, and `*.slx` files, there is also another important file type that is called P-code with a file extension of `*.p` that is very handy and recommended to create in some specific instances. The main reasons for creating a P-code file from an M-file are to prevent valuable files from being edited, to keep them secure, and to speed up the simulation time of scripts.

Specifically, here are some good reasons to create P-code:

- To speed up the process of simulation in the MATLAB platform.

- To keep valuable M-files secure to a certain extent; however, P-code should not be considered as a substitute for secure encryption.

- P-code/files provide a simple means of hiding proprietary algorithms.

- When you call an M-file function, MATLAB parses the M-code and stores the instructions as P-code in cache memory. P-code remains in memory until it is cleared using the `clear` command or until MATLAB quits.

- P-code is platform-independent pseudocode for a virtual MATLAB machine.

- Since P-files are in a binary format, their source code is hidden.

Here is how to create P-code:

– To create P-code from a given M-file residing in the current folder, e.g., `My_fileM.m`, type the following in the Command window:

```
>> pcode My_fileM.m
```

After running the previously shown command, a P-code of the M-file `My_fileM.m` will be created under the same file name but with the extension of `*.p` instead of `*.m`.

– To execute the created P-code called `My_fileM.p`, type in the Command window:

```
>> My_fileM.p
```

– Or type in the Command window:

```
>> run My_fileM
```

This executes the created P-code even if we have in our current directory our primary M-file called `My_fileM.m`.

Some Remarks on Scripts/M/MLX-Files

Note the followings:

- In general, any MATLAB commands can be executed from scripts or from the Command window. Which method is best depends on what tasks/computations or evaluations need to be performed. If there are simple computations composed in a one-step process and no repetitions are required, then there is no need to create M/MLX-files or scripts. On the other hand, when large computations with different operations including loop iterations and conditional statements are required, then writing M/MLX-files is the best option.

- The error and warning messages will not only help you locate errors in your scripts or models but also provide some hints on what the causes of errors are and how to eliminate them. In fact, warning messages will help you make your programs more robust and also inform a user about the ignored data in computed/plotted outputs.

- To write scripts and M-files, the M/MLX-file editors must be employed since they include a number of helpful tools. For instance, they automatically detect unused variables, the display of data sets, and the unclosed loops. They also enable you to work in cell and debug modes and help you detect errors and warnings about costly computations within loops, suggest memory allocation options, and do much more.

Display and Print Operators: display, sprintf, and fprintf

There are several commands and operators (built-in functions) to display computation results in the Command window or export them into external files compatible with MATLAB. They are `disp()`, `display()`, `sprint()`, and `fprintf()`. Out of these commands, `disp()` and `display()` are straightforward ways to display any comments, strings, or numerical values in the Command window without any additional formatting tools or characters. They are not robust enough to display any comments/strings and numerical values in various formats. They cannot write data into external files. On the other hand, `sprint()` and `fprintf()` can substitute all functions of `disp()` and `display()`. They can be used to print various data types in the Command window by using formatting operators and characters. Moreover, they can print textual and numerical data into external files. Let's look at several examples to demonstrate how to employ these display and print commands.

Example 1

Given $f(t) = \sin(t), t = \left[0, \dfrac{\pi}{4}, \dfrac{\pi}{2}, \dfrac{3\pi}{4}, \pi \right]$. Displaying computation results of the function

$f(t)$ and its argument t with short explanatory comments in the Command window is straightforward with `disp()` and `display()`.

```
t=[0, pi/4, pi/2, 3*pi/4, pi];
disp(['Sine @', num2str(t(1)),' is equal to: ', num2str(sin(t(1))) ])
disp(['Sine @', num2str(t(2)),' is equal to: ', num2str(sin(t(2))) ])
disp(['Sine @', num2str(t(3)),' is equal to: ', num2str(sin(t(3))) ])
```

```
disp(['Sine @', num2str(t(4)),' is equal to: ', num2str(sin(t(4))) ])
disp(['Sine @', num2str(t(5)),' is equal to: ', num2str(sin(t(5))) ])
```

These commands display the following in the Command window:

```
Sine @ 0 is equal to: 0
Sine @ 0.7854 is equal to: 0.70711
Sine @ 1.5708 is equal to: 1
Sine @ 2.3562 is equal to: 0.70711
Sine @ 3.1416 is equal to: 1.2246e-016
```

The displayed results are all correct and readable, but the demonstrated procedure is tedious. Note that there is a difference in the output from disp() and display(). The command disp() displays output without a variable name. By contrast, display() displays the variable name and its value like a simple calculation ending without a semicolon. Here's an example:

```
>> D1=sind(90);
>> display(D1)
D1 =
     1
>> disp(D1)
     1
>> D1
D1 =
     1
>> D1=sind(90)
D1 =
     1
```

Example 2

The command clock is a built-in command to show the current year/date/time according to a user's computer clock in a row matrix format. Let's display the current time in a more explicit way with some explanations. How do you do it with the disp() and display() commands?

```
>> format short G
>> TT=clock

TT =

  Columns 1 through 5

        2022                8                9                4                5

  Column 6

       37.261
>> display('This year is: '); disp(TT(1))
This year is:
        2022

>> display('This month is: '); disp(TT(2));
This month is:
     8
>> display('Day of this month: '); disp(TT(3));
Day of this month:
     9
>> disp('Current time is '); display(['hour: ', num2str(TT(4))]);
display(['minutes: ', num2str(TT(5))])
Current time is
hour: 4
minutes: 5
```

The output is legible and explicit, but the commands are too long and inefficient from a programming point of view.

fprintf( )

For the previous two examples, the fprintf() command can be employed much more simply. The general syntax of this command is as follows:

```
fprintf(format, A, ...)
```

For example 1, to display the sine function values, fprintf() gives a much simpler solution, as shown here:

```
fprintf('Sine @ %1.5f is equal to: %2.5f\n', t, sin(t));
```

135

Here are the output results from the command `fprintf()`:

```
Sine @ 0.00000 is equal to: 0.78540
Sine @ 1.57080 is equal to: 2.35619
Sine @ 3.14159 is equal to: 0.00000
Sine @ 0.70711 is equal to: 1.00000
Sine @ 0.70711 is equal to: 0.00000
```

The formatting specifiers, namely, %1.5f, %2.5f, and \n, used in this example define the field width and precision with the floating-point number format (%1.5f, %2.5f) and new line (\n). More details of formatting specifiers are given later in this section.

This is a much simpler and effortless method; it contains only one line of code.

For example 2, to display the time and date, `fprintf()` can be employed again as follows:

```
>> TT=clock;
fprintf('Year:%g; Month: %g; Day: %g; Hour: %g; Min passed: %g\n',TT(1),
TT(2), TT(3), TT(4), TT(5))

Year: 2022; Month: 8; Day: 13; Hour: 9; Min passed: 28
```

For such cases, `sprintf()` could be also an option. The general syntax of the command `sprintf()` is as follows:

```
str = sprintf(format, A, ...)
[str, errmsg] = sprintf(format, A, ...)
```

Note that to display outputs (`str` string messages) from `sprintf()`, you need to use either the `disp()` or `display()` command again. So, `sprintf()` is less flexible than `fprintf()`. Note that `sprintf()` always returns a string. To demonstrate how to employ `sprintf()` more explicitly and improve the script created to solve a quadratic equation, let's use the `sprintf()` command to display the computation results in the Command window with some additional information. Let's compute the roots of a quadratic equation and display some information in the Command window. Here is the solution script (`QuadEq1.m`):

```
% QuadEq1.m
% Solve quadratic equations based on coefficients of: a, b, & c
disp('Solve: ax^2+bx+c=0')
```

```
a=input('Enter value of a: ');
b=input('Enter value of b: ');
c=input('Enter value of c: ');
D=b^2-4*a*c;
[S, Errm]=sprintf('Discriminant of the equation is: %g', D); disp(S)
% Roots
x1=(-b+sqrt(D))/(2*a); x2=(-b-sqrt(D))/(2*a);
% Display roots in the Command window
[xr1, Errm1]=sprintf('Root1 of the equation is x1= %g', x1);
disp(xr1); display(x1)
[xr2, Errm2]=sprintf('Root2 of the equation is x2= %g', x2);
disp(xr2); display(x2)
```

Let's test the created M-file (QuadEq1.m) by executing it and providing the following entries (a=11, b=11, c=13) in the Command window:

```
Solve: ax^2+bx+c=0
Enter value of a: 11
Enter value of b: 12
Enter value of c: 13
Discriminant of the equation is: -428
Root1 of the equation is x1= -0.545455

x1 =

  -0.5455 + 0.9404i

Root2 of the equation is x2= -0.545455

x2 =

  -0.5455 - 0.9404i
```

Note that within `sprintf()` the numerical data is defined by the % sign followed by a formatting sign/letter g that is called a conversion specifier. Here is a command syntax of declaring data formats with `sprint()` and `fprintf()`:

```
[S, Errm]=sprintf('Discriminant of the equation is: %g', D);
```

Conversion specifications begin with the % character and contain these optional and required elements:

- Flags (optional)

- Width and precision fields (optional)

- A subtype specifier (optional)

- Conversion character (required)

These elements are used in the following order:

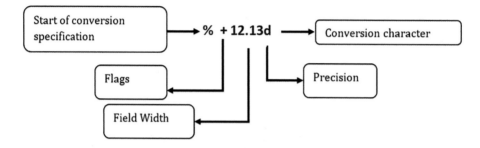

Flags are to control the alignment of the output, for instance, the - sign for the left justification of the output, the + sign for the right justification of the output, the space character for space before the value, and 0 to put a zero before the output. The field width is defined with a non-negative integer that specifies the number of digits or characters in the output, and the number (in precision) specifies the number of digits after the decimal point of the output; for example, %12.0 produces no decimal digits after the decimal digit number, and %12.13 produces 13 decimal digits after a decimal sign. Table 2-1 lists conversion characters, and Table 2-2 lists escape characters to specify nonprinting characters.[1]

---

[1] ANSI specification X3.159-1989: "Programming Language C," ANSI, 1430 Broadway, New York, NY 10018.

Table 2-1. *Conversion Characters*

Specifier	Description
%c	Single character
%d	Decimal notation (signed)
%e	Exponential notation (using a lowercase *e* as in 3.1415e+00) for floating point
%E	Exponential notation (using an uppercase *E* as in 3.1415E+00) for floating point
%f	Fixed-point notation for floating point
%g	The more compact of %e or %f, as defined in [*]. Insignificant zeros do not print for floating point
%G	Same as %g, but using an uppercase *G* for floating point
%i	Base 10 values for integer signed
%o	Base 8 octal notation (unsigned)
%s	String of characters
%u	Base 10 integer (unsigned)
%x	Base 16 hexadecimal notation (using lowercase letters a–f)
%X	Base 16 hexadecimal notation (using uppercase letters A–F)

Table 2-2. *Characters for Escape Formatting*

Symbol	Description
\b	Backspace
\f	Form feed
\n	New line
\r	Carriage return
\t	Horizontal tab
\\	Backslash
' or ' (quote)	Single quotation mark
%	Percent character

One of the most used escape formatting commands (Table 2-2) is \n. This is the escape character used to write the following data in a new line. Note that in the previous examples (examples 1 and 2), we have used \n within the `fprintf()` command.

Let's consider the previously discussed quadratic equation example to display and write its real and complex roots. Note that this exercise will demonstrate how to employ `sprintf()` and `fprintf()`.

There are several ways to display complex numbers (e.g., complex roots of the quadratic equation) in the form of $R + I * i$ explicitly with additional comments and then write/export them into an external file. They are to employ numerical conversions along with comments and to employ the string conversion character s.

Let's start with the first way, which requires adding several formatting items from Tables 2-1 and 2-2 to display complex numbers correctly. Now the existing M-file (QuadEq1.m) will be updated with the following by employing a conversion function `num2str()` to obtain the complex roots converted into strings. Now there is no need to separate out real and imaginary components of the complex roots of the equation.

```
[xr1, Errm1]=sprintf('Root1 of the equation is x1= %10s', num2str(x1));
disp(xr1); display(x1)
[xr2, Errm2]=sprintf('Root2 of the equation is x2= %10s', num2str(x2));
disp(xr2); display(x2)
% Or an alternative way:
disp(sprintf('Root2 of the equation is x1= %10s', num2str(x1)))
disp(sprintf('Root2 of the equation is x2= %10s', num2str(x2)))
```

After executing the updated script (QuadEq1.m), three input prompts to enter values for *a*, *b*, and *c* in a consecutive order are displayed; and after entering the following entries (a=11, b=12, c=13), the following display results of the given equation's complex roots are obtained:

```
Solve: ax^2+bx+c=0
Enter value of a: 11
Enter value of b: 12
Enter value of c: 13
Discriminant of the equation is: -428
Root1 of the equation is x1= -0.54545+0.94037i
x1 =

    -0.54545 +    0.94037i
```

Root2 of the equation is x2= -0.54545-0.94037i
x2 =

      -0.54545 -   0.94037i

Note that in this script `fprintf()` can be used instead of `sprintf()` in a similar manner that would make the code simpler. Thus, let's employ `fprintf()` to display the complex roots of the quadratic equation and write the computation results of the quadratic equation into an external file.

A general syntax for writing data into an external file with `fprintf()` is as follows:

```
FileID=fopen('filename.extension', 'permission')
fprintf(FileID, 'format', data);
...
fclose(FileID)
```

Note that `permission` can be w for writing, r for reading, r+ for reading and writing, and a for appending permissions. MATLAB supports a few common data file types, such as `*.txt`, `*.dat`, `*.xls`, `*.csv`, `*.jpeg`, etc. Via `format`, a type of data is defined, and `format` in `fprintf()` is defined like in `sprint()`. Using the general syntax of writing/ exporting data (existing data is discriminant and roots of the quadratic equation are saved in the workspace) into an external file, the following short script (`DataWrite.m`) writes the computed data (saved data in the workspace) into a file called `DataWrite.txt`:

```
% DataWrite.m - Write computed data into an external file
fid1=fopen('Results_QE.txt', 'w');
fprintf(fid1,'Discriminant of the equation is: %g\n', D)
fprintf(fid1,'Root1 of the equation is x1= %10s\n', num2str(x1))
fprintf(fid1,'Root2 of the equation is x2= %10s\n', num2str(x2))
fclose(fid1); open('Results_QE.txt')
```

Note that this script first creates a new external file called `Results_QE.txt` with the permission to write in it. This external file (`Results_QE.txt`) has a file ID called `fid1` (note that a file ID can be named freely) through which it can be called up to write in it assigned data, D, and the roots x1 and x2. Within the `fprintf()` command, the \n escape character is used to write the following data in a new line. Finally, the data-writing procedure ends with a file close command: `fclose(fid1)`. This is a standard procedure of exporting data of any size and of any format into an external file. This procedure of data export is covered more extensively via different examples in other chapters of the book. After executing the previously shown script, the following `*.txt` file content is displayed in the M-file Editor window:

```
Discriminant of the equation is: -428
Root1 of the equation is x1= -0.54545+0.94037i
Root2 of the equation is x2= -0.54545-0.94037i
```

There are many different format tools for data display and writing to external files that all can be employed more freely with respect to the given tasks. The demonstrated script can be improved further with data entries and by displaying/writing the calculated data. These will be discussed further with control statements in the following section.

Control Statements: [if, else, elseif, end], [switch, case, end]

The control statements [if ... elseif... else ... end] are also called *conditional statements* or *Boolean operators* and are employed to select logically which block of the code is to be executed while running the whole script according to the given input entries or computed results or defined outputs of the script. They are one of the few most crucial programming operators in all programming languages. These operators are employed in building and branching structures of programs (or codes). The general structure of these control statements follows:

```
if <logical statement A is true>
        DO X Operations
elseif <logical statement B is true>
        DO Y operations
...
else
        DO W operations
end
```

This indicates that if logical statement A is true, then X operations will be executed. If logical statement B is true, Y operations will be executed. If none of the previous logical statements is true, W operations will be executed.

Note that if there is only one logical statement, then `elseif` is not needed, and `else` is used directly instead of `elseif`.

Conditional operations can be expressed in two different ways: either on operational forms or on M/MLX-file forms. The conditional and logic statements given in Table 2-3 are applicable to all logical operations and conditional statements used in M/MLX-files/scripts.

Table 2-3. *Control Logical Operators and Their Descriptions*

Operation	M/MLX-file	Description
A<B	LT(A,B)	Less than
A>B	GT(A,B)	Greater than
A<=B	LE(A,B)	Less than or equal to
A>=B	GE(A,B)	Greater than or equal to
A~=B*	NE(A,B)	Not equal to
A==B	EQ(A,B)	Equality
A&B	AND(A,B)	Logical AND
A|B	OR(A,B)	Logical OR

The ~ tilde sign followed with = means "Not equal to."

Note   The operations < versus `lt()`, > versus `gt()`, <= versus `le()`, >= versus `ne()`, ~= versus `ne()`, == versus `eq()`, & versus `and()`, | versus `or()` can be used interchangeably and lead to no difference in code performance.

In various examples, we employ these operators and their syntaxes interchangeably. Let's consider several examples to demonstrate how to employ the control statements given in Table 2-3.

Example 1

The command clock displays the current year/date/time of a user computer in a row matrix form. Let's display the data in a readable format with some additional textual information. Here is a possible solution script (TimeDisp.m) of the given problem with the clock command:

```
% TimeDisp.m  Time Display
display(clock);
NOW=fix(clock);% Round to nearest integers
fprintf('This year is: %g \n', NOW(1));
if NOW(2)==1
    display(['Date is January ' num2str(NOW(2))]);
elseif NOW(2)==2
    display(['Date is February ' num2str(NOW(2))]);
elseif NOW(2)==3
    display(['Date is March ' num2str(NOW(2))]);
elseif NOW(2)==4
    display(['Date is April ' num2str(NOW(2))]);
elseif NOW(2)==5
    display(['Date is May ' num2str(NOW(2))]);
elseif eq(NOW(2),6)
    display(['Date is June ' num2str(NOW(2))]);
elseif eq(NOW(2),7)             % No difference between eq() and ==
    display(['Date is July ' num2str(NOW(2))]);
elseif eq(NOW(2),8)
    display(['Date is August ' num2str(NOW(2))]);
elseif eq(NOW(2),9)
    display(['Date is September ' num2str(NOW(2))]);
elseif eq(NOW(2),10)
    display(['Date is October ' num2str(NOW(2))]);
elseif eq(NOW(2),11)
    display(['Date is November ' num2str(NOW(2))]);
else
    display(['Date is December ' num2str(NOW(2))]);
end
```

```
if NOW(4)>=12
fprintf('Current time:  %g PM - %g min - %g sec\n', ...
        NOW(4), NOW(5), NOW(6))
else
fprintf('Current time:  %g AM - %g min - %g sec\n', ...
        NOW(4), NOW(5), NOW(6))
end
```

By running the script, the following outputs are displayed in the Command window:

```
ans =

  Columns 1 through 5

        2018              11            1            4            51

  Column 6

      34.932
```

This year is: 2018
Date is November 1
Current time: 4 AM - 51 min - 34 sec

Let's consider the previous example with a quadratic equation and correct the display of computed roots with disp, sprintf, fprintf, if, elseif, and else to demonstrate how to employ these programming operators.

Example 2

Here is one of the possible solution scripts (Quad_Eq2.m) of computing roots and displaying the roots of quadratic equations with additional comments:

```
% Quad_Eq2.m
% Solve quadratic equations based on coefficients of: a, b, & c
a=input('Enter value of a: ');
b=input('Enter value of b: ');
c=input('Enter value of c: ');
fprintf('Solve: (%3g)x^2+(%3g)x+(%3g)=0\n', a,b,c)
```

```
D=b^2-4*a*c;
% Roots
x1=(-b+sqrt(D))/(2*a); x2=(-b-sqrt(D))/(2*a);
if lt(D,0)
disp('This equation does not have real value roots!');
Dm=sprintf('Because discriminant is negative. D = %g', D); disp(Dm)
fid1=fopen('Results_QE.txt', 'w');
fprintf(fid1, 'This equation does not have real value \n');
fprintf(fid1, 'roots!\n');
fprintf(fid1,'Because discriminant is negative. D = %g\n', D);
fprintf(fid1,'Complex Root1: x1= %10s\n', num2str(x1));
fprintf(fid1,'Complex Root2; x2= %10s\n', num2str(x2));
fclose(fid1); open('Results_QE.txt');
elseif eq(D,0)
disp('This equation has one unique root! ');
disp('Because discriminant is zero: D = 0 ');
fid1=fopen('Results_QE.txt', 'w');
fprintf('This equation has one unique root! \n');
fprintf(fid1,'Because discriminant is zero: D = 0 \n');
fprintf(fid1,'Unique Root: x = %g \n', x1);
fclose(fid1); open('Results_QE.txt');
else
    disp('This equation has two roots! ');
    Dm =sprintf('Because discriminant is: D = %g ', D); disp(Dm);
    fid1=fopen('Results_QE.txt', 'w');
    fprintf(fid1,'This equation has two roots \n');
    fprintf(fid1,'Because discriminant is: D = %g \n', D);
    fprintf(fid1,'Real Root1 is: x1= %g \n', x1);
    fprintf(fid1,'Real Root2 is: x2= %g \n', x2);
    fclose(fid1); open('Results_QE.txt');
end
```

In this script, a few sets and combinations of [if, elseif, else, end] conditional statements are employed that provide nice outputs. Also, verbal versions of gt, lt, and eq of the $\leq, \geq, <, >, =$ logical operators are employed. Yet, the robustness of the script is still insufficient. For instance, the user may enter wrong entries by typing the value of c

for *b* or *a*, or vice versa, or skip entering any of the coefficients. As a result, the previous script produces the wrong output. To attain the full robustness of the script for different case scenarios and user entries including mistakes while entering input values, a loop control statement with [while ... end] will be employed. We discuss it in the following section.

Example 3

This example warns drivers about their driving speed based on where they are driving (school zone, residential/business district, unpaved town road, two and multilane highways, interstate highways). The speed limits in North Dakota are used in this example.

```
% Drive_Safe.m gives a warning sign what is the speed limit and
% how to behave in specific roads, highways, expressways, etc.
% E.g. CS is the only input variable
% Speed Limit [1] School area: 20 [mph]
% Speed Limit [2] Residential and business area: 25 [mph]
% Speed Limit [3] Town gravel roads: 55 [mph]
% Speed Limit [4] Two-lane highways: 65 [mph]
% Speed Limit [5] Multi-lane highways: 70 [mph]
% Speed Limit [6] Divided Interstate: 75 [mph]

CS = input('Enter your car"s current speed in [mph]:    ');
disp('Choose WHERE you are driving:  ')
disp('[1] School area;  [2] Residential and Business Area; ')
disp('[3] Town gravel roads; [4] Two-lane Highways; ')
disp('[5] Multi-lane Highways; [6] Divided Interstate Roads; ')
DZone = input('Enter your drive zone, e.g., 1, 2, 3, ... 6:  ');
if isempty(DZone) || DZone >7 || DZone<0
    warndlg('Not clear where you are driving!')
elseif CS ==0
    warndlg('Your car is not moving')
elseif CS<0
    warndlg('Your car is moving in a rear direction that is DANGEROUS! ')
else
```

```
    if DZone ==1 && CS > 0 && CS<10
        fprintf('Your speed = %g [mph] is too slow \n ', CS)
       fprintf('even if it is during the student arrival/departure time \n')
        warndlg('Speed UP to make your car speed around 15 mph!')
    elseif  DZone ==1 && CS > 10 && CS<20
        fprintf('Your speed = %g [mph] is OK in a school area \n', CS)
        fprintf('during the student arrival/departure time \n')
        warndlg('Keep your car speed around 15...20 mph!')
    elseif DZone ==1 && CS >20
    fprintf('Your speed = %g [mph] is TOO fast for a school area \n', CS)
    warndlg('DANGER! Slow DOWN up to 20 mph!')
    elseif DZone ==2 && CS > 15 && CS <20
fprintf('Your speed = %g [mph] is too slow for a residential area!\n', CS)
warndlg('Speed up! Speed limit for residential/business areas is 25 mph')
    elseif DZone == 2 && CS > 20 && CS < 28
    fprintf('Your speed = %g [mph] is adequate for a residential
    area!', CS)
    warndlg(' Do not speed over 25 mph!')
    elseif DZone == 2 && CS > 28
    fprintf('Your speed = %g [mph] is higher for a residential
    area!\n', CS)
    warndlg(' DANGER!  Slow DOWN and do not speed over 25 mph!')
    elseif DZone == 3 && CS < 48  && CS > 25
    fprintf('Your speed = %g [mph] is slower for gravel town roads!\n', CS)
    warndlg(' Speed up to 55 mph!')
    elseif DZone == 3 && CS > 48  && CS < 58
fprintf('Your speed = %g[mph] is within limits for gravel town
roads!\n',CS)
        warndlg(' Keep your speed around 55 mph!')
    elseif DZone == 3 && CS > 58
        fprintf('Your car speed = %g [mph] is beyond \n', CS)
        fprintf('the limits for a gravel town roads! \n')
        warndlg(' DANGER!  Slow DOWN up to 55 mph!')
    elseif DZone == 4 && CS > 55  && CS < 60
        fprintf('Your speed = %g [mph] is slower for two-lane HW!\n', CS)
```

```
        warndlg(' Speed up to 65 mph!')
    elseif DZone == 4 && CS > 60  && CS < 68
fprintf('Your speed = %g[mph] is within the limits for two-lane HW!
\n', CS)
        warndlg(' Keep your speed around 65 mph!')
    elseif DZone == 4 && CS > 68
        fprintf('Your speed = %g [mph] is beyond  \n', CS)
        fprintf('the limits for two-lane HW! \n')
        warndlg(' DANGER!  Slow DOWN up to 65 mph!')
    elseif DZone == 5 && CS > 55  && CS <65
        fprintf('Your speed = %g [mph] is slower than \n' , CS)
        frpintf('the speed limits for multi-lane lane HW! \n')
        warndlg(' Speed up to 70 mph!')
    elseif DZone == 5 && CS > 65  && CS <75
        fprintf('Your speed = %g [mph] is within \n', CS)
        fprintf('the speed limits for multi-lane lane HW! \n')
        warndlg(' Keep your speed around 70 mph!')
    elseif DZone == 5 && CS > 75
        fprintf('Your speed = %g [mph] is beyond \n', CS)
        frintf('the speed limits for multi-lane lane HW! \n')
        warndlg(' DANGER! Slow DOWN up to 70 mph!')
    elseif DZone == 6 && CS > 60  && CS <70
        fprintf('Your speed = %g [mph] is slower than \n', CS)
        fprintf('the speed limits for Divided Interstate in ND! \n')
        warndlg(' Speed up to 70 mph!')
    elseif DZone == 6 && CS > 70  && CS <78
        fprintf('Your speed = %g [mph] is within \n', CS)
        fprintf('the speed limits for Divided Interstate in ND! \n')
        warndlg(' Keep your speed around 75 mph!')
    else
        fprintf('Your speed = %g [mph] is beyond  \n', CS)
        fprintf('the speed limits for Divided Interstate in ND! \n')
        warndlg(' DANGER!  Slow DOWN up to 75 mph!')
    end
end
```

In this answer script to the given exercise, the conditional statements or Boolean operators [if ... elseif ... else ... end] are used twice. The first [if... elseif ...elseif... else ...] identifies according to the user entries whether the user has specified or not where they are driving (DZone) and what their car speed (CS) is. The second [if... elseif ... ... else ... end] starts only if the first [if...elseif ...] conditions are met or, in other words, the user has entered their driving speed and driving zone.

If the user information is valid for the predefined six driving zones, then under [elseif CS ==0], it verifies the given car is moving. If it is moving, [elseif CS<0] indicates in which direction (backward for negative speed values). The last [else] evaluates the positive values of car speed.

Beyond this first [else], all positive values (forward movement of a car) and a second loop of [if ... elseif ... ... else ... end] are executed. They define the speeding level and provide respective warning signals to the driver with respect to where (zones and roads) they are driving and what their speed is.

Let's test the script with different input speed values and the different driving zones.

Case 1

The given car is driving at 9 mph in a school zone.

```
Enter your car"s current speed in [mph]:    9
Choose WHERE you are driving:
[1] School area;   [2] for Residential and Business Area;
[3] Town gravel roads; [4] Two-lane Highways;
[5] Multi-lane Highways; [6] Divided Interstate Roads;
Enter your drive zone, e.g. 1, 2, 3, ... 6:   1
Your car speed = 9 [mph] is too slow
 even if it is during the student arrival/departure time
```

In addition, the following warning dialog box is displayed:

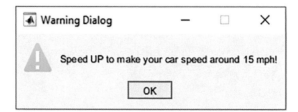

Case 2

The given car is driving at 72 mph on a two-lane highway.

```
Enter your car"s current speed in [mph]:     72
Choose WHERE you are driving:
[1] School area;  [2] for Residential and Business Area;
[3] Town gravel roads; [4] Two-lane Highways;
[5] Multi-lane Highways; [6] Divided Interstate Roads;
Enter your drive zone, e.g., 1, 2, 3, ... 6:  4
Your car speed = 72 [mph] is beyond
the limits for two-lane HW!
```

The following warning dialog box is displayed:

Case 3

The given car is driving at 77 mph on a divided interstate highway.

```
Enter your car"s current speed in [mph]:     77
Choose WHERE you are driving:
[1] School area;  [2] for Residential and Business Area;
[3] Town gravel roads; [4] Two-lane Highways;
[5] Multi-lane Highways; [6] Divided Interstate Roads;
Enter your drive zone, e.g., 1, 2, 3, ... 6:  6
Your car speed = 77 [mph] is within
the speed limits for Divided Interstate in ND!
```

The following warning dialog box is displayed:

Another set of control statements is [switch, case, otherwise, end]. These statements have one big advantage over [if, else, elseif, end]: they can handle many cases and simplify the code. They have the following general syntax structure:

```
switch expression of scalar or string (variable)
    case variable value1
   A statements
    case variable value2
    B statement
    case variable value3
    C statement
...
    otherwise
    X, ... W statements
end
```

Let's look at a simple example to understand how the conditional statements [switch, case, otherwise, end] operate.

Example 4

Determine whether the user-entered integer from 1 to 50 is odd or even or a prime number. The following script (Num50Type.m) defines the type of any integer from 1 to 50:

```
% Num50Type.m - Defines entered integer type
clear all; clearvars; clc
N=input('Enter ANY integer up to 50:  ');
if N<=50
```

```
fprintf('Your entry is: %d \n', N)
switch N
case mod(N,2)==0
 fprintf('you have entered %g which is EVEN number \n', N)
case {9,15,21,25,27,33,35,39,45,49}
fprintf('You have entered %g which is ODD number but not prime\n', N)
otherwise %N=[3,5,7,11,13,17,19,23,29,31,37,41,43,47]
 fprintf('You have entered %g which is PRIME number \n ', N)
end
else
disp('This code works with integers up to 50 to identify if they are even,
odd or prime!')
end
```

If you enter 23 as an input, you will get the following output in the Command window:

```
Enter ANY integer up to 50:   23
You have entered 23 which is PRIME number
```

This script correctly defines all entered numbers up to 50 (including 50); however, we may still improve its robustness by adding conditional statements and while loops to verify whether a user entry is correct and within the range of 1 to 50.

The robustness of the scripts with the control branching statements [if, elseif, else, end] can be improved by employing them within loop statements, namely, [while ... end] and [for ... end], which will be discussed via examples in the following section.

Loop Control Statements: while, for, continue, break, end

There are two types of loop control statements, namely, [while...end] and [for... end]. A first loop control statement is [while...end], and its general syntax is as follows:

```
while expression is NOT true
perform A, B, C,..., W operations
end
```

Unlike [if .. elseif ... else... end] statements, the [while...end] loop keeps executing the the following [A, B, C,..., W] operations until the expression behind the [while] operator becomes valid. Once the expression becomes true/valid, the execution process halts.

Example 1

To compute the sum of all odd numbers from 1 up to 100 using the [while ... end] loop control (Sum100.m), use this:

```
% Sum100.m sums of all odd numbers with while
Sum=0; N=1;
while ne(N, 100)
if ne(mod(N,2),0) % Verifies all odd numbers are added.
    Sum=Sum + N;
else                % Verifies if the number is even, NO sum.

end
    N=N+1;
end
disp(['Sum of all odd numbers 1...100 is:  ', num2str(Sum)]);
```

This loop-based script (Sum100.m) will keep running until the iteration process reaches the end value of 100. The [if ... else ... end] conditional statements verify if the number is odd or even with the help of mod(). The iteration loop is used to compute the remainder of N/2. If the number is odd, the sum will take place; otherwise, it will be ignored.

Note   In many summation types of exercises that use [while ... end] and [for ... end] iteration loops, users have difficulty understanding the computation algorithm and writing the code to perform the summation process. The summation algorithm does not appear to be intuitive for many beginners in programming. Their difficulty comes from the fact that they overlook the necessity of assigning 0 to the sum variable in the initial step of the summation process. This means that adding any number to 0 gives the same number back. Subsequently, within a [for ... end] or [while ... end] loop, this summation variable will get a new value in each iteration. For example, this is what it looks like for the previous example:

Step 1. Sum(1) = 0;

Step 2. Sum(2) = Sum(1)+N(1) = 0+1=1; when N(1)= 1 and ne(mod(N(1),2), 0) is valid;

Step 3. Sum(3)=Sum(2)=1; when N(2)= N(1)+1=2 and because N(2)/2 = 1 and remaining "0" ← ne(mod(N,2), 0) is not valid.

Step 4. Sum(4)=Sum(3)+N(3)=1+3 =4; when N(3)=N(2)+1=2+1=3 and ne(mod(N(3),2), 0) is valid;

Step 5. Sum(5)=Sum(4)=4; when N(4)=N(3)+1=3+1=4; because N(4)/2 = 2 and remaining "0" ← ne(mod(N(4),2), 0) is not valid.

Step 6. Sum(6)= Sum(5)+N(5)=4+5=9; when N(5)=N(4)+1=4+1=5 and ne(mod(N(5),2), 0) is valid;

And so on. The iteration process continues until N equals 100.

The final output of the script in the Command window is as follows:

```
Sum of all odd numbers 1...100 is:   2500
```

Example 2

To better understand how to employ the control loop statement [while ... end], let's look at the example of solving a quadratic equation: $ax^2 + bx + c = 0$.

Let's say the problem is that users can enter the values of a, b, c in the wrong order or mistype their values or enter wrong numbers.

To make a program more robust and resolve any wrong entries, the conditional operators [if ... elseif ... else ... end] are employed along with the [while ... end] loop iteration operators. The following is one of the possible solutions to the problem. The script (Quad_Eqn.m) is robust. It accepts different entries and computes all possible roots, real and complex, and then displays them correctly in the Command window. It resolves many possible wrong entries such as wrong values, wrong data sizes, or mixed-up values of a, b, c. Note that a first while loop statement in the script verifies the size of the entry row array containing the values of a, b, c, and a second one with the conditional operator [if ... else ... end] verifies the order correctness of the a, b, c entries.

```
% Quad_Eqn.m solves quadratic equations based on the user entries for:
% a, b, c
clearvars; clc
SIZE_chk = 0;
while SIZE_chk ~=1
    disp('Solution of: ax^2+bx+c=0')
    abc=input('Enter values as [a, b, c]:    ');
    SIZE_abc=numel(abc);
    if SIZE_abc ==3
        SIZE_chk=1;
        a=abc(1);
        b=abc(2);
        c=abc(3);
        CorrectEntry=0;
        while CorrectEntry~=1
            if SIZE_abc == 3
                CorrectEntry=1; CorrectABC=0;
                % Check the order of a, b, c
                while CorrectABC~=1
                    disp('Is this the equation you"d like to find roots?')
                    disp('   ')
                    Eqn=sprintf(' %g*x^2+(%g)*x + (%g)=0', a, b, c);
                    disp('   ')
                    disp(Eqn);
            CorrectABC=input('If yes, enter 1, otherwise any integer!  ');
                    disp('   ')
                    if CorrectABC~=1
                        abc=input('Enter values as [a, b, c]:    ');
                        a=abc(1); b=abc(2); c=abc(3);
                    end
                end
                D=b^2-4*a*c;
                % Roots
                x1=(-b+sqrt(D))/(2*a); x2=(-b-sqrt(D))/(2*a);
```

```
        if lt(D,0)
            disp('The equation has no real roots!');
            Dm=sprintf('Because discriminant <0. D = %g', D);
            disp(Dm)
            % Display the results
            fprintf(' %g*x^2+(%g)*x + (%g)=0\n', a, b, c);
            fprintf('It does not have real roots!\n');
            fprintf('Because its discriminant <0. D=%g\n',D);
fprintf('Complex Roots are: x1=%10s; x2=%10s\n', num2str(x1),num2str(x2));
        elseif eq(D,0)
            disp('This equation has one unique root! ');
            disp('Because discriminant is zero. D=0 ');
            fprintf('%g*x^2+(%g)*x + (%g)=0\n', a, b, c);
            fprintf('It has a unique root! \n');
            fprintf('Because discriminant is "0". D=0 \n');
            fprintf('Unique Root: x = %g \n', x1);
        else
            disp('This equation has two real roots! ');
    [Dm, Errm2]=sprintf('Because discriminant >0. D = %g ', D); disp(Dm);
            fprintf(' %g*x^2+(%g)*x + (%g)=0\n', a, b, c);
            fprintf('It has two real roots \n');
            fprintf('Because discriminant >0. D = %g\n', D);
    fprintf('Roots of the equation are x1= %g; x2= %g \n', x1, x2);
        end
    else
        CorrectEntry=0;
        disp(' a, b, c cannot have more than 1 element (value)');
        disp('Re-enter values of a, b, c');
        disp('    ')
        abc=input('Enter values as [a, b, c]:    ');
        a=abc(a); b=abc(2); c=abc(3);
        sa=numel(1); sb=numel(b); sc=numel(c);
    end
end
else
```

```
fprintf('Your entry is wrong \n')
fprintf('Enter the correct entries for [a, b, c] as a row vector in [ ]: ')
    end
end
```

The created script is tested for robustness with intentionally wrong entries for *a, b* and *c* as stated in cases 1 and 2.

```
Solution of: ax^2+bx+c=0
Enter values as [a, b, c]:    [12 1 1 ]
Your entry is wrong
Enter the correct entries for [a, b, c] as a row vector in [ ]:  Solution
of: ax^2+bx+c=0
Enter values as [a, b, c]:    [12 1 1 ]
Your entry is wrong
Enter the correct entries for [a, b, c] as a row vector in [ ]:  Solution
of: ax^2+bx+c=0
Enter values as [a, b, c]:    [1 2 1 1 ]
Your entry is wrong
Enter the correct entries for [a, b, c] as a row vector in [ ]:  Solution
of: ax^2+bx+c=0
Enter values as [a, b, c]:    [1 2 1  ]
Is this the equation you"d like to find roots?

 1*x^2+(2)*x + (1)=0
If yes, enter 1, otherwise any integer!  0

Enter values as [a, b, c]:    [1 1 2]
Is this the equation you"d like to find roots?

 1*x^2+(1)*x + (2)=0
If yes, enter 1, otherwise any integer!  1

The equation has no real roots!
Because discriminant <0. D = -7
 1*x^2+(1)*x + (2)=0
It does not have real roots!
Because its discriminant <0. D=-7
Complex Roots are: x1=-0.5+1.3229i;   x2=-0.5-1.3229i
```

From the previous simple example, it is clear that the [while ... end] loop operations have more power than the [if ... elseif ... else ... end] operators to enhance the robustness of scripts.

A second loop control statement is [for ... end] that has the following general syntax:

```
for i=startloop:stepsize:endloop
      Iteration 1
      Iteration 2
...
      Iteration N
end
```

The [for ... end] loop control statement works in the following fashion. It starts executing every statement of [1, 2, ..., N] starting from startloop for every step, with a step equal to stepsize until the loop iteration reaches the value of endloop, which corresponds to N number of iterations.

Note   If the step size is not specified when you're assigning vector spaces or index ranges in [for ... end] loops, the default step size is 1.

Example 3

Compute the sum of all odd numbers within [1, 20] with the [for ... end] loop (Sum20odd.m).

```
% Sum20odd.m sums odd numbers within 0...20 with for loop
clearvars;
S=0;
for N=1:20;
if rem(N,2)==0 %Verifies if the number is even, then no sum.
        S=S;
else            %All odd numbers are added:
        S=S+N;
end
end
fprintf('Sum of all odd numbers 1 to %g is equal to %g \n', N, S)
```

The [if ... else ... end] conditional statements determine whether the number is odd. If the number is odd, it is added to the summation variable (S). Otherwise, nothing will be added to the summation variable (S). The algorithm behind this script is the same one in Example 1, and this short script is an alternative solution to Example 1. The only difference here is that the number of odd numbers is 20.

Two alternative, simpler solutions of this example using [for ... end] and [while ... end] are as follows:

```
% Sum20odd_alt.m sums odd numbers within 0...20 with [for ... end] loop
clearvars;
S=0;
for N=1:2:20
    S=S+N;
end
fprintf('Sum of all odd numbers 1 to 20 is equal to %g \n', S)
% Sum of odd numbers within 0...20 with [while ... end] loop
clearvars;
S=0; jj=1;
while le(jj,20)
    S=S+jj; jj=jj+2;
end
fprintf('Sum of all odd numbers 1 to 20 is equal to %g \n', S)
```

The computation algorithm of this script [for ... end] is as follows:

1. $S(1) = 0$

2. $S(2)=S(1)+N(1)=0+1=1$ when $N(1)=1$

3. $S(3)=S(2)+N(2)=1+3=4$ when $N(2)=3$ because the step size is equal 2 (i.e., $N(2)=N(1)+2=1+2=3$)

4. $S(4)=S(3)+N(3)=4+5=9$ when $N(3)=5$ because ...
   $N(3)=N(2)+2=3+2=5$

5. And so on. The iteration loop runs until $N = 20$. In fact, in this case, N cannot be 20 because the step size is 2 and the maximum value of N will be 19.

Finally, the outputs of the scripts are identical:

```
Sum of all odd numbers 1 to 20 is equal to 100
```

As demonstrated, the [for ... end] loop can be easily substituted by the [while ... end] loop or vice versa; their efficiencies are similar. However, in some specific examples (e.g., Example 2), the robustness gained by using [while ... end] cannot be attained using [for ... end]. Theoretically, in some problems (e.g., Example 2), [while ... end] can be substituted by [for ... end] when the number of iterations is unknown.

Note   In many loop iteration-based computation problems, [for ... end] can be easily substituted by the [while ... end] loop and vice versa. Their computation efficiencies are similar. However, there are many specific cases, where [while ... end] operations are employed to attain the high robustness of the script, where [for ... end] cannot provide as much robustness and flexibility as [while ... end]. This is illustrated in Example 2.

It is recommended to use ii or jj for loop iterations for indices so as not to get confused with imaginary numbers, which use i and j as reserved in MATLAB by default. If these letters (i and j) are employed for indices, then sqrt(-1) can be used for imaginary numbers alternatively, or you can clear these variables from the workspace.

Example 4

Evaluate the square and cubic powers of even numbers (1 to 10) with the loop control statements for and end.

Here is one of the possible solution scripts (SqCube10.m) of the exercise. Note in this example startloop=2, stepsize=2, and endloop=10.

```
% SqCube10.m computes square and cube of 2...10
%% [for  ... end]
for k=2:2:10
    Square = k^2;
```

```
    Cube    = k^3;
    fprintf('Square of %g is %g \n', k, Square);
    fprintf('Cube of %g is %g \n'  , k, Cube);
end
%% Alternative way: [while ... end]
jj =2;
while le(jj, 10)
    Square = jj^2;
    Cube    = jj^3;
    fprintf('Square of %g is %g \n', jj, Square);
    fprintf('Cube of %g is %g \n'  , jj, Cube);
    jj =jj+2;
end
```

When the script is executed, the following will be displayed in the Command window:

```
Square of 2 is 4
Cube of 2 is 8
Square of 4 is 16
Cube of 4 is 64
Square of 6 is 36
Cube of 6 is 216
Square of 8 is 64
Cube of 8 is 512
Square of 10 is 100
Cube of 10 is 1000
```

Example 5

Evaluate the square and cubic powers of odd numbers (1 to N) with the loop control statements for and end.

Note that this exercise is similar to Example 4 with two major differences. It uses only odd numbers to compute squares and cubes and performs calculations until it reaches the user-defined N value.

Here is one of the possible solution scripts (SqCubeN.m) of the exercise. Note in this example startloop=1, stepsize=2, and endloop=N for the [for end] loop. For the [while end] loop, the starting value of the loop is 1 (jj = 1).

```
% SqCube_N.m computes square and cube of all odd numbers up to N
%% [for ... end]
N = input('Enter N =   ');
for ii=1:2:N
    Square = ii^2;
    Cube   = ii^3;
    fprintf('Square of %d is %d; Cube is %d \n', [ii, Square Cube])
end
%% Alternative way: [while ... end]
N = input('Enter N =   ');
jj =1;
while le(jj, N)
    Square = jj^2;
    Cube   = jj^3;
    fprintf('Square of %d is %d; Cube is %d \n', [jj, Square Cube])
    jj =jj+2;
end
```

When you run the script and enter 13 for N, the following will be displayed in the Command window:

```
Enter N =   13
Square of 1 is 1; Cube is 1
Square of 3 is 9; Cube is 27
Square of 5 is 25; Cube is 125
Square of 7 is 49; Cube is 343
Square of 9 is 81; Cube is 729
Square of 11 is 121; Cube is 1331
Square of 13 is 169; Cube is 2197
Enter N =   13
Square of 1 is 1; Cube is 1
Square of 3 is 9; Cube is 27
Square of 5 is 25; Cube is 125
```

```
Square of 7 is 49; Cube is 343
Square of 9 is 81; Cube is 729
Square of 11 is 121; Cube is 1331
Square of 13 is 169; Cube is 2197
```

Example 6

Compute the series $1 - \dfrac{1}{2} + \dfrac{1}{3} - \dfrac{1}{4} + \dfrac{1}{5} - \dfrac{1}{6} + \dfrac{1}{7} \ldots$ by using [for ... end] and [while ... end].

Here are five possible solution scripts of the given exercise gathered within one script called SumSeries.m:

```
% SumSeries.m computes a sum of Series:
%% Version A. [for ... end]
clearvars
sign=1;
S=0;
N=input('Enter number of series to compute:  ');
for n=1:N
    S=S+sign/n;
    sign=-sign;
end
fprintf('Sum of %g series is equal to %2.5f \n', n, S)
%% Version B. [for ... end]
clearvars S N sign
S=0; sign=1;
N = input('Enter number of series to compute:  ');
for k=1./(1:N)
    S=S+sign*k;
    sign=-sign;
end
fprintf('Sum of %g series is equal to %2.5f  \n', N, S)
%% Version C. [for ... end]
clearvars S N
S=1;
```

```
N = input('Enter number of series to compute:   ');
for k=2:N
    if eq(mod(k,2), 0)
        S=S-1/k;
    else
        S=S+1/k;
    end
end
fprintf('Sum of %g series is equal to %2.5f  \n', k, S)
%% Version D. [while ... end]
clearvars
sign=1;
S=0;
N = input('Enter number of series to compute:   ');
n = 1;
while n<N+1
    S=S+sign/n;
    n=n+1;
    sign=-sign;
end
fprintf('Sum of %g series is equal to %2.5f \n', n, S)
%% Version E. [while ... end]
clearvars S N
S=0;
N=input('Enter number if series to compute:   ');
ii=1;
while ii~=N+1
if mod(ii,2)==0
        S=S-1/ii;
else
        S=S+1/ii;
end
    ii=ii+1;
end
fprintf('Sum of %g series is equal to %2.5f \n', ii, S)
```

All of these solution methods with [for ... end] and [while ... end] result in identical solutions. It must be noted that the MATLAB function mod(N, 2) defines whether the number is even or odd, which works in this script to separate the even and odd components of ii and k.

In all of the previous examples, the last final values from the iteration processes are saved. What about saving the values from all iteration steps?

```
%% Version A. [for ... end]
clearvars
sign=1;
S=0;
N=input('Enter number of series to compute:  ');
for n=1:N
    S=S+sign/n;
    SS(n)=S;
    sign=-sign;
    fprintf('Sum of %g series is equal to %2.5f \n', n, SS(n))
end
%% Version B. [for ... end]
clearvars S N sign
S=0;
sign=1;
ii=1;
N = input('Enter number of series to compute:  ');
for k=1./(1:N)
    S=S+sign*k;
    SS(ii)=S;
    sign=-sign;
    fprintf('Sum of %g series is equal to %2.5f \n', ii, SS(ii))
    ii=ii+1;
end
%% Version C. [while ... end]
clearvars S N ii SS
S=0;
N=input('Enter number if series to compute:  ');
ii=1;
```

```
while ii~=N+1
    if mod(ii,2)==0
        S=S-1/ii;
    else
        S=S+1/ii;
    end
    SS(ii)=S;
    fprintf('Sum of %g series is equal to %2.5f \n', ii, SS(ii))
    ii=ii+1;
end
%% Version D. [while ... end]
clearvars n N S SS
sign=1;
S=0;
N = input('Enter number of series to compute:  ');
n = 1;
while n<N+1
    S=S+sign/n;
    SS(n)=S;
    fprintf('Sum of %g series is equal to %2.5f \n', n, SS(n))
    n=n+1;
    sign=-sign;
end

%% Version E. [while ... end]
clearvars SS ii S N
S=0;
N=input('Enter number if series to compute:  ');
ii=1;
while ii~=N+1
if mod(ii,2)==0
        S=S-1/ii;
        SS(ii)=S;
else
        S=S+1/ii;
        SS(ii)=S;
```

```
end
    fprintf('Sum of %g series is equal to %2.5f \n', ii, SS(ii))
    ii=ii+1;
end
```

One of the easiest and straightforward approaches to preserving all values from all iterations is to initiate a new series of variables, e.g., SS(ii) and SS(n)s. When you are saving all values of calculated outputs within loops, it is salient to interpret SS(ii) and SS(n) correctly. For example, SS(1) first takes the value of S=1, and S = S-1/2 leads to SS(2) = 1-1/2 = ½. In the next step (step 2), S=S +1/3 leads to SS(3) =1/2+1/3 =5/6, and so forth.

Example 7

Compute the expression $f(t) = e^{sin(3t)}$ $t = 0 : 0.001 : 6.28$; by employing $[\text{for} \ldots \text{end}]$ and $[\text{while} \ldots \text{end}]$ loop iteration operators and preserving all of the values from all iterations. From every iteration, one value of $f(t)$ is saved, corresponding to each value of t.

```
%% Ex7_FOR_WHILE.m
clearvars
% Ver 1
ii =1;
for t=0:.001:6.28
    f(ii)=exp(sin(t));
    ii=ii+1;
end
% Ver 2
t=0:.001:6.28;
for k=1:numel(t)
    f(k)=exp(sin(t(k)));
end
%% Ver 3
jj=0;
```

```
while jj~=numel(t)
    f(jj+1)=exp(sin(t(jj+1)));
    jj=jj+1;
end
% Ver 4
m=1;
while m~=numel(t)+1
    f(m)=exp(sin(t(m)));
    m=m+1;
end
```

Note   It is salient to use f(ii) or f(k) or f(jj+1) or f(k) while computing $f(t)$ within the [for ... end] and [while ... end] loops. This preserves all the values of $f(t)$ from the iteration process with respect to the values of t. If only f is used instead of f(ii) or f(k) or f(jj+1) or f(k), then only one very last value of $f(t)$ is saved.

One of the most common mistakes that users make is that while working with loop operators ([for ... end], [while ... end]), they overlook index (ii) or (k) or (jj) or (m) after the main variable. For example, f(ii), f(k), f(jj+1), and f(k) in the previous example collect all values from the whole iteration process.

Another most common mistake that users make is that while working with loop operators ([for ... end], [while ... end]), they assign wrong index values for (ii), (jj). For example, they use negative values or values starting with 0 or noninteger values or use mismatched sizes. In the previous example (version 3), we overlook assigning the value 0 to jj before the [while end] loop or assign jj=0 and at the same time assign f(jj). This is not acceptable. f(0) is meaningless for one important reason: a variable cannot have an index of 0. Indexes can be 1, 2,3, ... 10^{12}, ...$10^{25} + 1$, ... but not ..., -3,-2, -1, 0   or 0.12, 2.1, 3.35, 5.5, 7/8, 100/899, etc.

There is a good alternative option or approach in collecting every value from every computation within a loop, which is to assign a new variable with an index. For example, this can be attained via the following:

```matlab
%% Ex7_FOR_WHILE.m
clearvars
% Ver 1
ii =1;
for t=0:.001:6.28
    f=exp(sin(t));
    F1(ii)=f;
    ii=ii+1;
end
% Ver 2
t=0:.001:6.28;
for ii=1:numel(t)
    f=exp(sin(t(ii)));
    F2(ii)=f;
end
%% Ver 3
jj=0;
while jj~=numel(t)
    f=exp(sin(t(jj+1)));
    F3(jj+1)=f;
    jj=jj+1;
end
% Ver 4
jj=1;
while jj~=numel(t)+1
    f=exp(sin(t(jj)));
    F4(jj)=f;
    jj=jj+1;
end
```

It must be noted that in this example the best and most efficient computation approach is vectorization, as shown here:

```
t=0:.001:6.28
f=exp(sin(t));
```

Note   For the efficiency of computation processes or to improve your code performance, it is recommended you avoid using the [`for` ... `end`] and [`while` ...`end`] loop control statements whenever feasible. Instead of the loop iteration, it is recommended to employ the vectorization approach, as shown in Example 7: `t=0:.001:6.28; f=exp(sin(t))`.

Example 8

Let's look at the famous "wheat and chessboard problem" story in this example. The problem is defined by the following: pieces of grain are placed in each chessboard square in the order of one grain in the first square, two pieces of grain in the second, four pieces of grain in the third, eight pieces of grain in the fourth, 16 pieces of grain in the fifth, 32 pieces of grain in the sixth, and so on, by doubling the pieces of grain on each subsequent square of the chessboard.

The sum of all grains to be placed on the chessboard is 1+2+4+8+16+32 ... and so forth. This can be also expressed as the sum of powers of 2: $2^0 + 2^1 + 2^2 + 2^3 + 2^4 + 2^5 + ...2^{63}$ and is equal to $2^{64} - 1 = 18{,}446{,}744{,}073{,}709{,}551{,}615$. This is indeed a huge number of grains requested by the chess inventor for his ingenious invention of the chess game as a reward from a king. In simple calculations, if one grain weighs 0.025 grams, then the whole amount would weigh over 461 billion tons of grains. That would be a mountain of wheat grains bigger than Mount Everest. Let's compute the number of grains using [`for` ... `end`] and [`while` ... `end`] loops.

```
% ChessInventorReward.m
% Example: Grains of Wheat on the Chess Board
%% Version A. [while ... end] loop
clearvars
S=1; Ncell=64; jj=1;
```

```
while jj~=Ncell
        S=S+2^jj; jj=jj+1;
end
display(['Number of cells: ', num2str(Ncell)])
display('& total sum of grains is: '), disp(uint64(S))
% Now test our results with a simple solution: 2^64-1;
Error=S-2^64 %#ok
%% Version B. [for ... end] loop
clearvars
S=0; Ncell=64;
for ii=0:Ncell-1
        S=S+2^ii;
end
display(['Number of cells: ', num2str(Ncell)])
display('& total sum of grains is: '), disp(uint64(S))
% Now test our results with a simple solution: 2^64-1;
Error=S-2^64 %#ok
```

Both of the solution scripts output the same number of grains: 18,446,744,073,709,551,615 with 0 error. Note that $18{,}446{,}744{,}073{,}709{,}551{,}615 = 2^{64} - 1$ is the maximum length of any integer to be displayed correctly in MATLAB's 64-bit version installed on a 64-bit processor computer.

Note   In this example, the most efficient way of computing the number of grains on the chessboard is the vectorization approach:

n=0:63; N = 2.^n; S = sum(N); disp(uint64(S))

Example 9

Prove that $\sum_{k=1}^{\infty} \dfrac{1}{k^2} = \dfrac{\pi^2}{6}$. Compute this summation by using the [for ... end] and [while ... end] loops.

```matlab
% SumPi6.m
%% Example: sum(1/k^2)=pi^2/6
% Version A. [while...end] loop
clearvars
S=0; jj=1; k=input('Enter number of iterations to compute:  ');
while le(jj, k)
    S=S+1/jj^2; jj=jj+1;
end
display(['Number of cells: ', num2str(Ncell)])
display('& total sum of grains is: '), disp(uint64(S))
Error = (pi^2/6)-S; display(Error)
% Example: sum(1/k^2)=pi^2/6
% Version B. [for ... end] loop
clearvars -except k
S=0;
for jj=1:k
    S=S+1/jj^2;
end
display(['Number of cells: ', num2str(Ncell)])
display('& total sum of grains is: '), disp(uint64(S))
Error = (pi^2/6)-S; display(Error)
```

Example 10

Prove that $\sum_{n=0}^{\infty} \frac{(-1)^n 4}{2n-1} = \pi$. Compute this summation by using the [for ... end] and

[while ... end] loops.

```matlab
%% Series_PI.m
% [for .. end]
clearvars
N = input('Enter the number series to compute:   ');
S = 0;
for n = 1:N
    S = S+(4*(-1)^(n+1))/(2*n-1);
end
```

173

```
fprintf('Sum of n = %d is series:  %1.5f \n', n, S)
%%
% [while .. end]
clearvars
N = input('Enter the number series to compute:   ');
S = 0;
n = 1;
while n~=N+1
  S = S+(4*(-1)^(n+1))/(2*n-1);
  n=n+1;
end
fprintf('Sum of n = %d is series:  %1.5f \n', n, S)
```

Example 11

Prove that $\displaystyle\sum_{n=0}^{\infty}\frac{1}{n!}=e$. In this exercise, the sum of rational factorial series is equal to the natural logarithm base e. Compute the sum of the series $\dfrac{1}{0!}+\dfrac{1}{1!}+\dfrac{1}{2!}+\ldots\dfrac{1}{n!}\ldots=e$ by using the [for ... end] and [while ... end] loops.

```
%% Series_e.m
% [for .. end]
clearvars
N = input('Enter the number series to compute:   ');
S = 0;
for n = 1:N
    S = S+1/factorial(n);
end
fprintf('Sum of n = %d is series:  %1.5f \n', n, S)
%%
% [while .. end]
clearvars
N = input('Enter the number series to compute:   ');
S = 0;
n = 1;
```

```
while n~=N+1
  S = S+1/factorial(n);
  n=n+1;
end
fprintf('Sum of n = %d is series:  %1.5f \n', n, S)
```

Note that in this exercise the function `factorial()` is employed. The most efficient way of computing the sum of these series is the vectorization approach: `S = sum(1./factorial(1:N))`.

Memory Allocation

When all values from all iterations are saved, it is recommended to employ memory allocation methods to improve the computation efficiency. Let's analyze how the computation efficiency is enhanced by the memory allocation approaches. With the memory allocation technique, a user-specified memory is allocated to the computed variable with a user-specified size, and during the computation with a loop, all of the variable's values are recorded in the prespecified memory. This speeds up the whole computation process considerably. With the memory allocation technique, the exact size of the variable being computed is created before the loop iteration, via the standard matrix/array generators (e.g., `ones()` or `zeros()`).

The general syntax/pseudocode of the memory allocation is as follows:

```
% [FOR ... END]
M = zeros(1, n);
for ii = 1:n
M(ii) = [computation];
end
% [WHILE ... END]
MM = zeros(1, k); jj=1;
while jj ~=k
 MM(jj) = [computation];
 jj=jj+1;
end
```

This syntax can be applied to loops within loops. Let's consider the memory allocation technique in the following examples.

Note   It is important to employ memory allocation techniques while using the [for ... end] and [while ... end] loops with considerable computations. This will decrease computation time and save computing resources.

Example 12

In the 17th century, Leibnitz used the series expansion of *arctan(x)* to find an approximation of π. By using the Maclaurin series formula, we can write it as follows:

$$arctan(x) \approx 0 + (1)x + \frac{0}{2}x^2 - \frac{2}{3!}x^3 + \frac{0}{4!}x^4 + \frac{0}{5!}x^5 \ldots = x - \frac{x^3}{3} + \frac{x^5}{5} - \frac{x^7}{7} + \frac{x^9}{9} \ldots$$

Considering that $(1) = \frac{\pi}{4}$, we can substitute $x = 1$ into the previous expression and will get the following expression:

$$\frac{\pi}{4} = 1 - \frac{1}{3} + \frac{1}{5} - \frac{1}{7} + \frac{1}{9} - \ldots$$

That can be also written for *N* terms in this form:

$$\frac{\pi}{4} \cong \sum_{k=0}^{N} \frac{(-1)^k}{2k+1} = S_N$$

The task is to write a script that computes the sum (S_N) of N terms, approximates the value of $\frac{\pi}{4}$, and then plots the iteration process by computing the difference (error) $Error = \frac{\pi}{4} - S_N$ using the [for ... end] and [while ... end] loops.

Here are possible solution scripts collected under one script called LeibnitzSeries.m to compute the Leibnitz series for an approximation of $\frac{\pi}{4}$:

```
%% LeibnitzSeries.m
% Version A.
clearvars
N=input('Enter number of terms to approximate pi/4:  ');
```

```
S=0;                    % Initial value of summation
Err=zeros(1, N);    % Memory allocation
for k=0:N-1
    p=(-1)^k;
    S=S+p/(2*k+1);
    Err(k)=pi/4-S;  % Accumulates all of Errors from all iterations
end
plot(1:N, Err), grid on
title(['\pi/4 Approximation of ', num2str(N), ' terms'])
ylabel('Error'), xlabel('Number of terms')
%% Leibnitz series
% Version B
clearvars
n=input('Enter number of terms to approximate pi/4:   ');
S=0;  % Initial value of summation
Error=zeros(1, n);  % Memory allocation
k=0;
while k<=n-1
    p=(-1)^k;
    S=S+p/(2*k+1);
    Error(k+1)=pi/4-S; % Accumulates all of Errors from all iterations
    k=k+1;
end
plot(1:n, Error), grid on
title(['\pi/4 Approximation of ', num2str(n), ' terms'])
ylabel('Error'), xlabel('Number of terms')
```

If you execute the script with *N=200* terms, you get the plot shown in Figure 2-3, which displays the results of series approximation.

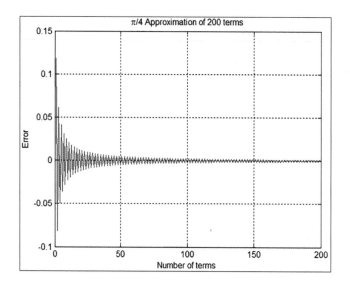

Figure 2-3. $\frac{\pi}{4}$ *approximation by 200 terms of the Leibnitz series*

It is also possible to have one loop within another. Let's look at several examples to generate loops within loops with [for ... end] and [while ... end].

Note   It is also possible to make multiple nested [for ... end] loops.

Example 13

Create elements of a 5-by-7 matrix with pseudorandom integer numbers of 1 to 10. Here is the solution script Pseudo_Randi.m:

```
% Pseudo_Randi.m
for k=1:5
    for n=1:7
        AB(k,n)=randi(10,1);
    end
end
display(AB)
```

By executing the script, the following results are obtained:

AB =

3	5	3	6	4	8	6
2	1	5	6	7	2	9
9	6	8	1	9	4	9
4	9	9	4	7	3	3
3	7	2	8	8	10	8

Note that every time when the script is executed, different pseudorandom integers ranging within [1, 10] are generated. As demonstrated in the previous examples, the [for, end] and [while, end] loop control statements can be employed interchangeably.

For this exercise, the following is the [while .. end] loop-based solution script:

```
% [while .. end]
k=1;
while k~=5
    n=1;
    while n~=7
        AB(k, n)=randi(10,1);
        n=n+1;
    end
    k=k+1;
end
display(AB)
```

Note where the second loop's iteration value n is assigned in the two nested [while .. end] loops.

Example 14

Compute the sum of the following series by using the [for ... end] and [while ... end] loop iterations: $f(b_n,x) = \sum_{n=1}^{\infty} \dfrac{b_n \sin\left((2n-1)\pi x\right)}{2L}$. Take $n = 1:11$, $b_n = [-5...5]$, $x = [0...L]$, and $L = 10$. For x take a step size of $\Delta x = \dfrac{L}{2000}$. Plot all computation results.

Here is the solution script with the [for ... end] loop (Sine_Series.m):

```
% Sine_Series.m
%% Ver A. [FOR ... END]
clearvars, close all
n=11; b = -5:5;x=0:.01:10; f=0; L=10;
f = zeros(n, numel(x)); % Memory allocation
for k=1:n
    f=f+(b(k)*sin((2*k-1)*pi*x(:))/(2*L));
    Fun(k,:)=f;
    plot(x(:), Fun(k,:)), label{k}=(['Iteration: ' num2str(k)]);
    legend(label{:}), hold all,
end
grid on, title('sum of series'), xlabel('x'), ylabel('f(b_n, x)')

%% Ver B.
clearvars
n=11; b = -5:5;x=0:.01:10; L=10;
f = zeros(n, numel(x)); % Memory allocation
for k=1:n
    f(k, :)=(b(k)*sin((2*k-1)*pi*x(:))/(2*L));
end
Fun = sum(f(:,:));  % Obtain the final values (total sum of all).
```

An alternative solution script with the [while ... end] loop (see Figure 2-4). Using f(k,:) saves all values of the vector $f(b_n, x)$ in the order of calculation inside the loop.

```
%% Ver C. [WHILE...END]
clearvars, close all
n=11; b = -5:5;x=0:.01:10; f=0; L=10; k=1;
Fun = zeros(n, numel(x)); % Memory allocation
while k<n+1
    f=f+(b(k)*sin((2*k-1)*pi*x(:))/(2*L));
    Fun(k,:)=f;
    plot(x(:), Fun(k,:)), label{k}=(['Iteration: ' num2str(k)]);
    legend(label{:}), hold all
```

```
    k=k+1;
end
grid on, title('sum of series'), xlabel('x'), ylabel('f(b_n, x)')
```

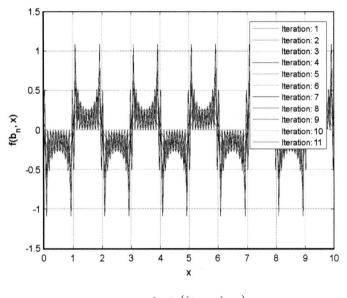

Figure 2-4. *Sum of series: f(b_n, x)=* $\sum_{n=1}^{\infty}\dfrac{b_n sin((2n-1)\pi x)}{2L}$

An alternative and best solution in terms of computational efficiency using vectorization approach with meshgrid() is as follows:

```
%% Ver D. Vectorization and No loop
clearvars, close all
n=11; b = -5:5;x=0:.01:10; L=10; k=1;
[bS, xS] = meshgrid(b,x);
[~, nS] = meshgrid(x, 1:n);
f = (bS.*sin(pi*(transpose(2*nS-1)).*xS)/(2*L));
plot(x, cumsum(f.'))
L = cell(n,1);
for ii=1:n
    L{ii}=strcat('n = ', num2str(ii));
end
legend(L)
grid on, title('sum of series'), xlabel('x'), ylabel('f(b_n, x)')
```

Note that the commands related to plot building, such as plot(), label{}, and legend(), do not have an effect on the accuracy of the calculations. But they considerably slow down the computation process. Therefore, it is advisable to avoid plotting the computed data within loops if possible.

Example 15

Compute the values of the function $F(\mu,t) = sin(2t) * \sqrt{sin\left(\mu t \sqrt{t + t^2 + t^3}\right)}$ for $t = 0...\pi$

($\Delta t = \pi/1000$) and $\mu = 1...7$ by using the [for ... end] and [while ... end] loop

operators.

Here is the plain script (F_Series.m) with two loop operators:

```
% F_Series.m
%% [FOR ... END]
mu=1:7; t=0:pi/1e3:pi;
F=zeros(numel(mu), numel(t)); % Memory allocation
for mu=1:7
    for ii=1:numel(t)
    F(mu, ii)=sin(2*t(ii))*sqrt(sin(mu*t(ii)*sqrt(t(ii)+t(ii)^2+t(ii)^3)));
    end
    plot(t, F(mu,:)), hold all
end
xlabel('t'), ylabel('F'), title('Sine waves'), grid on
%% [WHILE ... END]
mu=1:7; t=0:pi/1e3:pi;
F=zeros(numel(mu), numel(t)); % Memory allocation
MU = 1;
while ne(MU, 8)
    ii=1;
    while ii~=numel(t)+1
    F(MU, ii)=sin(2*t(ii))*sqrt(sin(MU*t(ii)*sqrt(t(ii)+t(ii)^2+t(ii)^3)));
    ii=ii+1;
    end
    plot(t, F(MU,:)), hold all
```

```
    MU=MU+1;
end
xlabel('t'), ylabel('F'), title('Sine waves'), grid on
```

Both iteration loops produce identical solutions and an identical plot figure, as shown in Figure 2-5. While simulating this script, there will be complex outputs from the square root of negative values as well. Therefore, some warning messages will be displayed in the Command window indicating omitted/ignored imaginary components in the plot shown in Figure 2-5.

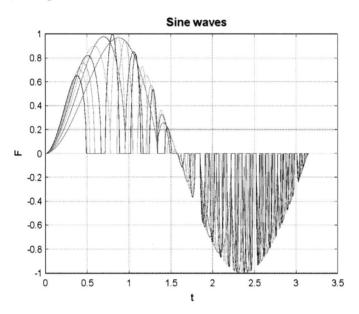

Figure 2-5. *Plot of* $F(\mu,t) = sin(2t) * \sqrt{sin\left(\mu t\sqrt{t+t^2+t^3}\right)}$ *for t = 0...π (Δt = π/1000),* $\mu = 1...7$

Example 16

Compute $f(t) = e^{\frac{-3t}{7}}$ function values for $t = 0...5$ with a time step of $\Delta t = 10^{-6}$ and $\Delta t = 10^{-3}$, and plot computed function values against time values. The very small time step $\Delta t = 10^{-6}$ is taken to demonstrate the importance of memory allocation and loop iteration versus vectorized calculations, and $\Delta t = 10^{-3}$ is taken to show the importance of where to put the `plot()` command while using loop iterations [`for ... end`] and [`while ... end`].

There are four different ways (combined into one script EXP_Calcs.m) of solving the given exercise with [for, end] and [while, end] loop control statements. All of the four ways produce identical results, but their computation (elapsed) time differs substantially. Note that in this exercise, [for, end] and [while, end] loop control statements are used for demonstration purposes only to explain how to employ loop statements and to improve their efficiency. By default, for this type of exercise or any other vector computations, the direct computation of vector values of $f(t)$ is much more efficient than using loop statements.

```
% EXP_Calcs.m computes Exp(-3t/7) Math function
% Solution method 1
clc; clear all; close all

tic
t=0:1e-6:5;
m_num=length(t);
for m=1:m_num
    f(m)=exp(-3*t(m)/7);
end
Tfor=toc; fprintf('[1] FOR loop computation TIME: %g \n', Tfor)
plot(t, f, 'ro'), xlabel('t'), ylabel('f(t)'), title('t vs. f(t)')
% % Solution method 2
clearvars, close all
tic
t=0:1e-3:5;
m_num=length(t);
for m=1:m_num
    f(m)=exp(-3*t(m)/7); plot(t(m), f(m), 'ro'), hold
end
TforP=toc;
xlabel('t'), ylabel('f(t)'), title('t vs. f(t)')

fprintf('[2] FOR loop computation TIME with plot() in it: %g \n', TforP)
% Solution method 3
clearvars, close all
tic
t=0:1e-3:5;
```

```
m_num=length(t); m=1;
while m<=m_num
    f(m)=exp(-3*t(m)/7); plot(t(m), f(m), 'ro'), on
    m=m+1;
end
Twhile=toc;
xlabel('t'), ylabel('f(t)'), title('t vs. f(t)')

fprintf('[3] WHILE loop computation TIME with plot() in it: %g \n', Twhile)
% Solution method 4. Improved by memory allocation.
clearvars, close all
tic
t=0:1e-6:5;
m_num=length(t); f=zeros(1,m_num-1);
for m=1:m_num
    f(m)=exp(-3*t(m)/7);
end
plot(t,f, 'bo'); TforIm=toc;
xlabel('t'), ylabel('f(t)'), title('t vs. f(t)')
fprintf('[4] Improved FOR loop TIME with memory allocation: %g\n', TforIm)
% Solution method 5. Vectorized method
clearvars, close all
t=0:1e-6:5;
tic
f=exp(-3*t/7);
Tvec=toc; plot(t, f, 'bo')
xlabel('t'), ylabel('f(t)'), title('t vs. f(t)')

fprintf('[5] Recommended VECTORIZED method time: %g\n', Tvec)
```

By executing the previous script (EXP_Calcs.m) in a personal computer (processor: Intel COREi7, RAM: 16GB, OS: Windows 10, in MATLAB R2022a), the following computation time results are obtained in the Command window. Note that the plot figure is not shown here.

```
[1] FOR loop computation TIME: 0.533504
[2] FOR loop computation TIME with plot() in it: 13.4151
[3] WHILE loop computation TIME with plot() in it: 13.1509
```

185

[4] Improved FOR loop TIME with memory allocation: 0.244569
[5] Recommended VECTORIZED method time: 0.0175538

From the simulation results, the influence of the plot() command on the computation time within [for, end] and [while, end] control loops is demonstrates once more. In this plain example, it is clear how much CPU time (~17.5 times more time) is spent on plotting the computed data in solutions 2 and 3 than in solutions 1 and 4 even though the step size $\Delta t = 10^{-3}$ is chosen to be 10^3 larger than the one ($\Delta t = 10^{-6}$) used with the three other ways. The simulation speed difference between the [for .. end] loop (way 1) and vectorization (method 5) is more than 30 times. In other words, the speed of the calculation with the [for ... end] loop is slower than the vectorization approach for more than 3000 percent.

Note   If it is possible, avoid using the plot() command within the [for... end] and [while ... end] loops to enhance the computation/simulation time efficiency.

The memory allocation is necessary whenever the use of [for, end] and [while, end] loop control statements is unavoidable. By using memory allocation techniques, you can improve the time efficiency of loop control statements substantially. In the previous simple example, the computation time has been improved more than 2.18 times by using a memory allocation technique (method 1 versus method 4). It must be noted that the time efficiency with a memory allocation technique can be observed if the number of iterations is considerably large, for instance, several thousand iterations or beyond. The vectorization approach is indeed superior. If method 4 with memory allocation is compared with the vectorized approach (method 5), the vectorization is more than 13 times faster than memory allocation method for computations of this example.

Note   If the computation/simulation problem requires using the [for...end] and [while ... end] loops, use the memory allocation technique to enhance the computation/simulation time efficiency.

To calculate elapsed (computation) time, stopwatch timer [tic ... toc] operators are recommended to track the computation time of a script. In fact, there are several other alternative operators as well in MATLAB that count elapsed (computation) of a script (code) or function file. They are [cputime] and [clock, etime]. These operators can be used as follows within M-files. This is the comparative analysis script (EXP_CT.m) of these operators with the previous studied example to compute $f(t) = e^{\frac{-3t}{7}}$ function values for $t = 0...5$ with a time step of $\Delta t = 0.0001$.

```
% EXP_CT.m
%% [tic ... toc]
clc; clear all, close all
T1=tic;
t=0:1e-4:5; m_num=length(t);
for m=1:m_num
    f(m)=exp(-3*t(m)/7);
end
Tfor1=toc(T1);
fprintf('FOR loop time [tic...toc]: %g \n', Tfor1)
%% [cputime ... cputime-Time]
clear all, close all
TT=cputime;
t=0:1e-4:5; m_num=length(t);
for m=1:m_num
    f(m)=exp(-3*t(m)/7);
end
Tfor2=cputime-TT;
fprintf('FOR loop time with [cputime]: %g \n', Tfor2)
%% [clock ... etime()]
clear all, close all
TT=clock; t=0:1e-4:5; m_num=length(t);
for m=1:m_num
    f(m)=exp(-3*t(m)/7);
end
Tfor3=etime(clock, TT);
fprintf('FOR loop time with [clock...etime]: %g \n', Tfor3)
```

187

By executing the previous script, the following results are obtained in the Command window:

```
FOR loop time [tic...toc]: 2.49076
FOR loop time with [cputime]: 2.46482
FOR loop time with [clock...etime]: 2.494
```

The simulation results show that these operators work similarly and produce very close numerical results. In addition to these operators, there is another operator called [timeit] that can be also employed to compute the overall computation time spent for all simulation processes of function files. This operator was introduced with MATLAB R2014b.

The [break] command is a powerful and handy operator that is used mainly within [while, end] and [for, end] loop control statements to halt the evaluation process when certain conditions are met. This is another technique to improve iteration time efficiency. Let's look at the [break] command's implementation in the following example.

Example 17

Let's compute the Fibonacci numbers. The Fibonacci sequence is defined by the following expressions: $f_1 = 1, f_2 = 1, f_3 = f_1 + f_2, ..., f_n = f_{n-2} + f_{n-1}$.

Thus, if $f_1 = 1, f_2 = 1, f_3 = 3, f_4 = 3 + 1, f_5 = 4 + 3, ...$, then do the following:

1) Show the first 13 elements of the Fibonacci sequence.

2) Compute the first 111 elements of the Fibonacci sequence or stop the iteration when the sequence has reached the value of 123456789.

Here is the solution script (Fibonacci.m). It computes the first 13 elements (series) with [for ... end] and [while ... end], and 111 elements (series) of the sequence or halts the computation if the sequence reaches to 12345789 by employing the [break] operator.

```
% Ver 1. Computation of 13 elements of the Fibonacci Numbers: [for ... end]
clearvars
f(1) = 1; f(2)=1;
f = [f(1), f(2), zeros(1, 11)]; % Memory allocation
```

```
fprintf('Element: %.0f  FN: %.0f \n', 1, f(1));
fprintf('Element: %.0f  FN: %.0f \n', 2, f(2));
for jj=3:13
    f(jj)=f(jj-2)+f(jj-1);
    fprintf('Element: %.0f  FN: %.0f \n', jj, f(jj));
end
%% Ver 2. Computation of 13 elements of the Fibonacci Numbers:
[while...end]
clearvars
f(1) = 1; f(2)=1; jj=3;
f = [f(1), f(2), zeros(1, 11)]; % Memory allocation

fprintf('Element: %.0f  FN: %.0f \n', 1, f(1));
fprintf('Element: %.0f  FN: %.0f \n', 2, f(2));
while jj<=13
    f(jj)=f(jj-2)+f(jj-1);
    fprintf('Element: %.0f  FN: %.0f \n', jj, f(jj));
    jj=jj+1;
end
%% [BREAK]. Iteration of the Fibonacci Numbers is controlled by [BREAK]
clearvars
F(1) = 1; F(2) = 1;
F = [F(1), F(2), zeros(1, 40)]; % Memory allocation

fprintf('Element: %.0f  FN: %.0f \n', 1, F(1));
fprintf('Element: %.0f  FN: %.0f \n', 2, F(2));
for ii=3:111
    F(ii)=F(ii-2)+F(ii-1);
    fprintf('Element: %.0f  FN: %.0f \n', ii, F(ii));
    if F(ii)>=123456789
    fprintf('Iteration is halted because \n ')
    fprintf('the last computed value is greater than 123456789 \n')
    break
    else
        continue
    end
end
```

189

This script produces the following output in the Command window:

```
Element: 1   FN: 1
Element: 2   FN: 1
Element: 3   FN: 2
Element: 4   FN: 3
Element: 5   FN: 5
Element: 6   FN: 8
Element: 7   FN: 13
Element: 8   FN: 21
Element: 9   FN: 34
Element: 10   FN: 55
Element: 11   FN: 89
Element: 12   FN: 144
Element: 13   FN: 233
```

The results from version 2 are identical to the results from version 1, and thus, they are not shown here. Here are the results from the previous section of the code with the [break] operator:

```
Element: 1   FN: 1
Element: 2   FN: 1
Element: 3   FN: 2
Element: 4   FN: 3
Element: 5   FN: 5
Element: 6   FN: 8
Element: 7   FN: 13
Element: 8   FN: 21
Element: 9   FN: 34
Element: 10   FN: 55
Element: 11   FN: 89
Element: 12   FN: 144
Element: 13   FN: 233
...
Element: 25   FN: 75025
Element: 26   FN: 121393
Element: 27   FN: 196418
...
```

```
Element: 39   FN: 63245986
Element: 40   FN: 102334155
Element: 41   FN: 165580141
Iteration is halted because
 the last computed value is greater than 123456789
```

Note that in the previous section with the [break] operator, the memory allocation operation F = [F(1), F(2), zeros(1, 40)] creates [1, 1, 40 elements of 0]. Therefore, after 41 iterations, the last element (42nd element) of F remains 0.

Example 18

Let's compute the values of the sine function $h(\theta) = sin(\theta)$ for $\theta = 0...2\pi$ with 1,000 incremental steps and stop computation when the value of the function $h(\theta) \approx 0.9999$.

The sine function fluctuates between 0 ... 1; 1...0; 0...-1; -1...0 for $\theta = 0...2\pi$; therefore, you should stop the computation loop when the function value reaches 0.9999. Here is the solution script (SINE.m) that computes and halts the computation process with a [break] statement when the preset condition is met:

```
% SINE.m computes sin(theta)
clearvars
theta=linspace(0, 2*pi, 1000); k=length(theta); h=ones(1,k-1);
for ii=1:k
    h(ii)=sin(theta(ii));
    data(ii,:)=[theta(ii);h(ii)];
if abs(h(ii))>=0.9999
    fprintf('Computation is halted after %g iterations\n', ii);
    fprintf('The function value is: %1.5f \n', h(ii))
fprintf('When theta is equal to % 1.5f degrees\n', theta(ii)*180/pi)
break
else
continue
end
end
plot(data(:,1), data(:,2), 'bd'), hold on
plot(data(ii,1),data(ii,2), 'p', 'markerfacecolor','c',  'markersize', 18)
grid on
```

When the script (SINE.m) is executed, the following output is displayed in the Command window:

```
Computation is halted after 249 iterations
The function value is: 0.99994
When theta is equal to  89.36937 degrees
```

Figure 2-6 shows the plot.

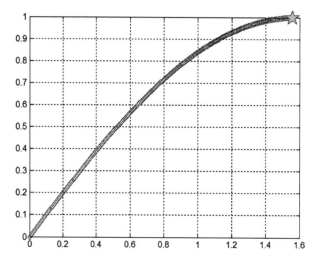

Figure 2-6. *Plot of h(θ) = sin (θ) for* $\theta = 0...\dfrac{\pi}{2}$ *and computation halted at h(θ)=0.9999*

This solution script can be also edited or rewritten with [while, end] in a similar way. That is left as an exercise for you to try for self-testing purposes.

Example 19

Write a script file with the [for ... end] loop and conditional statements [if, elseif, else, end] to compute the values of the function. Then plot the computed values.

$$f(t) = \begin{cases} e^{-2t_1} + 100t_1, & t_1 \le -13 \\ \cos(100t_2) - \sin(10t_2) - 3 \ge t_2 \ge -13 \\ \cos(100t_3) + e^{t_3}, & -3 \le t_3 \le 0 \\ 1, & t_4 > 0 \end{cases}$$

For the values of $t \in [-4\pi, 3\pi]$, take $\Delta t = 0.0001$.

Here is the solution script (Fun_Sets.m):

```matlab
%% Fun_Sets.m
clc, clearvars
t=-4*pi:.0001:3*pi;  % Whole t series
F=ones(size(t));     % Memory allocation
tic;
for ii = 1:numel(t)
    if t(ii)<=-2*pi
        F(ii)=exp(sin(2*t(ii)))+exp(cos(100*t(ii)));
    elseif t(ii)>-2*pi && t(ii)<=pi
        F(ii) = sin(2*t(ii))+cos(100*t(ii));
    elseif t(ii)>pi && t(ii)<=2*pi
        F(ii)=exp(sin(100*t(ii)))+exp(cos(2*t(ii)));
    else
        F(ii)=1;
    end
end
    Tma=toc;
    plot(t, F), grid on;
    plot(t, F), grid on, xlabel('\it t'), ylabel('\it F(t)'),
    title('Plot of function values w.r.t value ranges of t'), shg
    fprintf('Computation Time with memory allocation:  %2.6g \n', Tma)
```

Here are the outputs in the Command window and plot figure of the script (Fun_Sets.m). Figure 2-7 shows the plot figure of the script.

```
Computation Time with memory allocation:  0.0286769
```

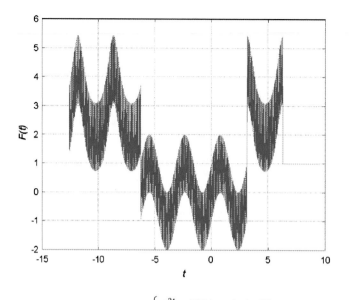

Figure 2-7. *Plot of the function* $f(t) = \begin{cases} e^{-2t_1} + 100t_1, & t_1 \leq -13 \\ cos(100t_2) - sin(10t_2) - 3 \geq t_2 \geq -13 \\ cos(100t_3) + e^{t_3}, & -3 \leq t_3 \leq 0 \\ 1, & t_4 > 0 \end{cases}$

One of the tricky points in the previous example is separating out the four value ranges of t according to the problem statement and then computing four separate function values. The loop operation [for ... end] before the conditional statements [if... elseif... elseif... elseif ... else ... end] perform the whole computation with respect to the indexes of t, whereas [if... elseif... elseif... elseif ... else ... end] will separate out the values of t by comparing every value of t(ii) according to the given conditions. In this case, the index (ii) is mandatory.

Symbol References in Programming

In MATLAB, there are dozens of important symbols used for various purposes in different contexts of code. The symbol references are discussed in the following sections.

Asterisk

In general, the symbol * is used for the multiplication operator, but in other instances, it can be used as a wildcard for file name identification. For instance, if you need to delete all files (MATLAB's autosave files of M/MLX-files) with a file extension of *.asv, the following command performs the task:

```
>> delete *.asv
```

In this way, you can delete all the files with the file extension of *.asv located in the current directory of MATLAB. Another example is if you need to locate an M-file called midpoint_rule.m, then you can type part of the file name as so:

```
>> dir('midp*.m')
midpoint_rule.m
```

Also, you can use * to get information on variables residing in the workspace by typing part of their names or delete them by typing part of their names, for instance:

```
>> whos
>> who
Your variables are:
H   N    NaH  NbN  NN  Nn  NCN  NkN
>> clear N*   % All variables whose names start with N will be deleted
from the Workspace
>> whos           % Check which variables left in the Workspace
  Name      Size              Bytes  Class      Attributes
   H        5x1                3525  struct
```

At Sign

The @ sign/operator is mostly used to construct a function handle. Here's an example:

```
F=@(argument lists)expression
```

Moreover, the @ operator is used to assign a new MATLAB class directory designator used to override MATLAB files. Here's an example:

```
\@myclass\myfun.m
```

Note that MATLAB classes can be used to define object-oriented designs, which are not discussed in this book.

Colon

The colon (:) operator is used to generate a sequence of numbers or elements of arrays or assign indices. Here's an example:

```
>>M=1:3:300;
>>N=1:200;
>>K=[N(1:10); M(1:10)];
>>Knew=K(1,:)
Knew =

        1    2    3    4    5    6    7    8    9    10
```

Note that when you use the colon to generate a sequence of numbers by specifying the step size, the last element of the being generated sequence may not match with the specified last element. For example, A = 1:3:20 generates A = [1 4 7 10 13 16 19], and B = 1:3:21 generates B = [1 4 7 10 13 16 19].

Comma

The comma (,) operator is used to separate variables or elements of arrays and indices, function input and output arguments, commands, or statements. Here's an example:

```
>> M, N,K
>> K(1,2)
>> G=@(t, x) sin(x*t)
>> for ii=1:2:20, Knewm(:)=K(1, ii)*2, end
```

Curly Brackets

The { } are used to assign or construct cell arrays and write fractions, superscripts, and subscripts via the LaTeX interpreter in plot figure titles and axis labels. Here's an example:

```
>> KnM=[{Knew}, {K}, {N},{M}];
```

```
>> title('$sin(\frac{2\phi}{5})$','interpreter', 'latex');
>> xlabel('t_{at}');
>> ylabel('F^{dot}')
```

Dollar Sign

The $ is used mostly in expressing fractions via the LaTeX interpreter in plot figure titles and axis labels. Here's an example:

```
>> ylabel('$\Omega(\frac{2\beta}{\alpha})$', 'interpreter', 'latex')
```

Dot

The . operator is used to add fields to an existing structure field, table variable name specifier, and object method specifier. It is used for decimal digits and element-wise matrix operations. Here's an example:

```
>> MH=H(1,1).bytes
>> P = [-3.1416, -1.5708, 0, 1.5708, 3.1416];
>> A = P.^2;
```

Dot-Dot

The two dots (..) operator is used to refer in sequence to the current directory. For instance, the following commands jump up one and two directories (if they already exist) from the current one:

```
>>cd ..\Fax
>>cd ..\..\Desktop
```

Dot-Dot-Dot (Ellipsis)

The three dots operator (...) is a line continuation operator. Here's an example:

```
>>   KmN=[1,2; ...
3,2; ]

KmN =
```

```
     1     2
     3     2
>> fprintf('This is what you have entered: %f\n', ...
KmN)
This is what you have entered: 1.000000
This is what you have entered: 3.000000
This is what you have entered: 2.000000
This is what you have entered: 2.000000
```

Parentheses

The () are mostly used for indexing of array elements or specifying arguments passed to a called function. Here's an example:

```
>> K(3, 4)
>> F=@(x, y)sqrt(x.^2+y.^2)
>> [m, n]=myFUN(x, y, k)
```

Percent

The % operator is one of the most commonly used reference symbols in MATLAB programming. It is used for five main purposes, of which four are for commenting purposes:

- To assign comments within M-files (%)

- To create cell modes with double percent signs (%%) and then leave a space

- To assign conversion specifiers when used within fprintf() and sprintf()

- To assign a block of comments when used with curly braces, %{, opened in one line and, %}, closed in another

- To remove (%#ok) a warning sign to display computation results in the Command window

Let's look at several examples for the % sign:

```
% This script is used for ...
% Created by ...
%% New cell mode starts from here ...
fprintf('you have entered %g which is EVEN number \n', N)
sprintf('you have entered %g which is EVEN number \n', N)
X1=sprintf('Root2 of the equation is x1= %g*i', x1);
%{
 Start comments and remarks...
 This script is used for....
 Created by ...
 Define entered integer type
 End comments and remarks...
%}
K=randi(100,3);
M=K*2.2          %#ok
```

Note   When using % with { } to insert comments in your script, the %{ open bracket must be on a separate line before the starting comments, and the %} closing must be at the end of the comments block on a separate line.

Semicolon

The ; is mainly used to construct arrays, suppress output display in the Command window, and separate commands entered on one line. Here's an example:

```
>>K=[1 2; 3,4];
>> clear all; close all; clc
```

Single Quotes

Single quotes (' ') are one of the most commonly used reference symbols of MATLAB. They are used as character and string constructors and with the inline function. Here's an example:

```
>> a=input('Enter value of a: ');
>> Hurray = 'Today is a great DAY ! '
>> F_1 =inline('2*x^2-3*x+13')
```

Note the single quote (') is also used for a matrix transpose operation:

```
>> A = [1, 2; 3, 4]
A =
       1      2
       3      4
>> A'
ans =
       1      3
       2      4
```

Slash and Backslash

The / \ characters are used to separate elements of a path or directory string. Note that these characters are also used for division (\ for left matrix division) operations, and \ is used to write Greek letters in the LaTeX format. Here's an example:

```
>> dir(['..\Circle13.m']) % Path or directory
>> title(' \alpha(\theta) vs. \Omega(\theta)')  % LaTeX interpretation
>> C = [2, 3, 4]; D=A/2;         % Division
>> A = [2 3; 3 4]; b = [1;2]; x = A\b;  % Backslash or left matrix divide
```

Square Brackets

The square brackets ([ ]) are used to construct, declare, and concatenate arrays, and to declare and capture values returned by functions. They are also used to indicate text and labels in plot figures. Here's an example:

```
>> K=[1 2; 3,4];
>> B=[K, K*2; rand(2,2), eye(2)];
>> [xout yout] = myfunction(z, t);
>> gtext(['End point: ' num2str(yout(end)) ' [m]'], 'background', 'y');
```

For more information on MATLAB symbol references, see the appendix.

Function Files

Function files are a remarkable strength of the MATLAB package for efficient code generation and computations. Once the function files are saved, they will have the symbol 📲; M-files will have the symbol 📄, and MLX-files will have symbol 📄. The way the function files are executed and the way their simulation results are obtained are substantially different from M- and MLX-files.

First, let's address why function files are necessary. The main reasons for using function files is that they are faster, are computationally more efficient, and consume less space in the workspace. Moreover, they can be called and executed within any M or MLX-files and the Command window. Function files are powerful and flexible and can be used for various purposes. They can be used not only to evaluate mathematical functions or perform calculations, but also to evaluate, compare and assess already evaluated data inputs as arguments. The general syntax and structure of the function file can be expressed in the following way:

```
function [A, B, ..., W] = FCN_name(a, b, ..., w)
% Help. ....
....
end
```

where [A, B, ..., W] are output variables and (a, b, ..., w) are input variables, also called input arguments. Note that function files have to start with the word function and the file name of the function file has to match with the word FCN_name in the previously shown example. In other words, the previous function file has to be named FCN_name.m. Otherwise, you cannot execute it. MATLAB automatically recognizes the function file as soon as you type in the word *function* and prompts you for the function file name when you save it the first time. Moreover, it is advised and required to end the function file with the keyword end. Naming the function files is like naming any M- or MLX-files. They have to start with a letter and must not contain any empty spaces or symbols except for an underscore, _.

In addition, some additional comments (help comments explaining how to use the function file) can be added to the existing function files by opening and editing them.

Note   One of the most common mistakes while recalling the function files is misspelling the name of the file.

In recent versions of MATLAB, function files can have different file names and function names. For example, a function file can be called `MY_fun`, but its script may be written as: `function [Out1, Out2, ...] = fun(In1, In2, ...)`. This will not cause any error. However, it is recommended to have the same file name and function name.

In more recent versions of MATLAB, the function file syntax (called *subfunctions/member functions*) can be implemented/embedded within M-files and called/executed as part of M-files.

The function files can be created for various purposes, for instance, to perform computations or display simulation results. The function files can be created/written in several ways.

- By using M-file or MLX-file editors or also using the Notepad text editor.

- By using the main menu [New ⊕] and selecting [*fx* Function] or [*fx* Live Function] from the drop-down option. The latter option of Live Function is available only starting from MATLAB 2018b.

- By using the Symbolic Math[2] toolbox's symbolic object identifier, `syms`.

When using the first way, you type in the function file and save a precisely matched function name with a file extension of `*.m` or `*.mlx`. It will be recognized by MATLAB automatically as a function file. When using the second and third ways, the editors will automatically generate the standard syntax components of the function file. The file can be edited without having to type in standard parts.

Let's look at several examples to demonstrate how to create function files and work with them.

---

[2] Symbolic Math is a registered trademark of MathWorks Inc.

Example 1

Write a function file to compute $f(t) = \dfrac{3}{e^{\sin(t)}} + \dfrac{5}{e^{\cos(t)}}$ and plot the computed values. This

can be implemented as shown in the following script called myfunction.m. It uses the
first way of creating the function file.

```
function f=myfunction(t)
% HELP: this function file (myfunction.m) computes values of the
expression:
%{
f(t)=3./exp(sind(t))+5./exp(cosd(t)) for given values of t, saves the
computed values of f(t) in the workspace, and plots computed resultsthem.
Note that [t] is input argument that has to be in degrees not in radians,
e.g., numerical array, e.g., >> t = -360:720360;
To execute this file: >> t = -720:10:720;  f=myfunction(t); that assigns
output values to a variable f and shows the plot.
To execute this file: >> t = 0:1:180;  myfunction(t);  that assigns output
values to a variable ans shows the plot.
To execute this file: >> FF=myfunction(-180:2:360);  that assigns output
values to a variable FF shows the plot.
%}

f=3./exp(sind(t))+5./exp(cosd(t));
plot(t, f, 'ro-', 'linewidth', 1.5), grid on
title('Plot of myfunction.m file')
end
```

It is good practice to have some comments in the function files specifying the
functionality of the function file, specifics about input arguments, help hints how to use
it with example input data, and so forth. When there are help comments in function files,
users can search for help about function files from the Command window. They do so by
recalling a function file name with the help command, as shown here:

```
>> help myfunction
  HELP: this function file (myfunction.m) computes values of the expression
f(t) = 3./exp(sind(t))+5./exp(cosd(t)) for given values of t, saves the
computed values of f(t) in the workspace, and plots computed results them.
```

Note that [t] is input argument that has to be in degrees not in radians, e.g.numerical array, e.g. >> t = -360:720360;
To execute this file: >> t = -720:10:720; f=myfunction(t); % that assigns output values to a variable f   % and shows the plot.
To execute this file: >> t = 0:1:180;   myfunction(t); % that assigns output values to a variable ans % shows the plot.
To execute this file: >> FF=myfunction(-180:2:360); % that assigns output values to a variable FF   % shows the plot.

Note that to call and use the created function files, the function files must reside in the current (active) MATLAB path directory. Or, you must add the path (directory) where the function file is residing to the search directory of MATLAB. You can do that using the addpath(' ') command and specifying the directory inside (' '). If the function file's directory (path) is not in the current directory or is not added to the list of the MATLAB paths, it cannot be recalled and executed. Subsequently, you'll get error messages indicating an undefined function or variable.

Simulation of the previous function file (called myfunction.m) can be done from the Command window or within another M-file with an input variable of *t*:

```
>> % If the output variable is not specified, "ans" will be the default
variable name.
>> % That has all values of "f" of the function file.
>> myfunction(0:2:720);  % The input argument "t" values are defined directly.
>> t=0:2:720; f=myfunction(t); % Input and output variables are defined
and given.
```

A second way of creating a function file is by clicking the main menu New and selecting *fx* Function or *fx* Live Function . The following function file will be opened with a default name of untitled.

This file is opened with the *fx* Function button in the M-file editor.

```
function [outputArg1,outputArg2] = untitled(inputArg1,inputArg2)
%UNTITLED6 Summary of this function goes here
%   Detailed explanation goes here
outputArg1 = inputArg1;
outputArg2 = inputArg2;
end
```

204

This one is opened with the [fx Live Function] button in the MLX-file editor.

```
Brief summary of this function.
Detailed explanation of this function.
function z = Untitled(x, y)
z = x + y;
end
```

The following example explains how to employ the third way of creating the function files.

Example 2

Given: $f(u_1,u_2)=\begin{cases} u_2-e^{2u_1} \\ 1.25u_1-e^{3u_2} \end{cases}$

A third way of creating a function file is to employ Symbolic Math[3] toolbox's symbolic object identifier syms. The MATLAB tool matlabFunction creates an anonymous function file to compute $f(u_1, u_2)$ with input arguments of u_1, u_2.

```
syms u1 u2          % Define variables (input data names)
% Formulate the given function equations
f=[u2-exp(2*u1), 1.25*u1-exp(3*u2)];
% Define a function file name, e.g. myFUN
matlabFunction(f, 'file', 'myFUN');
```

When you execute these commands, a function file called myFUN.m is created in the current MATLAB directory.

```
function f = myFUN(u1, u2)
%MYFUN
% F = MYFUN(U1,U2)

% This function was generated by the Symbolic Math Toolbox version 5.9
%      15-Apr-2022 14:20:01

f = [u2-exp(u1.*2.0),u1.*(5.0./4.0)-exp(u2.*3.0)];
```

---

[3] Symbolic Math is a registered trademark of MathWorks Inc.

Note that once a function is created with matlabFunction, you should not reuse this function tool to re-create the already created function file. If you try to do so, error messages will pop up. Another salient point in employing the matlabFunction tool is that this tool cannot generate function files containing variables with indexes, and if there are variables with indexes, the matlabFunction tool prompts you with errors.

Certain operations, commands, computations, or execution of several function files can be performed within the function file without any input arguments or outputs. A general syntax of such function files without any input arguments and output variables is very simple.

```
function functionNAME
% ...
```

Let's now look at the following example.

Example 3

Create a function file that clears all previously typed in commands from the Command window and displays benchmarking results of your computer. This getREADY.m function file performs the assigned tasks without any input arguments and any output data:

```
function getREADY
% HELP. getREADY.m
% This function file cleans up command history of MATLAB  and
% performs benchmarking calculations of several MATLAB functions (LU, FFT,
% ODE, etc.) installed computer against some of the computer configurations
of the date.

clc; close all; bench
end
```

Note   There are two cases when the function files can be executed with the Run

button from the M/MLX-file editor. (1) If the function file has no input arguments, (2) if the function file has a varying number of input arguments.

Example 4

Create a function file that computes the roots (x_1, x_2) and discriminant (D) of the quadratic equation $ax^2 + bx + c = 0$ with respect to the input arguments of $a, b, c,$ which are scalars. This function file (called QUAD.m) will produce three output variables according to the three input arguments.

```
function [x1, x2, D]=QUAD(a, b, c)
% QUAD.m
% Solves quadratic equations based on coefficients of: a, b, c
% Note that a, b, c need to be scalars.
fprintf('Solve: (%g)x^2+(%g)x+(%g)=0\n', a,b,c)
D=b^2-4*a*c;
% Roots
x1=(-b+sqrt(D))/(2*a); x2=(-b-sqrt(D))/(2*a);
if lt(D,0)
    fprintf('This equation does not have real value roots!\n');
    fprintf('Because discriminant is negative. D = %g\n', D);
    fprintf('Complex Root1: x1= %10s\n', num2str(x1));
    fprintf('Complex Root2: x2= %10s\n', num2str(x2));
elseif eq(D,0)
    disp('This equation has one unique root! ');
    disp('Because discriminant is zero. D=0 ');
    fprintf('This equation has one unique root! \n');
    fprintf('Because discriminant is zero. D=0 \n');
    fprintf('Unique Root: x = %g \n', x1);
else
    fprintf('This equation has two roots \n');
    fprintf('Because discriminant is: D = %g \n', D);
    fprintf('Root1 of the equation is: x1= %g \n', x1);
    fprintf('Root2 of the equation is: x2= %g \n', x2);
end
```

Let's test this function file with different input arguments and obtain the output variables from the Command window.

```
>> a=1; b=2; c=3;
>> [x1, x2, D]=QUAD_01(a, b, c)       % Case 1. The values of a, b, c are
predefined
Solve: (1)x^2+(2)x+(3)=0
This equation does not have real value roots!
Because discriminant is negative. D = -8
Complex Root1: x1= -1+1.4142i
Complex Root2: x2= -1-1.4142i
x1 =
  -1.0000 + 1.4142i
x2 =
  -1.0000 - 1.4142i
D =
     -8
>> [x1, x2, D]=QUAD_01(1, 2, 3)    % Case 2. The values of a, b, c are
entered directly
Solve: (1)x^2+(2)x+(3)=0
This equation does not have real value roots!
Because discriminant is negative. D = -8
Complex Root1: x1= -1+1.4142i
Complex Root2: x2= -1-1.4142i
x1 =
  -1.0000 + 1.4142i
x2 =
  -1.0000 - 1.4142i
D =
     -8
>> [x1, x2, D]=QUAD_01(b, c, a)  % Case 3. The values of a, b, c are
mixed up
Solve: (2)x^2+(3)x+(1)=0
This equation has two roots
Because discriminant is: D = 1
Root1 of the equation is: x1= -0.5
Root2 of the equation is: x2= -1
```

```
x1 =
   -0.5000
x2 =
    -1
D =
    1
>> QUAD_01(b, c, a);   % Case 4. The values of a, b, c are mixed up and
outputs are not specified
Solve: (2)x^2+(3)x+(1)=0
This equation has two roots
Because discriminant is: D = 1
Root1 of the equation is: x1= -0.5
Root2 of the equation is: x2= -1
>> a1=1; a2=5; a3=8;
>> [X, Y, Dis]=QUAD_01(a1, a2, a3) % Case 5. The output variable names
are changed
Solve: (1)x^2+(5)x+(8)=0
This equation does not have real value roots!
Because discriminant is negative. D = -7
Complex Root1: x1= -2.5+1.3229i
Complex Root2: x2= -2.5-1.3229i
X =
  -2.5000 + 1.3229i
Y =
  -2.5000 - 1.3229i
Dis =
    -7
```

From the five simulations, it is clear that while calling and executing the function files, the names of the input arguments and output variables are not essential. That means users can change the output variables (instead of x1, x2, D) and use different input arguments (other than a, b, c). Furthermore, the function files take the input arguments according to their order (case 2 versus case 3 and case 4). The output variables can be omitted, and the function file will still run, but there will be only one output corresponding to the first output variable specified in the function file's context.

Most Common Errors with the Function Files

There are a few common mistakes made while working with the function files. They are as follows:

- Misspelling function names while calling them.

- Providing the wrong number of input arguments than actually defined in the function file.

- Calling the wrong number of output variables than actually defined in the function file.

- Incorrectly using input arguments, such as mismatched size, variable type, etc.

- Defining the input arguments in the function file and providing their values within the function file context. In other words, defining the input arguments twice without any preconditions of missing arguments and use of alternative (default) values.

Let's test some of the most common errors that occur when working with the function files in the case of Example 1.

```
>> t=0:2:720;
>> % Error:  Misspelled function name with "M" instead of "m"
>> f=Myfunction(t);
Cannot find an exact (case-sensitive) match for 'Myfunction'

The closest match is: myfunction in
C:\ ...\Documents\MATLAB\myfunction.m
Did you mean:
>> f=myfunction(t);   % MATLAB's automatically detected
similar  function name

>>  t1 = [0:2:720]; t2= -720:2:0;
>> % Error: Two input arguments are inserted instead of one as defined in
the function file
>> f=myfunction(t1, t2);
Error using myfunction
Too many input arguments.
```

```
>> t2= -720:2:0;   % Input argument name can be altered that is not
an ERROR!
>> % Error: Two output variables are requested instead of one as defined in
the function file
>> [f1, f2]=myfunction(t2);
Error using myfunction
Too many output arguments.

>> % Error: Wrong type of input arguments
>> t3 =['a', 'b', 'c', 'd', 'f', '-a'  '-b' '-c' '-d' '-f'];  % t3 is a
character array and not numerical array
>> f=myfunction(t3);
Undefined function 'sind' for input arguments of type 'char'.
Error in myfunction (line 4)
f=3./exp(sind(t))+5./exp(cosd(t));
```

Let's analyze the case whereby the input argument is defined within the function file contexts and is called as the input argument while executing the function file. Let's make some changes to the function file (myfunction.m).

```
function f=myfunction(t)
% HELP: this function file (myfunction.m) computes values of the expression
% f(t) = 3./exp(sind(t))+5./exp(cosd(t)) for
% given values of t, saves the computed values of f(t) in the workspace,
% and plots computed results.
% Note that [t] is input argument that has to be in degrees not in radians,
% e.g.numerical array, e.g. >> t = -360:720360;
% To execute this file: >> t = -720:10:720;  f=myfunction(t); that
% assigns output values to a variable f     % and shows the plot.
% To execute this file: >> t = 0:1:180;       myfunction(t); that assigns
% output values to a variable ans % shows the plot.
% To execute this file: >> FF=myfunction(-180:2:360);  that assigns output
% values to a variable FF
t = 0:720;  % Input variable t is assigned internally
f=3./exp(sind(t))+5./exp(cosd(t));
plot(t,f, 'ro-', 'linewidth', 1.5), grid on
title('Plot of myfunction.m file')
```

211

In this case, the function file takes the input argument values and ignores the input arguments for *t* when it is called/executed. Here's an example: `>> t =0:1:180; f=myfunction(t);`.

The example outputs the computed values of f(t) for *t* = 0:720 by ignoring our specified input variable *t* =0:1:180.

Varying Number of Inputs and Outputs

One of the remarkable advantages of the function files over M or MLX-files is that they may have a varying number of input arguments and output variables. This flexibility of the function files can be attained via several MATLAB's built-in tools, commands, and functions. To vary the number of input arguments, you use `varargin`, and to vary the output variables, you use `varargout`. To make these functions more useful with respect to the given problem, there are two other MATLAB built-in tools, commands, and functions, namely, `nargin` and `nargout`, which determine how many input arguments and output variables are called while executing the function file. The varying inputs and outputs of the function files can be used in a few different combinations while declaring function input arguments and output variables. Here's an example:

1. `function [Out1, Out2] = function_name(varargin)`

2. `function varargout = function_name(Input1, Input2, Input3)`

3. `function varargout = function_name(varargin)`

To find out more detailed information on these tools, type the following into the Command window:

```
>> help varargin
>> help varargout
>> help narargin
>> help narargout
>> doc varargin
>> doc varargout
```

Let's look at two different examples dealing with varying input arguments and output variables.

1) Varying number of inputs and three output variables

2) Varying inputs and varying number of output variables

Example 5

Create a function file (Quad_Var.m) to solve the quadratic equation ($ax^2 + bx + c = 0$) with a varying number of input variables (a, b, c) and with three outputs (x_1, x_2, D).

```
function [x1, x2, D]=Quad_Var(varargin)
% Quad_Var.m computes roots of the quadratic equation with varying number
% of input arguments for a, b, c. Gives three output variables: x1, x2, D.
% There are four cases considered:
% Case 1. No Input arguments: a =1; b=2; c=3; values are taken as inputs.
% Case 2. One Input argument given: a, and b=2, c=3 are taken as inputs.
% Case 3. Two Input arguments given: a and b, c=3 is taken as a 3rd input
% Case 4. Three Input arguments given: a, b and c.
% E.g.
% Run: [x1, x2, D] = Quad_Var();        % No Input
% Run: [x1, x2, D] = Quad_Var(1);       % One Input: a=1;
% Run: [x1, x2, D] = Quad_Var(1, 2);    % Two Inputs: a=1; b=2;
% Run: [x1, x2, D] = Quad_Var(1, 2, 3); % Three Inputs: a=1; b=2; c=3;

if nargin==0          % Case 1. No input arguments
    a=1; b=2; c=3;
elseif nargin==1      % Case 2. One input argument only
    a=varargin{1}; b=2; c=3;
elseif nargin==2      % Case 3. Two input arguments
    a=varargin{1}; b=varargin{2}; c=3;
else                  % Case 4. Three input arguments
    a=varargin{1}; b=varargin{2}; c=varargin{3};
end
D=b^2-4*a*c;x1=(-b+sqrt(D))/(2*a); x2=(-b-sqrt(D))/(2*a);
end
```

In the function file Quad_Var.m the MATLAB's built-in command/tool nargin automatically counts how many inputs are given while calling the function file. The varargin{1} represents the first input argument, varargin{2} represents the second input argument, and likewise varargin{3} represents the third variable. In this order, n varying number of a function file can be created and called.

Let's test the function file (Quad_Var.m) in four different scenarios:

- *No inputs*: The values of *a, b, c* are taken from the function file context *a=1, b=2, c=3*.

- *One input*: *a* is given (e.g., *a = 1* or any scalar to be assigned to *a*), and the two values (*b, c*) are taken from the context of the function file: *b =2, c=3*.

- *Two input arguments*: *a, b* are given (e.g., *a = 1, b = 2* or any scalars to be assigned to *a, b*), and one input (*c*) is taken from the context of the function file: *c = 3*.

- *Three input arguments*: *a, b, c* are given (e.g., *a = 1, b = 2, c=3* or any scalars to be assigned to *a, b, c*).

```
>> [x1, x2, D]=Quad_Var          % Case 1. No Input
x1 =
  -1.0000 + 1.4142i
x2 =
  -1.0000 - 1.4142i
D =
     -8
>> [x1, x2, D]=Quad_Var(1)       % Case 2. One Input
x1 =
  -1.0000 + 1.4142i
x2 =
  -1.0000 - 1.4142i
D =
     -8
>> [x1, x2, D]=Quad_Var(1,2)     % Case 3. Two Inputs
x1 =
  -1.0000 + 1.4142i
```

```
x2 =
  -1.0000 - 1.4142i
D =
    -8
>> [x1, x2, D]=Quad_Var(1,2, 3) % Case 4. Three Inputs
x1 =
  -1.0000 + 1.4142i
x2 =
  -1.0000 - 1.4142i
D =
    -8
```

Let's look at the following example to analyze the varying number of outputs and specific (nonvarying) input arguments (two input arguments).

Example 6

Write a function file to compute the Leibnitz series expansion of *arctan(x)* to find an approximation of π by using the sum of the Maclaurin series formula for N terms in the following form:

$$\frac{\pi}{4} \cong \sum_{k=0}^{N} \frac{(-1)^k}{2k+1} = S_N$$

Write a script that computes the sum (S_N) of N terms, which is the approximation of $\frac{\pi}{4}$. The difference (error) is $Error = \frac{\pi}{4} - S_N$ and *Nhalt* is the number when the iteration is halted due to the user-specified error tolerance (E_tol) is attained. Use the [for ... end] and [while ... end] loops.

```
function [varargout] = Leibnitz_VarOut(varargin)
% Leibnitz_VarOut.m computes and plot the Leibnitz series expansion of
%          arctan(1) to find an approximation of PI.
%          Input arguments:
%          N (number of terms, e.g. 10000)
%          E_tol (Error tolerances, e.g. 0.0001)
%          OUTPUT variables are:
%          Out1 = Error;
```

215

```
%           Out2 = SN (Sum of series);
%           Out3 = Nhalt (terms when simulation is halted).
if nargin==0
    N =1000; E_tol = 0.001;
elseif nargin<2
    E_tol = 0.001;
    N = varargin{1};
else
    N = varargin{1}; E_tol=varargin{2};
end

SN=0;
for ii=0:N
    SN = SN+((-1)^ii)/(2*ii+1);
    Error=abs(pi/4-SN);
    if abs(Error)<=E_tol
        Nhalt=ii;
        break
    else
        Nhalt=ii;
        continue
    end
end
if nargout == 0       % No Output specified: NO outputs
    disp('NO Outputs!')
elseif     nargout ==1  % One Output: Error
    varargout{1}=Error;
elseif nargout ==2     % Two Outputs: Error and Sum of series
    varargout{1}=Error; varargout{2}=SN;
else      % Three Outputs: Error, Sum of Series, Number of Iterations,
    varargout{1}=Error; varargout{2}=SN;varargout{3}= Nhalt;
end
```

If no input arguments are given, then the output will be an error and two input arguments. For example, the number of iterations and error tolerance will be 1000 and 0.001, respectively. If there is one input, then that will be the number of iterations, and the error tolerance will take the value of 0.001.

216

While executing this function file, if the output variable is not specified, then no outputs will be obtained but only a short note (no outputs!) displayed in the Command window. If one output variable is called (with any variable name), then the achieved last error (Error) value will be the output variable's value. If two output variables (names) are specified while calling/executing the function file, then the error (Error) and sum of the series (S_N) will be output. If three output variables are called, then the Nhalt, S_N, and *Error* values will be obtained. It must be noted that from the specified conditions if the given (*N* input) number of iterations cannot produce the specified input (*E_tol*) error tolerance, then Nhalt (the output variable) will be equal to *N* (the input argument).

```
[EE, SM]=Leibnitz_VarOut()
EE =
    1.0000e-03
SM =
     0.7844
>> [EE, SM]=Leibnitz_VarOut(10000)
EE =
    1.0000e-03
SM =
     0.7844
>> Leibnitz_VarOut(1e5, 1e-4)    % No output variable specified and thus,
No output
NO Outputs!
>> E = Leibnitz_VarOut(1e5, 1e-4) % One output specified and thus, Error
displayed
E =
    1.0000e-04
>> [ Ee Ss]  = Leibnitz_VarOut(1e5, 1e-4) % Two outputs: Error and Sum
of series
Ee =
    1.0000e-04
Ss =
     0.7853
>> [ Ee Ss Nn ]  = Leibnitz_VarOut(1e5, 1e-4) % Three outputs: Error, Sum
of series, N Iterations
Ee =
```

```
    1.0000e-04
Ss =
    0.7853
Nn =
        2499
```

Let's look at another example with some computations to demonstrate and analyze the function files with a varying number of input arguments and output variables.

Example 7

Compute the values of this function:

$$ff(t) = \{e^{-2t_1} + 100t_1, \ t_1 \leq -13 \quad \cos\cos(100t_2) - \sin(10t_2) - 3 \geq t_2 \geq -13$$
$$\cos\cos(100t_3) + e^{t_3}, \ -3 \leq t_3 \leq 01, \quad t_4 > 0$$

The function $f(t)$ will have four input arguments (t_1, t_2, t_3, t_4), whose default set values are $t = [t_1, t_2, t_3, t_4] \in [-20, 6]$, $t_1 = -20 : 0.001 : -13$; $t_2 = -12.999 : 0.001 : -3$; $t_3 = -2.999 : 0.001 : 3$; and $t_4 = 3.001 : 0.001 : 6$.

Task 1

Simulate a system within the values of $t = [t_1, t_2, t_3, t_4]$, $t \in [-20, 6]$ with $\Delta t = 0.001$. Note that this function file may have the following:

1) No input argument and as default values of the function file: $t_1 = -20 : 0.001 : -13$; $t_2 = -12.999 : 0.001 : -3$; $t_3 = -2.999 : 0.001 : 3$; $t_4 = 3.001 : 0.001 : 6$ will be taken.

2) One input argument that has to represent t as a vector space of data points within $[-20, 6]$.

3) Two or three input arguments, which will be ignored, and the default set values of $t_1 = -20 : 0.001 : -13$; $t_2 = -12.999 : 0.001 : -3$; $t_3 = -2.999 : 0.001 : 3$; and $t_4 = 3.001 : 0.001 : 6$ will be taken instead.

4) Four input arguments that have to represent t_1, t_2, t_3, t_4 in consecutive order.

5) More than four input arguments; then all entries need to be ignored, and some warning messages need to be displayed. Instead of the entries, the default set values of the function file, $t_1 = -20:0.001:-13;$ $t_2 = -12.999:0.001:-3;$ $t_3 = -2.999:0.001:3;$ $t_4 = 3.001:0.001:6,$ will be taken.

Task 2

The function file (Ex7Var.m) has to produce any of the following underlined outputs with respect to the call commands of the function file:

1) No outputs and plot of the computed values of $f(t)$

2) One output: $f_1(t) = e^{-2t_1} + 100t_1$

3) Two outputs:

$$f_1(t) = e^{-2t_1} + 100t_1, f_2(t) = cos(100t_2) - sin(10t_2)$$

4) Three outputs:

$$f_1(t) = e^{-2t_1} + 100t_1, f_2(t) = cos(100t_2) - sin(10t_2),$$
$$f_3(t) = 100t_3 + e^{t_3}$$

5) Four outputs:

$$f_1(t) = e^{-2t_1} + 100t_1, f_2(t) = cos(100t_2) - sin(10t_2),$$
$$f_3(t) = 100t_3 + e^{t_3}, f_4(t) = 1$$

6) Five outputs:

$$f_1(t) = e^{-2t_1} + 100t_1, f_2(t) = cos(100t_2) - sin(10t_2),$$
$$f_3(t) = 100t_3 + e^{t_3}, f_4(t) = 1, t = t,$$

7) Six outputs:

$$f_1(t) = e^{-2t_1} + 100t_1, f_2(t) = cos(100t_2) - sin(10t_2),$$
$$f_3(t) = 100t_3 + e^{t_3}, f_4(t) = 1, t = t, t_1$$

8) Seven outputs:

$$f_1(t) = e^{-2t_1} + 100t_1, f_2(t) = cos(100t_2) - sin(10t_2),$$
$$f_3(t) = 100t_3 + e^{t_3}, f_4(t) = 1, t = t, t_1, t_2$$

9) Eight outputs:

$$f_1(t)=e^{-2t_1}+100t_1, f_2(t)=cos(100t_2)-sin(10t_2),$$

$$f_3(t)=100t_3+e^{t_3}, f_4(t)=1, t=t, t_1, t_2, t_3$$

10) Nine outputs:

$$f_1(t)=e^{-2t_1}+100t_1, f_2(t)=cos(100t_2)-sin(10t_2),$$

$$f_3(t)=100t_3+e^{t_3}, f_4(t)=1, t, t_1, t_2, t_3, t_4$$

If more than nine outputs are requested, then the function file will prompt you with a warning message and will note that this function file can produce a maximum of nine output variables.

The final solution script is called Ex7Var.m and has incorporated all the points in tasks 1 and 2 concerning the varying number of input arguments and output variables.

```
function [f1, f2, f3, f4, t, t1, t2, t3, t4, varargout
] = Ex7Var(varargin)
% HELP: Ex7Var.m is a function file to compute a complex function whose
% computation function components differ w.r.t the values of t
% It may have no, one, two, three, or four input arguments for t
% [f1, f2, f3, f4, t] = Ex7Var(t1, t2, t3, t4)
% [f1, f2, f3, f4, TT, t1, t2, t3, t4]=Ex7Var(linspace(-20, 6, 10000));
if eq(nargin,0)  % No input argument
t1 = -20:.001:-13; t2=-12.999:.001:-3; t3=-2.999:.001:3; t4 = 3.001:1e-3:6;
t = [t1,t2, t3, t4];
f1=exp(sin(2*t1))+exp(cos(100*t1)); f2=cos(2*t2)+sin(100*t2);
f3=exp(sin(100*t3))+exp(cos(2*t3))+sin(100*t3)+cos(2*t3);
f4=ones(size(t4));
elseif eq(nargin,1)&&min(varargin{1})<=-13&&max(varargin{1})>=6 % One Input
    t = varargin{1};
    for ii=1:numel(t)
        if le(t(ii),-13)
            t1(ii)=t(ii);
        elseif gt(t(ii),-13) && le(t(ii),-3)
            t2(ii)=t(ii);
        elseif gt(t(ii),-3) && le(t(ii),3)
```

```
            t3(ii)=t(ii);
        else
            t4(ii)=t(ii);
        end
    end
                      f1=exp(sin(2*t1))+exp(cos(100*t1));
  t2=t2(find(t2~=0)); f2=cos(2*t2)+sin(100*t2);
  t3=t3(find(t3~=0));f3=exp(sin(100*t3))+exp(cos(t3))+sin(100*t3)+
  cos(2*t3);
  t4=t4(find(t4~=0)); f4=ones(size(t4));
elseif gt(nargin,1) && lt(nargin,4) % Two or Three Inputs
    warndlg('t series need to be in four separate ranges or one
    united !!!')
    warndlg('Your entries are ignored and example data taken instead!!!')
    t1=-20:.001:-13; t2=-13.001:.001:-3; t3=-3.001:.001:3; t4=3.001:1e-3:6;
    t = [t1,t2, t3, t4];
    f1=exp(sin(2*t1))+exp(cos(100*t1)); f2=cos(2*t2)+sin(100*t2);
    f3=exp(sin(100*t3))+exp(cos(2*t3))+sin(100*t3)+cos(2*t3);
    f4=ones(size(t4));
elseif eq(nargin, 4)                 % Four Inputs
    t1 = varargin{1}; t2=varargin{2}; t3=varargin{3}; t4 = varargin{4};
    t = [t1,t2, t3, t4];
    f1=exp(sin(2*t1))+exp(cos(100*t1)); f2=cos(2*t2)+sin(100*t2);
    f3=exp(sin(100*t3))+exp(cos(2*t3))+sin(100*t3)+cos(2*t3);
    f4=ones(size(t4));
else                                 % More than Four Inputs
warndlg('Check your entries: input arguments for [t] or [t1, t2, t3, t4]')
    warndlg('Your entries are ignored and default data taken instead!!!')
    t1=-20:.001:-13;t2=-12.999:.001:-3; t3=-2.999:.001:3;
    t4 = 3.001:1e-3:6;
    t = [t1,t2, t3, t4];
    f1=exp(sin(2*t1))+exp(cos(100*t1)); f2=cos(2*t2)+sin(100*t2);
    f3=exp(sin(100*t3))+exp(cos(2*t3))+sin(100*t3)+cos(2*t3);
    f4=ones(size(t4));
end
```

```
plot(t1, f1, 'r', t2, f2, 'b', t3, f3, 'm',t4, f4, 'go--'), grid on; shg;
legend('toggle')

% Set number of output variables is verified
MIN_outs = 0; MAX_outs=9;
if nargout>MAX_outs
fprintf('Asked %3g outputs that are more than set outputs!!! \n', nargout);
warndlg('This fucntion file is assigned to have max. of 9 outputs!!!')
fprintf('Asked %3g outputs are beyond the set outputs !!! \n', nargout);

end
% NARGOUTCHK; Checks and prompts error if the number of outputs requested
% by the user is beyond 9!
% nargoutchk(MIN_outs, MAX_outs) % Can be also employed

end
```

It should be noted that in the Ex7Var.m function file, the nargin function will check how many input arguments are specified while calling this function file for task 1. The function's input arguments are taken for simulations depending on the number of the input arguments verified within the [if... elseif... elsefif ... elseif ... else ... end] conditional operators, addressing all points in task 1 for input arguments.

Moreover, the [for ... end] loop operator with another internal [if ...elseif ... elseif...else...end] conditional operator set after a first ([elseif ...]) splits up the given one input argument (t) values for four separate sets of values for t_1, t_2, t_3, t_4 according to the given exercise statements.

The logical indexing operations of find( ) with ~= 0 define which elements (taken from t within [for ... end] loop and [if ...elseif ... elseif...else...end] operations) of t_2, t_3, t_4 are to be taken for simulations to skip overlapping points. When the user enters two or three input arguments, nargin and [elseif ...] verify the entry. Two warning message dialogs will pop up informing the user that all entries are ignored and the default values are taken instead.

In the case of four input arguments, all input arguments are considered in the order of t_1, t_2, t_3, t_4, and simulations are performed. If the number of input arguments exceeds four, then again two warning message dialogs will pop up informing the user that all entries are ignored and the default values are taken instead. The plot( ) command plots all of the simulation results. This completes all verifications.

The specified MIN_outs = 0; MAX_outs=9; along with nargout verify the number of requested outputs. At this area of the script, the nargoutchk function can be also employed to detect the wrong number of requested output variables. If the requested number of outputs is more than nine, the warning message is displayed in the Command window. A plot figure is displayed, a warning dialog is displayed, and no output variables are obtained. The conditional statements [if ... elseif... ... end] with nargout determines which output corresponds to which simulation output.

Let's test the script (Ex8Var.m) for a different number of input arguments and output variables.

1) No input and no output

```
>> Ex7Var();
No outputs!
```

There will be no computation results except for the plot figure shown in Figure 2-8.

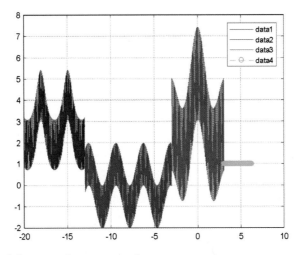

Figure 2-8. *Plot of the simulation results*

$$f(t) = \{e^{-2t_1} + 100t_1, \quad t_1 \leq -13 \; \cos(100t_2) - \sin(10t_2) - 3 \geq t_2 \geq -13$$
$$\cos(100t_3) + e^{t_3}, \quad -3 \leq t_3 \leq 01, \quad t_4 > 0$$

2) One input *t* is given, and four outputs are requested:

```
>> t=linspace(-20, 6, 1000);
>> [f1, f2, f3, f4]=Ex7Var(t);
>> whos
```

Name	Size	Bytes	Class	Attributes
f1	1x269	2152	double	
f2	1x385	3080	double	
f3	1x230	1840	double	
f4	1x116	928	double	
t	1x1000	8000	double	

Four outputs (f1, f2, f3, f4) are saved in the workspace and the plot figure (shown in Figure 2-8).

3) No input and nine output variables are requested:

```
>> clearvars;
>> clearvars; [f1, f2, f3, f4, t_all, t1, t2, t3, t4]=Ex7Var();
>> whos
```

Name	Size	Bytes	Class	Attributes
f1	1x7001	56008	double	
f2	1x10000	80000	double	
f3	1x6000	48000	double	
f4	1x3000	24000	double	
t1	1x7001	56008	double	
t2	1x10000	80000	double	
t3	1x6000	48000	double	
t4	1x3000	24000	double	
t_all	1x26001	208008	double	

Nine outputs (f1, f2, f3, f4, t1, t2, t3, t4, t_all) are obtained in the workspace.

4) Three input arguments and no output variables are requested:

```
>> clearvars; t=-20:.005:6; t3=-2.999:.002:3; t4 = 3.001:1e-2:6;
>> Ex7Var(t, t3, t4);
No outputs!
>> whos
```

Name	Size	Bytes	Class	Attributes
t	1x5201	41608	double	
t3	1x3000	24000	double	
t4	1x300	2400	double	

There are no outputs in the workspace from the simulations except for entries (input arguments, t, t3, t4), and there is a plot figure in Figure 2-8. In addition, the following two warning dialog boxes are displayed, stating the mismatch of the entries with the necessary t series, and the input entries are ignored:

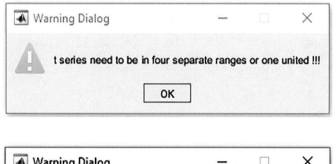

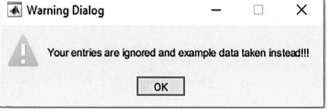

5) Three input arguments and three output variables are requested:

```
>>clearvars; t_all=-20:.001:6; In2=-2.999:.001:3; In3 = 3.001:1e-3:6;
>> [Out1, Out2, Out3]=Ex7Var(t_all, In2, In3);
>> whos
```

Name	Size	Bytes	Class	Attributes
Out1	1x7001	56008	double	
Out2	1x10000	80000	double	
Out3	1x6000	48000	double	
t_all	1x26001	208008	double	
In2	1x6000	48000	double	
In3	1x3000	24000	double	

Three outputs (Out1, Out2, and Out3 representing f1, f2, and f3, respectively) from the simulations and entries (input arguments: t_all, In2, In3) are obtained in the workspace. There is a plot figure as well (Figure 2-8). In addition, there are two warning dialog boxes displayed as shown in the previous case.

Note   While calling/executing the function files, output variable names and input argument names can be altered. That would not cause any errors. This is one of the useful attributes of the function files over M/MLX-files.

---

6)   Four input arguments representing t1, t2, t3, t4 and nine output variables representing f1, f2, f3, f4, t_all, t1, t2, t3, t4 are requested:

```
>> clearvars
t1 = -20:.0025:-13; t2=-12.999:.005:-3; t3=-2.999:.002:3; t4 =
3.001:5e-3:6;
>> [f1, f2, f3, f4, T_all, time1, time2, time3, time4]=Ex8Var(t1, t2,
t3, t4);
>> whos
```

Name	Size	Bytes	Class	Attributes
T_all	1x8401	67208	double	
f1	1x2801	22408	double	
f2	1x2000	16000	double	
f3	1x3000	24000	double	
f4	1x600	4800	double	
t1	1x2801	22408	double	
t2	1x2000	16000	double	
t3	1x3000	24000	double	
t4	1x600	4800	double	
time1	1x2801	22408	double	
time2	1x2000	16000	double	
time3	1x3000	24000	double	
time4	1x600	4800	double	

All requested output variables are obtained in the workspace along with the plot figure (Figure 2-8). Moreover, it should be noted that the input entries (arguments) t1, t2, t3, and t4 match with the output variables time1, time2, time3, and time4, respectively.

7) Nine output variables representing f1, f2, f3, f4, t_all, t1, t2, t3, t4 are requested without any input arguments:

```
>> clearvars
>> [f1, f2, f3, f4, T_all, time1, time2, time3, time4]=Ex7Var();
>> whos
  Name        Size                Bytes  Class     Attributes

  T_all       1x26001            208008  double
  f1          1x7001              56008  double
  f2          1x10000             80000  double
  f3          1x6000              48000  double
  f4          1x3000              24000  double
  time1       1x7001              56008  double
  time2       1x10000             80000  double
  time3       1x6000              48000  double
  time4       1x3000              24000  double
```

With no input arguments and requesting only nine output variables, the function file produces nine output variables with the assigned output variable names in the workspace by using the default set values for t_1, t_2, t_3, t_4 and the plot figure (Figure 2-8).

8) With one input for t and 10 outputs:

```
>> clearvars
>> tall=-20:.001:6;
>> [f1, f2, f3, f4, T_all, time1, time2, time3, time4, T_all]=Ex7Var(tall);
Asked  10 outputs are beyond the set outputs !!!
One or more output arguments not assigned during call to
"varargout". >> whos
  Name        Size                Bytes  Class     Attributes

  tall        1x26001            208008  double
```

Warning Dialog — ☐ ✕

⚠ This function file is assigned to have max. of 9 outputs!!!

OK

The 10th output variable is beyond the number of assigned output variables. Therefore, the function file does not produce any output variables in the workspace except for two warning messages, warning dialogs, and the plot figure (Figure 2-8).

There are many other possible scenarios to test this function file for robustness. By default the robustness of this function file is considerably high within the predefined conditions of the exercise for input arguments in particular. But there are a few other cases in which the outputs of this function file may not be accurate due to inaccurately chosen entries for t, or t_1, t_2, t_3, t_4. For example, one input for t might contain just two elements ([-20, 6]) of a row or column vector. Or similarly, the entries for t_1, t_2, t_3, t_4 might also be [-20, -13], [-12.999, -3], [-2.999, 3], [3.001, 6]. To improve the robustness of this function file, you can use MATLAB's built-in tool/operator nargchk to verify the allowable number of input arguments along with nargin. Likewise, another built-in tool/operator nargoutchk can be used to verify the number of defined output variables along with nargout.

Nested and Subfunctions of Function Files

The function files may contain more one, two, three, or more nested (subfunctions) functions to perform specific computations and analyses. The general syntax of such function files is as follows:

```
function [Output1, Output2, Output3, ...] = MAIN(In1, In2, In3, ...)
% MAIN.m contains several nested sub-functions
Output1 = Nest1;
Output2 = Nest2;
Output3 = Nest3;
...
function Out1Nest1()
Out1 = [do sth]
end
```

```
function Out2=Nest2(In2)
Out2 = [do sth]
end

function Out3=Nest3(In3)
Out3 = [do sth]
end Code
```

Let's look at several examples how to employ nested (subfunctions) functions in a function file.

Example 8

This example generates three different square matrices (of Pascal, Cauchy, Krylov) within three nested functions with respect to the main function's input arguments (only integers). Note that input arguments (integers) define the size of the output magic matrices.

Here is the solution script; it is a function file called Ex8_MAIN:

```
function [Output1,Output2, Output3] = Ex8_MAIN(In1,In2, In3)
% Ex8_MAIN.m generates three square matrices within three nested functions
%    Input arguments: In1, In2, In3 are integers defining the size of the
%    output matrices.
%    Nest1 generates the square matrix of Pascal
%    Nest2 generates the square matrix of Cauchy
%    Nest3 generates the square matrix of Krylov
disp(['This is ' num2str(In1)  '-by-' num2str(In1) ' Pascal matrix'])
Output1 = Nest1; disp(Output1)
disp(['This is ' num2str(In2)  '-by-' num2str(In2) ' Cauchy matrix'])
Output2 = Nest2; disp(Output2)
disp(['This is ' num2str(In3)  '-by-' num2str(In3) ' Krylov matrix'])
Output3 = Nest3; disp(Output3)

    function Out1=Nest1()
%    Nest1 generates the square matrix of Pascal
        Out1 = pascal(In1);
    end
```

```
    function Out2=Nest2()
%   Nest2 generates the square matrix of Cauchy
        Out2 = gallery('cauchy',In2);
    end
    function Out3=Nest3()
%   Nest3 generates the square matrix of Krylov
        Out3 = gallery('krylov',In3);
    end
end
```

Now, let's test the function file with three nested functions:

```
>> Input1 = 2; Input2=3; Input3=4;
>> [Output1,Output2, Output3] = Ex8_MAIN(Input1,Input2, Input3);
This is 2-by-2 Pascal matrix
     1     1
     1     2
This is 3-by-3 Cauchy matrix
     0.5000    0.3333    0.2500
     0.3333    0.2500    0.2000
     0.2500    0.2000    0.1667
This is 4-by-4 Krylov matrix
     1.0000    0.6731    3.8494   -2.7613
     1.0000    1.7986    2.3591    4.8361
     1.0000   -0.8049    1.0833   -0.5926
     1.0000    1.7760   -1.1635    1.2842
>> whos
  Name         Size          Bytes  Class      Attributes

  Input1       1x1               8  double
  Input2       1x1               8  double
  Input3       1x1               8  double
  Output1      2x2              32  double
  Output2      3x3              72  double
  Output3      4x4             128  double
```

Function Files Within M-Files

In recent versions of MATLAB, the function files (including nested and subfunctions) can be employed within M-files. This makes the computation process more efficient and saves memory space in the workspace. Let's take a look at the following example.

Example 9

Given: $y(x, t) = y_0 \sin (kx - \omega t)$ that is the solution of the wave equation. Here, $k = \dfrac{2\pi}{\lambda}$ is the wave number, $\omega = 2\pi f$ is the angular frequency, and λ is the wavelength.

Write an M-file with two nested functions to compute the values of $y(x, t)$ with four input arguments of k, ω, x, t.

$$\text{Given } y_0 = \frac{1}{2}; \lambda = 495*10^{-9}; k = \frac{2\pi}{\lambda}; f = 5000; \omega = 2\pi f; x = [0,1], t = [0,5]; \ N = 10^6$$

(number of data points for x and t).

The solution script is called Ex9_wFUN.m that has two embedded functions follows:

```
y0=1/2;                  % Magnitude of the wave
lambda=495e-9;           % Visible light wave length
k=2*pi/lambda;           % Wave number
f=5000;                  % Frequency
omega=2*pi*f;            % Angular frequency
x=[0, 1e-5];             % Length
t=[0, 1e-3];             % Time length
N=1e5;                   % Number of data points to be computed & simulated
xs=linspace(x(1), x(2), N);     % Wave Length series
time=linspace(t(1), t(2), N);   % Time series

y1 = F1(y0, k, xs, omega, time);
y2 = F2(y0, k, time, omega, xs); % NB: it is vital the order of the
varaibales: xs vs. time && V5 vs. V3
plot(xs, y1, 'bo', xs, y2, 'rx-'), grid on, hold on
title('Nested Function files within an M-file')
ORG0 = y0*sin(k*xs-omega*time);
plot(xs, ORG0, 'k', 'linewidth', 2)
legend('Fun1','Fun2','Original'), shg
```

```
function out1 = F1(var1, var2, var3, var4, var5)
% M-file nested function 1 called: F1
out1=var1*sin(var2*var3-var4*var5);
end

function out2 = F2(V1, V2, V3, V4, V5)
% M-file nested function 2 called: F2
% NB: it is vital the order of the variables: xs vs. time && V5 vs. V3
out2=V1*sin(V2*V5-V4*V3);
end
```

This is an M-file that can be executed by clicking the Run ▷ᴿᵘⁿ button in the menu panel or can be called from another M-file or from the Command window. After executing the file, you would get the following output:

```
>> clearvars
>> Ex9_wFUN
>> whos
  Name        Size                   Bytes  Class      Attributes

  N           1x1                        8  double
  ORGO        1x100000              800000  double
  f           1x1                        8  double
  k           1x1                        8  double
  lambda      1x1                        8  double
  omega       1x1                        8  double
  t           1x2                       16  double
  time        1x100000              800000  double
  x           1x2                       16  double
  xs          1x100000              800000  double
  y0          1x1                        8  double
  y1          1x100000              800000  double
  y2          1x100000              800000  double
```

In addition, the plot in Figure 2-9 is created.

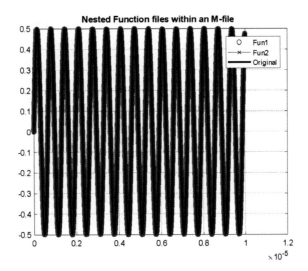

Figure 2-9. *Comparing the simulations obtained from the nested functions within the M-file*

The simulation results show that the M-file with embedded functions works like any other M-file. Like with the function files, the order of the input arguments is vital for the embedded functions within M-files. If the order is mixed up, then the outputs will be incorrect because one variable's values will be used for another.

Summary of Scripts and Function Files

Warning and error messages included in M-files and function files are of great help not only to users but also to developers. Thus, when you are writing scripts (programs), it is important to add as warning and error messages along with additional hints (as comments).

Consider these essential hints on how to write robust and efficient M-files and function files:

- Do as much checking (of input and output arguments, e.g., size, data type) as possible before executing a whole script to avoid the heavy burden of time-consuming calculations that may result in incorrect outputs.

233

- – Always start with most general checks (obvious ones) and then move to more complicated and complex ones. Add warning and error messages wherever necessary or appropriate. For example, add a unique text for display (e.g., `disp('This is ...')`) during the debugging process when you have varying numbers of input arguments and output variables. These messages not only help developers to spot errors but also go through all the anticipated scenarios in the code/script.

- – Explain how to correct errors when they occur using a warning message, e.g., `warning('Do something about this and that ...'`, A, B, C) and an error message, e.g., `error('There is an error on this and that ...'`, `Num(1)`, `Den(3)`, `u(1))`.

- – Add warning message identifiers so users can turn warnings on and off.

- – Add numerical examples to test your script with known correct solutions.

- – Use the M-file profiler to check the efficiency of your created M-files. To use the profiler, type >> `profile on; My_function; profile viewer` in the Command window.

Inline Functions

The MATLAB command `inline` lets users develop one or more analytical expressions with one or more input variables and assign that expression to a variable. A general syntax of creating an inline function expression is: f=inline('[expression1; expression2, ...]','arg1','arg2','arg3', ...);

In this command syntax, `expression1`, `expression2`, and so on, are inline functions and `arg1`, `arg2`, `arg3`, and so on, are input variables.

Note   An inline function is not recommended; instead, a function file is recommended.

Example 1

$f(t) = \dfrac{sin(2t)}{e^{2t}}$ can be expressed via a straightforward inline function without any

argument definition, since there is one variable that is identified automatically.

```
>> f = inline('sin(2*t)/exp(2*t)');
```

Example 2

The inline function of $f(t, \theta) = [e^{-2t\theta}; sin(2t)]$ should contain two input arguments, namely, t, θ. Therefore, the given function expressions can be expressed via the following inline function:

```
>> F = inline('[exp(2*t.*theta); sin(2*t)]', 't', 'theta');
```

That can be tested with predefined arguments, t and theta.

```
>> t=0:pi/20:pi; theta=linspace(0,1,length(t));
```

You can get the calculation results from the inline function F simply by recalling it with predefined input arguments.

```
>> fcalc=F(t, theta);
```

In addition, it is possible to perform computations and plot computed data from the inline function F simultaneously.

```
>> plot(t, log(F(t, theta)), 'linewidth', 1.5);
```

Example 3

Given: a second-order differential equation $\ddot{y} + \dot{y} + 3\,y = t^2$ that can be expressed by two first-order differential equations.

$$\{\dot{y}_1 = y_2 \ \dot{y}_2 = t^2 - (y_2 + 3y_1)$$

These two first-order differential equations can be defined by the following inline function:

```
f =inline('[y(2); t^2-(y(2)+3*y(1))]', 't', 'y');
```

Based on the previously shown examples, you can see that it is easy and straightforward to use the inline function tool not only in the Command window but also within scripts to solve various computation problems.

Note   The `inline` command will be removed in future releases of MATLAB, and therefore, it is recommended that you use anonymous functions instead.

Anonymous Functions with Handles

An anonymous function is a function that is not stored in a program file. It is associated with a variable and is called with a function handle (@). In general, the function handles accept input variables and return output variables similar to function files (as demonstrated previously). One major difference between a function file and an anonymous function expression is that an anonymous function with a handle (@) contains a single executable statement. The general syntax of expression anonymous functions with handles is as follows:

```
Fun_handle=@(arg1, arg2, arg3, ...)([expression1; expression2;...])
```

Example 1

$$f(t) = \frac{sin(2t)}{e^{2t}}$$

The given function $f(t)$ can be expressed via a function handle with the following command:

```
>> f_handle = @(t)(sin(2*t)./exp(2*t));
>> t=linspace(0,2*pi, 100);
>> f=f_handle(t); plot(t, f, 'o-')
```

It is also feasible to use function handles within loop iterations of scripts, for instance, Fun_handle.m.

```
% Fun_handle.m
f_handle = @(t)(sin(2*t)./exp(2*t));
```

```
t=linspace(0,2*pi, 100);
for ii=1:length(t)
    f(ii)=f_handle(t(ii));
end
plot(t, f, 'o-')
title('plot: f(t)=(sin(2*t)./exp(2*t)')
```

Example 2

Given $f(t, \theta) = [e^{-2t\theta}; sin\,(2t)\,]$.

The given function $f(t, \theta)$ can be expressed via one function handle.

```
>> F_handle = @(t, theta)([exp(2*t.*theta); sin(2*t)]);
```

Let's specify data values for input arguments of t and theta.

```
>> t=0:pi/20:pi; theta=linspace(0,1,length(t));
```

To obtain computation results from the function handle F_handle, the function handle has to be recalled with input arguments of t and theta.

```
>> Fvalues=F_handle(t, theta);
```

Example 3

Given: $\ddot{y} + \dot{y} + 3\,y = t^2$

It can be expressed by two first-order differential equations in a state-space form.

$$\{\dot{y}_1 = y_2 \; \dot{y}_2 = t^2 - \left(y_2 + 3y_1\right)$$

The previously derived system of two first-order differential equations can be defined via the following function handle (@):

```
f =@(t, y)([y(2); t^2-(y(2)+3*y(1))]);
```

The function handle (@) tool of an anonymous function can be also implemented like an inline function within scripts to solve many different kinds of problems including differential equations.

Up to this point, in a few examples, we have demonstrated how to employ function files, inline functions, and anonymous function tools. There are some advantages and disadvantages of employing anonymous functions defined with handles (@) and inline functions in comparison to function files. Anonymous functions are easy to implement and design, and there is a very small chance of getting confused with the function name and input arguments. On the other hand, one of the main advantages of function files over function handles and inline functions is that they are more efficient and flexible in terms of computation time. This is important when numerical simulation space is large. Moreover, function files can be employed within Simulink models.

Summary

In this chapter, we covered all the basic programming essentials, including different operators and symbols used in programming and how to employ them efficiently. Moreover, we highlighted the most frequently occurring potential programming errors and pitfalls and how to avoid them.

In summary, to speed up simulation/computation processes, there are a few procedures to follow while coding. When you are dealing with large data sets, you may face memory problems. To prevent memory problems or errors in MATLAB, or to just improve computation efficiency, the following tips are recommended:

- Avoid large temporary variables, break variables into smaller components, and clear variables (using `clearvars`) when they are no longer needed.

- Use lower-precision data types, if possible, and whenever there is no need for higher precision.

- Pre-allocate memory for arrays of fixed size to reduce defragmentation by using standard matrices such as `zeros()` or `ones()`.

- Use the function `pack()` to defragment memory.

- Consider writing data to a disk periodically.

- Increase the size of a swap file. This can be done via My Computer ➤ properties ➤ Advanced - Performance Settings.. Note that it is recommended to set the swap file size to twice the size of your computer's RAM.

Here are some other techniques to speed up the simulation processes:

- Functions (function files) are faster than scripts.

- Load and save are faster than file I/O functions when you need to import or export data sets.

- Avoid large background processes.

- Replace argument checking with `try-catch`.

- Use `switch-case` over `if-elseif-else`.

- Use sparse arrays for sparse data.

Exercises for Self-Testing

Exercise 1

Write an M-file using the [`while`, `end`] loop control statement to compute the values of the cosine function $g(\theta) = cos(\theta)$ for $\theta = 0...\pi$ with 2,000 incremental steps in this range and stop computation when the value of the function $g(\theta) \approx 0.99999$. Also, display how many steps it takes to get through the computation process and plot your simulation results. Also, display the end value of 0.99999 in the same plot.

Exercise 2

Compute the area of a circle, square, and rectangle with regard to these user entries:

```
W = input('Width of a rectangle:        ');
L = input('Length of a rectangle:       ');
R = input('Radius of a circle:          ');
S = input('Side length of a square:     ');
```

There are several scenarios to consider in your script: (1) if a user enters two dimensions for the width (W) and length (L) of a rectangle, your code has to compute the area of a rectangle and display it with comments; (2) if a user enters two dimensions, such as radius (R) and side (S) of a square, it has to compute the area of a circle and square, respectively, and display it with comments; (3) if a user enters all the required entries, it has to compute three areas and display with adequate comments; (4) if a user enters not all entries or misses any of the required entries (W, L, R, S), your code has to display "You need to ENTER all dimensions!". The input(), isempty(), exist(), length(), numel(), size(), and fprintf() commands can be employed.

Exercise 3

Write a script to display student grades based on their earned points. Student grades have to be defined according to the following scales:

$$
\begin{aligned}
F &= 0...65; \\
D &= 66 ... 70; \\
C &= 71 ... 81; \\
B &= 82 ... 87; \\
A &= 88 ... 100;
\end{aligned}
$$

Use input(), disp(), fprintf(), [if, elseif, else ... end].

Test your script with these example points: [70, 82, 35, 90, 99, 56, 81, 89, 66, 87, 88, 83, 71, 69, 55].

Exercise 4

These exercises cover the [while, end] and [for, end] loop control operators:

1. Compute the series $\dfrac{\pi^2}{8} = 1 + \dfrac{1}{3^2} + \dfrac{1}{5^2} + \dfrac{1}{7^2} + ... = \sum_{n=0}^{\infty} \dfrac{1}{(2n+1)^2}$ using

    the [while, end] and [for, end] loop control statements. Take $n = 101$ and plot error values over iterations.

2. Compute the sum of these series: $1^2, 2^2, 3^2, ...n^2$. Take $n = 500$ and

    compute the total sum $\sum_{n=1}^{500} n^2$ using the [while, end] and [for,

    end] loop control statements. Plot the sum as a function of n.

3. Compute the sum of these series: $1 * 2 + 2 * 3 + 3 * 4 + 4 * 5 + ...$

    This can be rewritten as a sum $\sum\limits_{k=1}^{N} k(k+1)$. To compute the total

    sum, use the [while, end] and [for, end] loop control
    statements. Plot the sum as a function of k.

4. Compute these series (discovered by Euler in 1735):

    $\dfrac{\pi^2}{6} = 1 + \dfrac{1}{2^2} + \dfrac{1}{3^2} + \dfrac{1}{4^2} + ... = \sum\limits_{n=1}^{\infty} \dfrac{1}{n^2}$ using the [while, end] and

    [for, end] loop control statements. Take $n = 1...99$ and plot error
    values over iterations. Run the exhaustive computation and halt
    the computation process when the error is smaller than 10^{-13}.

5. Compute the total sum of this series of odd numbers, namely, 1, 3,
    5, 7, ..., that can be written in the form of this sequence $2n + 1$.

    Take $N = 111111$ and compute the total sum $\sum\limits_{n=0}^{N} (2n+1)$ using the

    [while, end] and [for, end] loop control statements. Halt the
    iteration process when the sum is larger than 30869136. Display
    the final number as an integer. Plot the summing iteration process.

Exercise 5

Write a script file with the conditional statements [if, elseif, elseif, else, end]
and the function handle (@) to compute the values of the function:

$$g(a,b,t) = \{ e^{-at} + bt, \ t \le -13 \quad \cos(at) - si(bt), \ -3 \ge t \ge -13$$
$$\cos at + e^{bt}, \ -3 \le t \le 01, \ t > 0$$

$$a = 3, \ b = 2.$$

Exercise 6

Here are some exercises for the [while, end] and [for, end] loop control operators:

1. Compute this series: $f(x) = \displaystyle\sum_{n=1}^{\infty} \frac{b_n sin\big((2n-1)\pi x\big)}{2L}$ using the [while ... end] and [for ... end] loop control statements. Take $n = 11$, $b_n = [-5...5]$, $x = [0...L]$, and $L = 10$. For x take a step

   size of $\Delta x = \dfrac{L}{100}$. Plot all computation results.

2. Compute $f(t) = cos\,cos\,(20t) - sin\,(10t)$ for $t \in [-\pi,\ \pi]$ with $\Delta t = \pi/50$ using the [while ... end] and [for ... end] loops.

Exercise 7

These exercises cover the [while ... end] and [for ... end] loop operators:

1. Compute the total sum of these series: $\dfrac{3}{4} + \dfrac{4}{5} + \dfrac{5}{6} + \dfrac{6}{7}... = \displaystyle\sum_{m=3}^{55} \dfrac{m}{m+1}$.

   Compute the total sum using the [while, end] and [for, end] loop control statements. Plot the sum as a function of m.

2. Compute the total sum of these series:

   $10 + 30 + 90 + 270 + 810 + 2430 +... = \displaystyle\sum_{m=0}^{65}\big(10*3^k\big)$. Compute the

   total sum using the [while, end] and [for, end] loop control statements, and halt the computation process when the sum is larger than 18446744073709551615. Display the final sum number as an integer. Plot the sum as a function of k.

3. Compute the sequence 4, 1, 0.5, 0.25, ... until it is smaller than 0.000005 by using the [while, end] and [for, end] loop control statements. Note that the sequence can be formulated as $4*(0.5)^{n-1}$. Find n that makes the sequence smaller than 0.000005.

4. Compute the total sum of the series $\dfrac{1}{2} + \dfrac{1}{4} + \dfrac{1}{8} + \dfrac{1}{16} + \cdots$ until it is

    equal to or larger than 0.999 999 999. Note that the sum of this

    series can be expressed with $\sum\limits_{k=0}^{n} \dfrac{1}{2}\left(\dfrac{1}{2}\right)^{k}$.

5. Compute the following total sum of $\sum\limits_{n=0}^{m} \dfrac{(-1)^{n}\, x^{2n+1}}{(2n+1)!}$ for $m = 202$

    and $x = 2.5$ by using the [while, end] and [for, end] loop
    control statements.

6. Compute the following total sum of $\sum\limits_{n=1}^{m} \dfrac{(-1)^{n+1}(x-1)^{n}}{n}$ for $m = 1001$

    and $x = 2$ by using the [while, end] and [for, end] loop control
    statements.

Exercise 8

Fix two errors in the following given script:

```
%% Find out if the entry is a Scalar or NOT.
% Prepare your entry data that MUST be in array
% or matrix format of any size: 1-by-1, 2-by-2, 2-by-3, etc, etc!
% Your entry can be also any standard array generating functions!
ABC=input('Enter ANY numerical entry of any size surrounded with square
brackets [  ]:   ');
if isnumeric(ABC) && isscalar(ABC)
    fprintf('This is a scalar: %20g \n', ABC);
else
    format short
    fprintf('Your entry is not scalar, but an array \n', ABC);
    fprintf(ABC);
end
```

1. After fixing the errors, the code has to produce the following
    outputs in the Command window with the user entry of $[1, 2; 3, -1]$:

```
Enter ANY numerical entry of any size within [  ]:    [1, 2; 3,-1]
Your entry is not scalar, but an array
     1     2
     3    -1
```

2. The code has to produce the following outputs in the Command
    window with the user entry of ['ab' 'bc'; 'cd' 'ef']:

```
Enter ANY numerical entry of any size within [  ]:    ['ab' 'bc';
'cd'  'ef']
Your entry is not scalar, but an array
abbc
cdef
```

3. The code has to produce the following outputs in the Command
    window with the user entry of 13.12:

```
Enter ANY numerical entry of any size within [  ]:    13.12
This is a scalar:                13.12
```

Exercise 9

Fix the three errors in the following script:

```
%% Find out whether the array is square and if it is, show its size.
% Prepare your entry data that MUST be in array
% or matrix format of any size: 1-by-1, 2-by-2, 2-by-3, etc, etc!
% Your entry can be also any standard array generating functions!!!
ABC=input('Enter ANY numerical entry of any size within [  ]:   ');
[Rows, Cols]=size(ABC);
if isnumeric(ABC) && Rows==Rows
    fprintf('This is a square ARRAY! ');
    fprintf('Your entry is of %5g -by- %5g  square ARRAY \n', Cols, Rows);
else
    format short
    fprintf('Your entry is NOT a square array \n')
    fprintf('BUT an ARRAY of size %5g - by - %5g \n', Cols, Cols);
end
```

After fixing the errors, the code has to produce the following outputs in the Command window, given these entries:

1. Entry: [magic(5)]

```
Enter ANY numerical entry of any size within [  ]:    [magic(5)]
This is a square ARRAY!
Your entry is of      5 -by-      5  square ARRAY
```

2. Entry: [rand(3,5)]

```
Enter ANY numerical entry of any size within [  ]:    [rand(3,5)]
Your entry is NOT a square array
BUT an ARRAY of size      3 - by -      5
```

Exercise 10

Fix the two errors in the following script:

```
%% Find out: the user entry is scalar or not. If it is, display it.
% otherwise, show the variable type.
ABC=input('Enter ANY numerical entry of any size within [  ]:    ');
if isnumeric(ABC)
    fprintf('This is a Scalar! \n');
    fprintf('Your entry is a scalar:  %5g  \n', ABC);
else
    class(ABC, 1)
end
```

After fixing the two errors, the following outputs are obtained with respective inputs.
For entry 1:

```
Enter ANY numerical entry of any size within [  ]:    123
This is a Scalar!
Your entry is a scalar:    123
```

For entry 2:

```
Enter ANY numerical entry of any size within [   ]:    '1001011'
ans =
    'char'
```

For entry 3:

```
Enter ANY numerical entry of any size within [   ]:    [1 3 -2]
ans =
    'double'
```

Exercise 11

Fix two errors in the following script:

```
%% Find out: the array is real and square. If it is, display it;
% otherwise, show its size and type.
% NB: size(), display(), class() can be used.
% Prepare your entry data that MUST be in array
% or matrix format of any size: 1-by-1, 2-by-2, 2-by-3, etc.
% Your entry can be also any standard array generating functions!

ABC=input('Enter ANY numerical entry of any size within [   ]:    ');
[Rs, Cs]=size(ABC);
if ischar(ABC) && Rs==Cs
    fprintf('This is a square array! \n');
    disp(ABC);
elseif
format short
fprintf('This is not a square array & its size: %5g-by-%5g \n', Rs, Cs);
disp(num2str(ABC));
end
```

After fixing the two errors, the following outputs should be obtained with respective inputs.

Entry 1:

```
Enter ANY numerical entry of any size within [ ]:   magic(3)
This is a square array!
     8     1     6
     3     5     7
     4     9     2
```

Entry 2:

```
Prepare your entry data that MUST be in array or matrix format of any size
1-by-1, 2-by-2, 2-by-3, etc., etc.! with real value elements
Your entry can be also any standard array generating functions!!!
Enter ANY numerical entry of any size within [ ]:   [1, 2/0; 0/0, 1]
This is a square array!
     1    Inf
   NaN      1
```

Entry 3:

```
Your entry can be also any standard array generating functions!!!
Enter ANY numerical entry of any size within [ ]:   [1, 2/0; 0/0, 1; 1, 0]
This is not a square array and its size:    3 - by -    2
   1  Inf
 NaN    1
 1    0
```

Exercise 12

Fix the five errors in the following script:

```
%% Q7. Computing area of a circle, square and rectangle w.r.t the user
entries:
```

```
W = input('Width of a rectangle:          ', 's');
L = input('Length of a rectangle:         ', 's');
R = input('Radius of a circle:            ', 's');
S = input('Side length of a square:       ', 's');
if isempty(R) && isempty(S)
    A1=W*L;
    fprintf('Area of a rectangle:  A1 =  %5g \n', A1);
elseif isempty(W) && isempty(L) && exist('R','var') && exist('S', 'var')
    A2 = pi*R^2; A3 = S^2;
    fprintf('Area of a circle:  A2 =  %5g \n', A2);
    fprintf('Area of a square:  A3 =  %5g \n', A3);
elseif isempty(W) && isempty(L) && isempty(R)
    A3 = S^2;
    fprintf('Area of a square:  A3 =  %5g \n', A3);
elseif isempty(S) && isempty(W) && isempty(L)
    A2 = pi*R^2;
    fprintf('Area of a circle:  A2 =  %5g \n', A2);
else exist('W','var') && exist('L','var') && exist('R','var') &&
exist('S','var')
    A1=W*L; A2 = pi*R^2; A3 = S^2;
    fprintf('Area of a rectangle:  A1 =  %5g \n', A1);
    fprintf('Area of a circle:  A2 =  %5g \n', A2);
    fprintf('Area of a square:  A3 =  %5g \n', A3);
else
    fprintf('You need to ENTER some dimensions! \n')
end
```

After fixing the errors, with the following entries, these outputs are displayed in the Command window:

```
Width of a rectangle:        1
Length of a rectangle:       1
Radius of a circle:          1
Side length of a square:     1
```

```
Area of a rectangle:  A1 =      1
Area of a circle:  A2 =  3.14159
Area of a square:  A3 =      1
```

Exercise 13

Fix the five errors in the following script:

```
%% Assessing the student performances
clc; clearvars
SP =input('Enter the student grade:  ');
if SP <65
    disp('Student Grade is  F ')
elseif SP>=66 && SP<=71
disp('Student Grade is  D ')
elseif SP>71 && SP<=81
    disp('Student Grade is  C ')
elseif SP>82 && SP<87
    disp('Student Grade is  B ')
else
    disp('Student Grade is  A ')
end
```

After fixing the errors and running the script with inputs [65, 71, 81, 82, 87 87.5] in sequential order, the following outputs should be shown in the Command window:

```
Enter the student grade:  65
Student Grade is  F
Enter the student grade:  71
Student Grade is  C
Enter the student grade:  81
Student Grade is  C
Enter the student grade:  82
Student Grade is  B
Enter the student grade:  87
Student Grade is  B
```

```
Enter the student grade:   87.5
Student Grade is   A
```

Exercise 14

Write a script (program) that computes all solutions:

> (1) But displays only real solutions of these third-order polynomial equations for any values of a, b, c, and f: i) $x^3 + bx^2 + cx = 0$; ii) $ax^3 + f = 0$; iii) $ax^3 + cx = 0$; iv) $x^2 + bx + c = 0$.

> (2) But displays only complex solutions of these equations for any values of a, b, c, and f: i) $ax^2 + c = 0$; ii) $ax^3 + f = 0$; iii) $ax^2 + bx + c = 0$

Exercise 15

Write a script (program) that computes the volume and weight of the model that may have a form of cube, cylinder, and rectangular prism. Users need to enter (via input prompt) the necessary geometric dimensions of the model and enter or select material properties (density) from the given data (aluminum, copper, and steel) in your script. Your script has to write all computed and user input data sets into an external file called RESULTS.txt with explanatory comments in it along with numerical data.

Exercise 16

Edit and correct the following given script to display the current date and time correctly in the Command window:

```
Format short e
T=clock;
fprintf('This year is: %n4 \n', T(1))
if T(2)==1
 sprintf('It is: %f4 -st month of the year:  %n4 \n', T(2),T(1))
elseif T(2)==2
 sprintf('It is: %f4 -nd month of the year:  %n4 \n', T(2),T(1))
```

```
elseif T(3)==3
 sprintf('It is: %f4 -rd month of the year:  %n4 \n', T(2),T(1))
else
 sprintf('It is: %f4 -th month of the year:  %n4 \n', T(2),T(1))
end
sprintf('current time is: %lo o"clock %lo min \n', T(4), T(5))
sprintf('and %s secs \n', (T(6)))
```

Your corrected script should display the current date and time in the following format:

```
It is: 11 - day of the 6-th month of the year:   2014
current time is: 15 o"clock 3 min  and 17.136 secs
```

Exercise 17

Given: $y(x, t) = y_0 \sin(kx - \omega t)$ is the solution of the wave equation where $k = \dfrac{2\pi}{\lambda}$ is the wave number, $\omega = 2\pi f$ is the angular frequency, and λ is the wave length.

- Write an anonymous function with a function handle to compute the values of $y(x, t)$ with input arguments of k, ω,x, t.

- Write an inline function to compute the values of $y(x, t)$ with input arguments of k, ω,x, t.

- Write a function file to compute the values of $y(x, t)$ with input arguments of k, ω,x, t.

Exercise 18

The equation for a power factor of a series resistor-capacitor (RC) circuit with no inductance is $\cos \delta = \dfrac{RC\omega}{\sqrt{1+(RC\omega)^2}}$.

- Write an anonymous function with the function handle to compute the values of δ with the input arguments of R, C, and ω.

- Write an inline function to compute the values of δ with the input arguments of R, C, and ω.

- Write a function file to compute the values of δ with the input arguments of R, C, and ω, and plot δ versus ω.

Exercise 19

The equation for charge in a resistor-inductance-capacitor (RLC) circuit in a series is determined by Kirchhoff's law: $L\ddot{q} + R\dot{q} + \dfrac{q}{C} = E_{max} \cos\cos \omega t$.

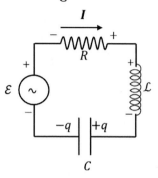

1. Write an anonymous function with a function handle for solving the given second-order differential equation for $q(t)$ with the input arguments of R, L, C, ω, and t.

2. Write an inline function necessary for solving the given second-order differential equation for $q(t)$ with the input arguments of R, L, C, ω, and t.

3. Write a function file necessary for solving the given second-order differential equation for $q(t)$ with the input arguments of R, L, C, ω, and t.

4. Create a Simulink model to simulate the given RLC system.

Exercise 20

The acceleration of a skydiver is determined by the following:

$$a = g\left(1 - \frac{v^2}{3600}\right)$$

where $g = 9.81 \ m/s^2$.

1. Write an anonymous function to compute a as a function of speed v.

2. Write a function file to compute a as a function of speed v and plot a versus v.

3. Create a Simulink model to simulate an acceleration of a skydiver.

4. Compute the terminal speed for a skydiver.

Exercise 21

A truck of mass m is accelerated from rest at $t = 0$ with constant power P along a level road. The speed of the truck as a function of time is given by $v(t) = \left(\dfrac{2P}{m}\right)^{\frac{1}{2}} \sqrt{t}$.

If $x = 0$ at time $t = 0$, the position function $x(t)$ is given by $x(t) = \left(\dfrac{8P}{9m}\right)^{\frac{1}{2}} \sqrt{t^3}$, where $P = 550 \ kW$ and m = 15000 kg.

1. Write a function handle to compute the speed of the truck $v(t)$ as a function of time t.

2. Write an inline function to compute the position of the truck from the function $x(t)$ as a function of time t.

3. Build plots of $x(t)$ versus t, $v(t)$ versus t in two separate plot figures.

Exercise 22

In a crash test, a car traveling 100 km/h (28 m/sec) hits an immovable concrete wall. We can treat this problem in general with the approximation that the car body is one piece, despite that different parts of the car when it hits the wall will accelerate differently. In fact, upon impact, the center of the car moves forward less than half of its length. Let's assume that the stopping distance of the car upon hitting the wall until full stop is

0.72 m. Time to full stop will be equal to $\Delta t = \dfrac{\Delta x}{v_{ave}} = \dfrac{0.72}{0.5v_0 + 0.5v} = \dfrac{0.72}{0.5v} = \dfrac{0.72}{14} = 0.051 \, \text{sec}.$

The average acceleration of the car until full rest is

equal to $a = \dfrac{\Delta v}{\Delta t} = \dfrac{v_0 - v}{\Delta t} = -\dfrac{v}{\Delta t} = -544.44 \, m/sec^2$

Note that this is about 55g, which means very high acceleration (deceleration) that takes over very large amount of energy from inertia forces and converts it into heat. Write a function file to compute a as a function of Δx (for different front bumper types) and v (for all cases, take $v_0 = 0$) for a crash test of any type of cars with different traveling speeds of v.

Exercise 23

Create a function file (called Ex23.m) taking one input argument (the planted year of the maple tree that is four-digit integer: 0 … 2018) and the two output variables AGE (calculated age of the maple tree) and NOTE (with regard to the age of the maple tree), as well as a warning/message box with notes.

1. If the planted year of the tree is before 1000, the outputs are NOTE = "CANNOT be TRUE", AGE = [ ] (empty) and also an error box with the note "Check Your Entry."

2. If the planted year of a tree is in between [1000...1918], the outputs are NOTE = "NEED to be PROTECTED", AGE = [∀] (computed age with respect to the current year) and a warning dialog box with the note "Maple Tree is to be Under Protection."

3. If the planted year of a tree is in between [1919 ...2000], the outputs are NOTE = "WELL fit", AGE = [∀] (computed age with respect to the current year) and a message box with the note "GOOD one for a timber";.

4. If the planted year of a tree is in-between [2001 ...2018], the outputs are NOTE = "TOO young", AGE = [∀] (computed age with respect to the current year) and a message box with the note "TOO young for a timber."

Exercise 24

The spiral of Archimedes (also called the Archimedean spiral) is a spiral curve (named after the 3rd-century BC Greek mathematician Archimedes). The equation of the spiral is written in the polar coordinate system by the following equation: $r = a + b\theta$, where r is the distance from the origin and θ is the angle of that point in radians with respect to the origin. The parameters a, b in the equation are real numbers that control the spiral and the distance between successive spiral turnings, respectively.

1. Compute the spiral of Archimedes for

   $\theta = 0 : \dfrac{\pi}{10000} : 13\pi, a = -1.25, b = -2.25$ with a vectorization method.

2. Compute the spiral of Archimedes for

   $\theta = 0 : \dfrac{\pi}{10000} : 13\pi, a = -1.25, b = -2.25$ with the [for ... end] and

   [while ... end] loops without memory allocation.

3. Compute the spiral of Archimedes for

   $\theta = 0 : \dfrac{\pi}{10000} : 13\pi, a = -1.25, b = -2.25$ with the [for ... end] and

   [while ... end] loops with memory allocation.

4. Compare the computation efficiencies in (2) and (3) against (1) using [tic ... toc].

Exercise 25

Create a function file (called Ex25.m) taking four input variables (n, b_n, Δx, L) and three output variables: $f(x)$, *ALL _ cell*, *ALL _ struct*.

Compute this series: $f(x) = \sum\limits_{n=1}^{\infty} \dfrac{b_n \sin\left((2n-1)\pi x\right)}{2L}$ using [for, end] loop control

statements. Take $n = 1 : 11$, $b_n = [-5...5]$, $x = [0...L]$, $L = 10$. $\Delta x = \dfrac{L}{2000}$.

ALL _ cell is a cell array composed of $f(x)$, x, b_n.

ALL _ struct is a structure array composed of $f(x)$, x, b_n, *ALL _ cell*.

Exercise 26

Create a function file (called Ex26.m) taking one input variable (k) and one output variable (S).

Compute a total sum of these series: $10+35+65+110+\ldots=\sum_{m=0}^{55}\left(10*2.5^{k}\right)$. Compute the total sum using the [while, end] and [for, end] loop control statements.

1. Halt the computation process when the sum is larger than $2.5*10^{18}$ and display in the Command window at what iteration step the computation is halted and the computed sum of the series.

2. Display the difference between the computed sum of series and $2.5*10^{18}$.

3. Display the final sum and the iteration number as integers, but collect all of the sums (S) from every step.

Exercise 27

Write a script file to compute the solutions of the third-order polynomial $a_3x^3 + a_2x^2 + a_1x = 0$ for any given values of its coefficients a_1, a_2, a_3 with a user prompt input and by using the conditional statements [if, elseif, else, end]. Include in your script to print out solutions of the polynomial in the external *.dat file in the correct data formats including complex values.

CHAPTER 3

Graphical User Interface Model Development

MATLAB has a number of graphical user interface (GUI) tools and functions that can be employed while building GUI models, either by using the GUI tools or by writing scripts. In this chapter, we cover how to build GUI models with the GUI development environment (GUIDE) and how to write scripts to generate pop-up and dialog boxes.

GUIDE

There are three principal groups of elements necessary to build a MATLAB GUI model. They are components, figures, and callbacks.

- MATLAB GUI components are graphical controls (push buttons, edit boxes, lists, sliders, and pop-up menus), static elements (frames, text strings), menus, and axes. Graphical control elements and static elements can be created by using the `uicontrol` function. Menus can be generated by using the `uimenu` and `uicontextmenu` functions. The `axes` function is used to display graphical data.

- The `figure` function is used to create figures that accommodate various combinations of components.

- Callbacks or callback functions are used to perform actions (simulations and displays) with respect to user entries, such as clicks with a cursor (mouse or touch screen) or entries from the keyboard.

© Sulaymon Eshkabilov 2022
S. Eshkabilov, *Beginning MATLAB and Simulink*, https://doi.org/10.1007/978-1-4842-8748-4_3

All of these components and figures and their properties and behaviors can be created by writing scripts, functions, and callback functions using MATLAB's built-in functionality tools. However, this approach is rather cumbersome and long, and it requires a considerable amount of code and programming. An alternative and rather straightforward approach to designing a GUI model is MATLAB's user-friendly GUI development environment tool.

GUIDE contains templates, including GUI with UI controls, GUI with axes and menus, and the modal question dialog. By using this tool, you can rather easily create a GUI model layout with all the necessary components, such as push buttons, menus, static elements, text, and so forth, and adjust their properties (size, color, type, etc.) and location with respect to the GUI model design. This can be done by using the mouse and keyboard, without any programming.

Note that MATLAB (starting from MATLAB R2016a) introduced a user-friendly drag-and-drop application development environment called App Designer. It's similar to GUIDE. In the future releases of MATLAB, App Designer and GUIDE will be integrated, and only App Designer will be supported. Many tools of App Designer and GUIDE are the same or similar. Here we demonstrate how to use GUIDE tools to create GUIs.

Many GUI controls in GUIDE are quite straightforward, and building a GUI model by employing GUI control tools is not difficult. We first launch the GUIDE tool and become familiar with its components and tools. Then we move on to working with GUI's Property Inspector tools and modifying the operational behavior of the GUI blocks used in our model. Using this approach, after completing GUI modeling with GUI blocks, we save the model. MATLAB will automatically generate a script function file (M-file) that we modify according to our given GUI model requirements.

You can launch the GUIDE application by typing `guide` in the Command window.

Subsequently, the GUIDE Quick Start window appears, as shown in Figure 3-1. It is straightforward to migrate the developed models and apps from GUIDE into App Designer, as shown in Figure 3-2, by downloading the Migration App from MathWorks and installing it. The apps developed in GUIDE can be exported into App Designer using the Export options, as shown in Figure 3-3.

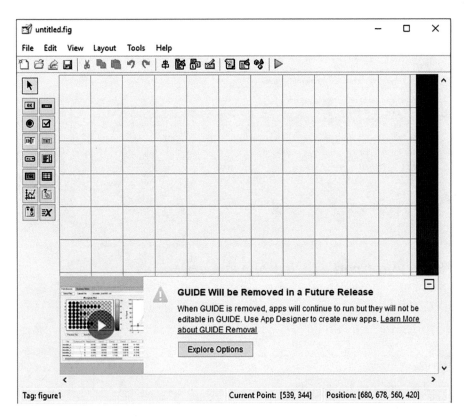

Figure 3-1. *GUIDE Quick Start window*

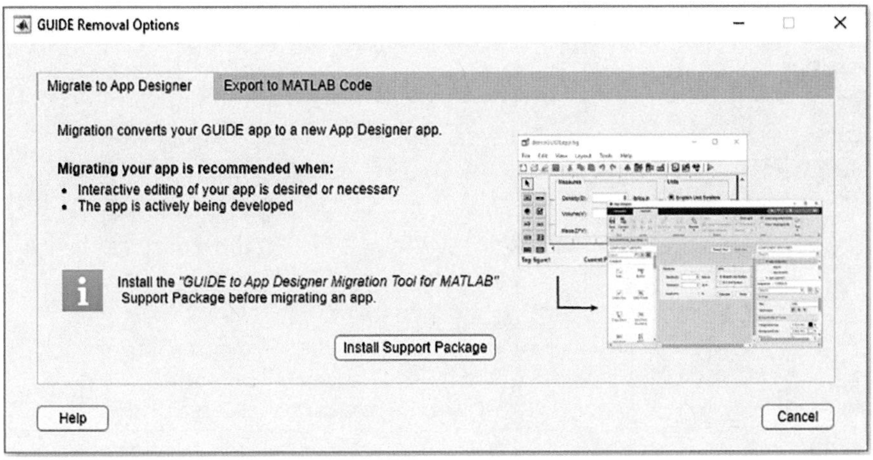

Figure 3-2. *GUIDE's Explore Options showing how to migrate from GUIDE to App Designer*

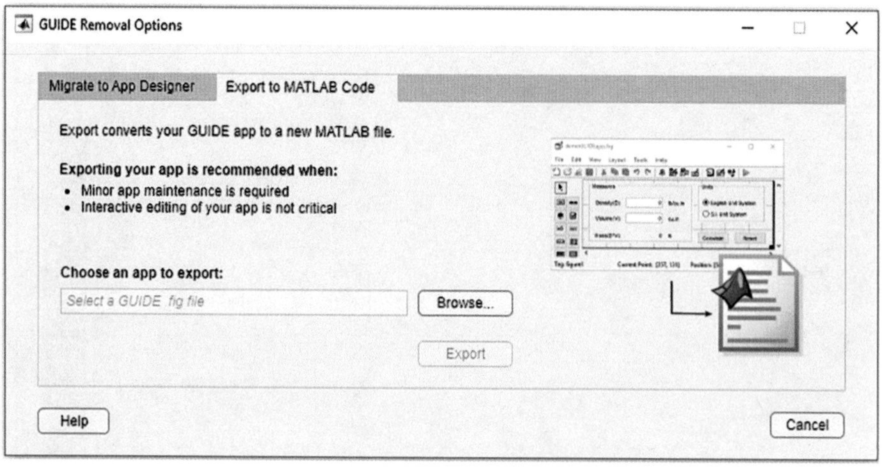

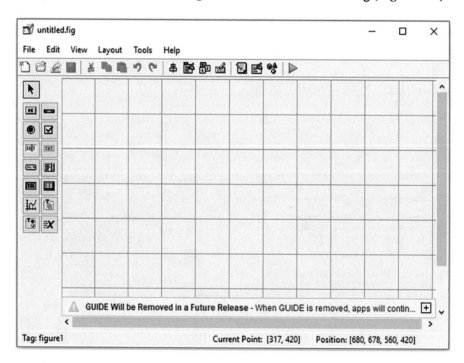

Figure 3-3. *How to export a GUIDE app to App Designer*

A blank GUI (the default) has UI controls, axes, and a menu bar, as shown in Figure 3-4. By default, the initial GUI figure is named untitled.fig (Figure 3-4).

Figure 3-4. *GUI figure window components without names*

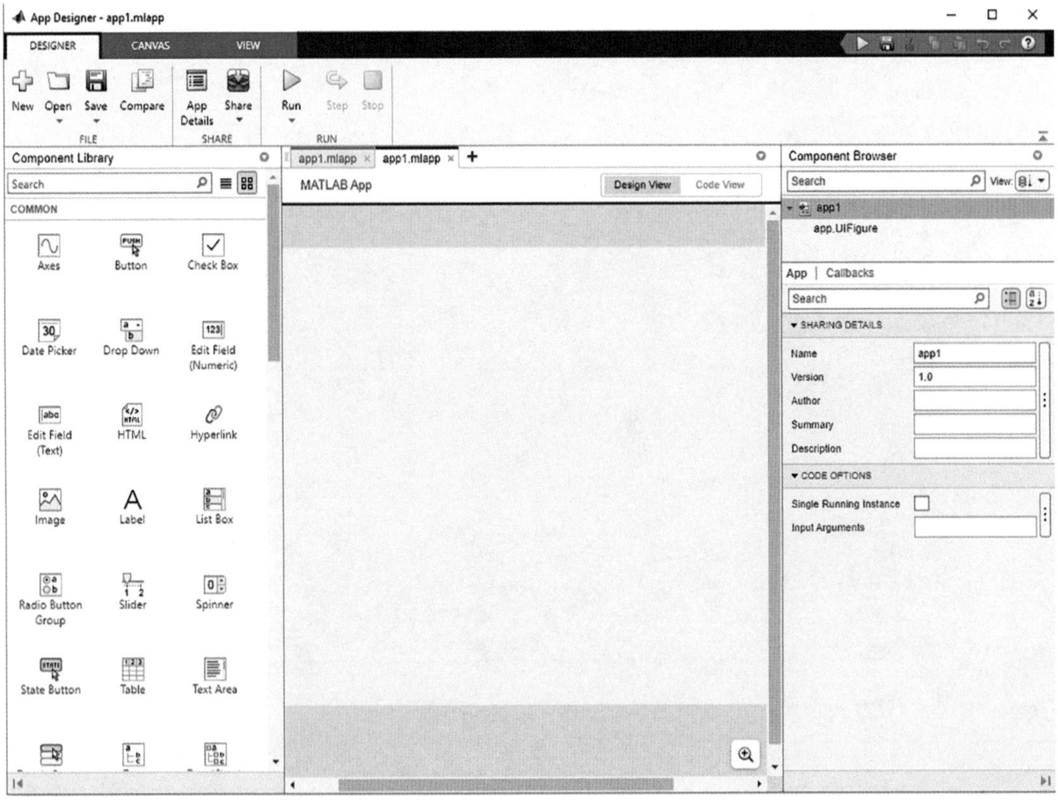

Figure 3-5. *Blank App Designer window*

Note   In the future releases of MATLAB, App Designer and GUIDE will be integrated, and only App Designer will be supported.

You can see from the tools in a blank GUIDE window (Figure 3-2) and in App Designer (Figure 3-3) that they have many common tools. In addition, App Designer has many other tools in its library such as Instrumentation, Aerospace, and Simulink Real-Time.

In the GUI layout window you can start working and add the name tags of the GUI components and tools by selecting Show Names on the Component palette from the MATLAB GUIDE Preferences. With this adjustment, the components will appear with their name tags, as shown in Figure 3-6.

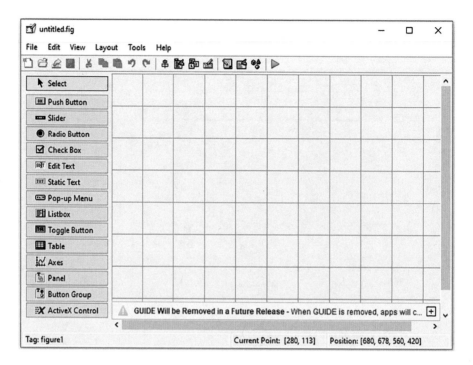

Figure 3-6. *GUI components with name tags displayed*

The GUI window shown in Figure 3-7 is composed of the GUI components, the design area, and the menu tools called Align Objects, Menu Editor, Tab Order Editor, etc. Once the GUI window is open, we can start building our GUI model by dragging the desired GUI components and dropping them into the design area (see Figure 3-7). You can resize and align the GUI components in our design area by clicking and dragging with a mouse or by using the right mouse button options or the Property Inspector, Align Objects, and other menu bar tools. Moreover, you can edit properties of the GUI components by clicking the Editor and adding menu tools to the GUI by clicking the Toolbar Editor.

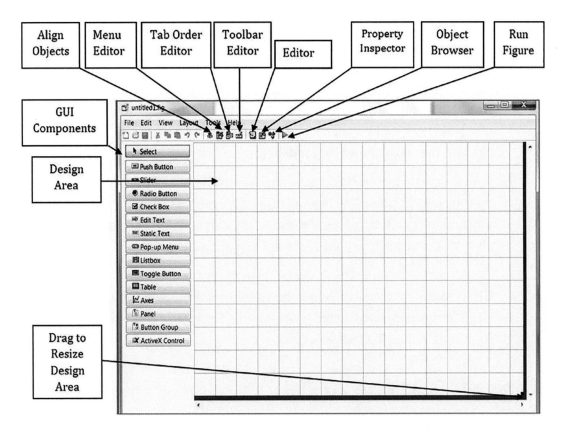

Figure 3-7. *The GUI GUIDE tool window and its menu bar*

There are several basic steps required to build a GUI model:

1. Clearly define the specific functions that the anticipated GUI model should have and which tools and components you need. As always, it is best to start with a draft sketch of the GUI model on a piece of paper.

2. Drag and drop all the necessary GUI components into the design area and then modify and align them according to your needs and project requirements. Once every component is in place and adjusted with respect to each other, you can pull them into a button group (there is a component called Button Group), which gives you good flexibility for moving and aligning objects in the design area. Note that grouping components into a button group is optional.

263

3. Rename or provide recognizable name "tags" to all components used in the model so you can easily identify them in later stages while editing scripts and M-files. Moreover, you may need to adjust and modify the color, font, and size of the components. Note that changing the color, font size, text type, size, and position of components in the model will not affect the execution/ simulation results of the GUI model.

4. Save the created work. It will be saved in two different file formats, one of which is `*.fig` (the Figure file) and the other is `.m` (the function file), containing a complete script of the model, including all nested callback functions for each GUI component.

5. Write/edit a script that implements the performance of each GUI component in association with the other components. This step requires some additional programming work, which will be discussed in the following examples.

Example 1: Building a 2D Plot

Let's look at a simple example of building a 2D plot of a cardinal sine function for a user's input value ranges: $sinc(\pi x) = \dfrac{sinsin(\pi x)}{\pi x}$.

By dragging and dropping the Push Button, Static Text, Edit Text, and Axes blocks, we start building the interface of the GUI model, as shown in Figure 3-8. After adjusting all the blocks in the GUI model with respect to each other, we can edit their properties by double-clicking each block or using the Property Inspector option via the right mouse. Note that Static Text serves as information to show the input entry names: Xmin and Xmax. The values of Xmin and Xmax represent boundary values of the variable and can be defined in the Edit Text blocks.

Note   There is one major difference between static text and edit text. The user can edit an entry in an edit text block but cannot do so in a static text block.

In addition, you add an Axes block that displays a plot of computed values of the function $sinc(x)$ according to our entries for x_{min} and x_{max} in the Edit Text blocks. You add a Push Button block that makes the built model (see Figure 3-8) compute the values of the function and display the results in the plot area.

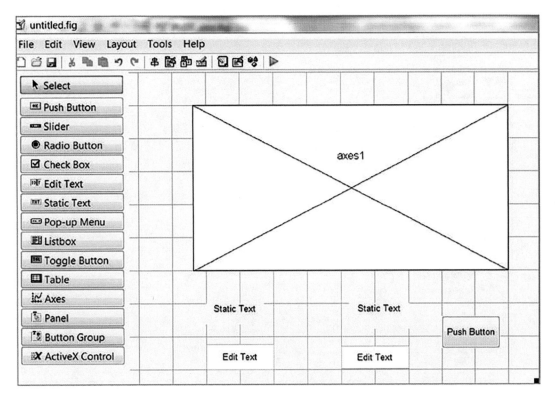

Figure 3-8. *Necessary blocks selected and placed onto the GUI model area*

To make your GUI model more informative, legible, and user friendly, you can edit each block by altering the font size, color, background color, and so forth.

There are many properties for each individual block that can be altered. For instance, you can change the front color, background color, font type, size and color, position, size, string, name, and a few other properties (see Figure 3-9). Among these properties, there are two main ones (see Figure 3-9) that you have to pay close attention to.

The first one is tag, and it's the name of a block and has to be a distinctive name. The second property is string, and it's used to display fixed text or an empty space. The string is displayed in the GUI model.

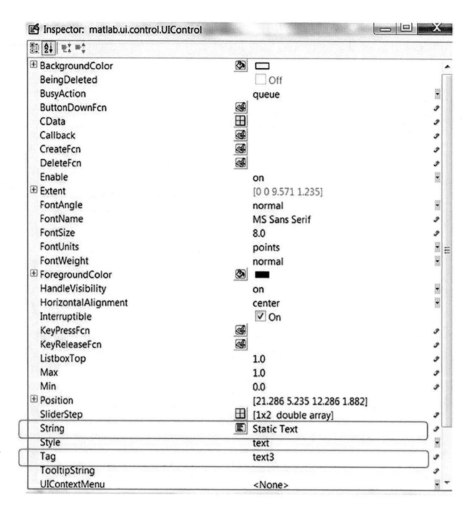

Figure 3-9. *Property Inspector window to edit properties of blocks*

Let's make some changes to our GUI model blocks by altering their background color, size, font size, color, and type, and by adjusting the positions of the blocks. Note that in the Static Text blocks, we edit the strings to be "Xmin" and "Xmax" in bold, 12.0 font size, and red and blue. In the Edit Text blocks, we remove their strings and make them empty. Moreover, we change their font size to 12.0, make them bold, and have a yellow background color.

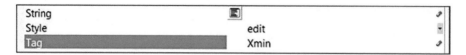

Note   For the edit text blocks to have predefined values (e.g., -6, 0, 3.14, and 100), you need to insert these values into the "string" value of each edit text block.

The Push Button's background color, font size, type, string, and tag names are modified. Its string is altered to "PLOT", and its tag is renamed to PLOTsinc. Figure 3-10 shows the completed GUI model.

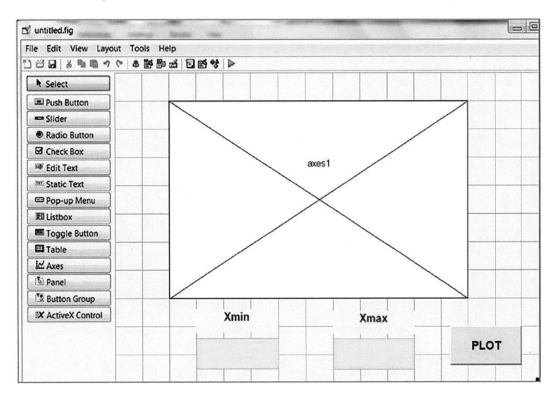

Figure 3-10. *The completed GUI model with all the modified buttons and components*

Save the model and click the Run button or press Ctrl+T on the keyboard. The new model, as shown in Figure 3-11, does not reflect any results yet. There is one important step that is still missing in the model's coding part.

Note   While building a GUI model and editing the properties of blocks, it is important to rename each block distinctively. The blocks' tags are required when modifying the callback functions of the GUI model.

To change the properties of several blocks simultaneously, select them all and then go to the Property Inspector to make changes.

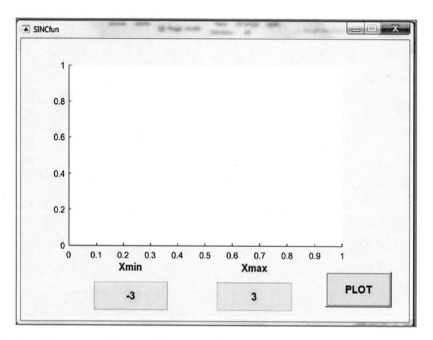

Figure 3-11. *The created SINC function plot GUI*

MATLAB saves the model in two file formats—SINC_fun.m (the function file) and SINC_fun.fig (the GUI figure). All the callback functions of the buttons used in SINCfun.m are shown here:

```
function varargout = SINCfun(varargin)
% SINCFUN MATLAB code for SINCfun.fig
% SINCFUN, by itself, creates a new SINCFUN or raises the existing
% singleton*.

...

% Begin initialization code - DO NOT EDIT gui_Singleton = 1;
```

```
gui_State = struct('gui_Name', mfilename, ... 'gui_Singleton', gui_
Singleton, ... 'gui_OpeningFcn', @SINCfun_OpeningFcn, ... 'gui_OutputFcn',
@SINCfun_OutputFcn, ... 'gui_LayoutFcn', [] , ... 'gui_Callback', []);

    if nargin && ischar(varargin{1})
    gui_State.gui_Callback = str2func(varargin{1});
    end

    if nargout
       [varargout{1:nargout}] = gui_mainfcn(gui_State, varargin{:});
    else
       gui_mainfcn(gui_State, varargin{:});
    end
    % End initialization code - DO NOT EDIT
    % --- Executes just before SINCfun is made visible.
    function SINCfun_OpeningFcn(hObject, eventdata, handles, varargin)
    % This function has no output args, see OutputFcn.
    ...

    function Xmax_Callback(hObject, eventdata, handles)
    % hObject handle to Xmax (see GCBO)
    ...

    % --- Executes on button press in PLOTsinc.
    function PLOTsinc_Callback(hObject, eventdata, handles)
    % hObject handle to PLOTsinc (see GCBO)
    ...
```

Note that most of this script (SINC_fun.m is automatically generated by the GUI model when it is saved) won't be changed, and only one callback function will be edited. The callback functions have to be edited to make the GUI model perform the anticipated computations and display a plot figure. Note that every callback function has three parameter handles—hObject, evendata, and handles.

While editing/writing callback functions, you can look up the properties of handles using the get() function and assign or change any property of the handles by using the set() function. Note that in the SINCfun.m model, there is only one callback function that has to be edited—PLOTsinc_Callback(hObject, eventdata, handles). The PLOTsinc_Callback callback function invokes the PLOT button, which is named with a tag name of PLOTsinc. The PLOTsinc tag name is given to the PLOT button via the Property Inspector while building the GUI model. This callback function computes $sinc(\pi x)$ by taking user entries for *xmin* and *xmax* and plotting the computed results.

To edit the PLOTsinc_Callback(hObject, eventdata, handles) callback function, you first convert the entries for Xmin and Xmax from the string format into the double format in the following way:

```
Xmin=str2double(get(handles.Xmin, 'string')); Xmax=str2double(get(handles.
Xmax, 'string'));
```

The get() command collects data from handles.Xmin and handles.Xmax as a string, and then str2double() converts them into a numerical format. Double-formatted values of Xmin and Xmax are taken to generate equally spaced values of the variable x, which are then taken to compute f(x) = $sinc(\pi x)$ and plot the results using the following commands:

```
x=linspace(Xmin, Xmax, 200); f=sinc(x*pi); plot(x, f, 'b'); grid on
title('sinc(\pi*x) = sin(\pi*x)/(\pi*x) ')
xlabel('x'),ylabel('f(x)=sinc(\pi*x)'),
axis([Xmin, Xmax, -.25, 1.1])
```

Finally, the edited callback function PLOTsinc_Callback(hObject, eventdata, handles) contains the following:

```
function PLOTsinc_Callback(hObject, eventdata, handles);
Xmin=str2double(get(handles.Xmin, 'string')); Xmax=str2double(get
(handles.Xmax, 'string'));

x=linspace(Xmin, Xmax, 200);
f=sinc(x*pi);
plot(x, f, 'b'); grid on title('sinc(\pi*x) = sin(\pi*x)/(\pi*x) ')
xlabel('x'),ylabel('f(x)=sinc(\pi*x)'),
axis([Xmin, Xmax, -.25, 1.1])
```

As noted, this is the only part of the M-file SINCfun.m that we have edited in order to make the model compute and plot the $sinc(\pi x)$ function with a single click in the GUI model. We save the SINCfun.m file and execute it by using ![Run] in the M-file editor or by pressing F5 from the keyboard. When we run SINCfun.m, the SINCfun.fig GUI model pops up. We enter two entries—$X_{min} = -5$ and $X_{max} = 6$—and click the PLOT button.

The results of the function are computed and plotted, as shown in Figure 3-9. Note that the value of Xmax must be larger than the value of Xmin; otherwise, there will be an error.

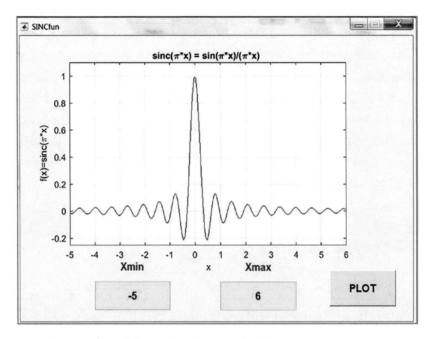

Figure 3-12. *The GUI model to plot the sinc(πx) function*

This simple example shows how easy it is to use GUIDE tools to build GUI models. Once the GUI model interface layout is complete and the components' properties have been edited and saved, the M-file's (function file's) callback functions are generated.

Subsequently, by editing only the required callback function or functions, we obtain the GUI model.

Example 2: Adding Functionality

Let's look at four more options for our GUI model (Figure 3-12):

- Creating an Exit/Quit button to close a GUI model window

- Playing a sound by recalling and executing another M-file (called SOUND_hear.m)

- Displaying an image (called GC2011.jpg) in *.jpg/*.jpeg format

- Displaying a message box with a message of "All Done Well Done!!!"

There are several ways to accomplish these tasks, one of which is to add a Push Button block and edit its callback function. You first go back to our GUI model editor and add one Push Button block. Change the Push Button's properties (background color, font size and type, and string and tag) as you did previously. Now you rename the Push Button block to QUIT by changing its string; its name tag is renamed QUIT_button. Before saving the updated GUI model SINCfun.fig, you need to alter the name of the window. To do that, you need to double-click the design area (the area outside of any blocks that opens the Property Inspector window) and then change the name tag from figure1 (the default) to GUI_window.

String	QUIT	
Style	pushbutton	
Tag	QUIT_button	

You can save the GUI model and execute it. After that, edit the last automatically added callback function called QUIT_button_Callback(hObject, eventdata, handles) by adding the following to it. The built-in MATLAB function called delete() halts the whole process and shuts the window.

```
function QUIT_button_Callback(hObject, eventdata,
handles) delete(handles.GUI_window)
run('SOUND_hear.m'); A=imread('GC2011.jpg');
image(A); msgbox('All Done Well Done!!!')
```

Figure 3-13 shows the updated GUI model. Test it with two numerical entries for Xmin and Xmax and then click the QUIT button. After pressing QUIT, the GUI window SINCfun.fig is closed.

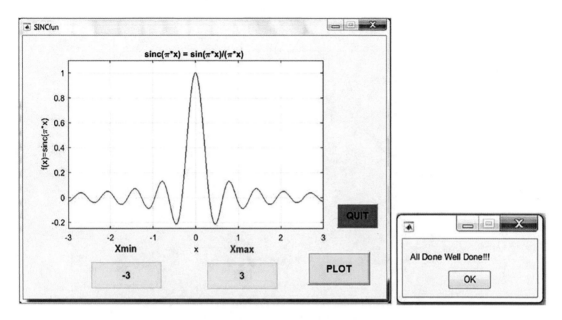

Figure 3-13. *The GUI model to display the sinc(πx) and quit options*

Subsequently, the built-in function run() executes the SOUND_hear.m file, which plays a sound wave. Another built-in function called image() displays the GC2011.jpg image, as shown in Figure 3-14. Finally, the msgbox() function displays the message: All Done Well Done!!!.

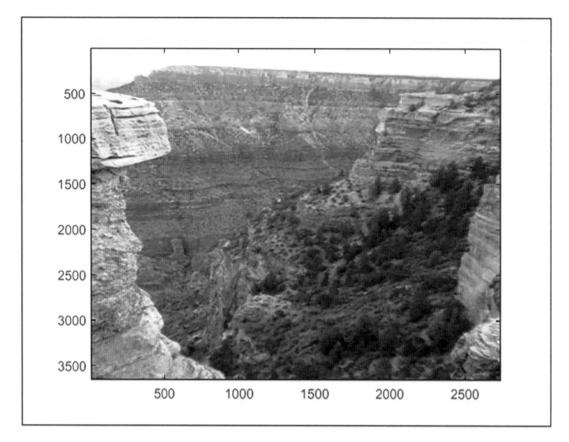

Figure 3-14.   *The image file GC2011.jpg is displayed*

The separate M-file called SOUND_hear.m contains the following scripts:

```
% SOUND_hear.m
% Three CHIRP signals to make a sound t=0:1/1e4:10;
D0=chirp(t, 0, 10, 3000, 'quadratic');
D1=chirp(t, 0, 10, 4000, 'q',[],'convex');
D2=chirp(t, .001, 10, 5000, 'logarithmic'); y=[D0, D1, D2]; sound(y);
```

In this simple exercise, we have demonstrated how to associate GUI model buttons with other GUI blocks and M-files. Similarly, Simulink's files, as well as any compatible applications with a MATLAB package, can be linked to GUI models.

Exporting the GUIDE GUI into App Designer

Let's export the created GUIDE's GUI model called SINCfun.fig into App Designer by following the options shown in Figure 3-3. To do that, you should select the file SINCfun.fig, and then click Open and Export. This creates a new M-file that is called SINCfun_export.m by default. Subsequently, the exported M-file called SINCfun_export.m opens automatically and can be executed by clicking the ▷ button in the M-file editor. After executing SINCfun_export.m, the exported GUI model shown in Figure 3-15 pops up.

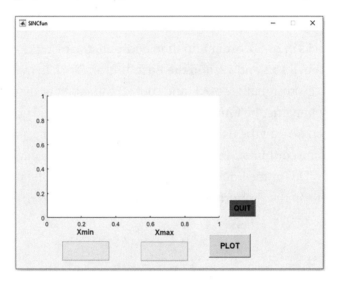

Figure 3-15. *The exported GUIDE App into App Designer to simulate and display the sinc(πx) plot*

After entering the values of -3 and 3 for Xmin and Xmax, respectively, and clicking the PLOT button, you will obtain the same simulation results as shown in Figure 3-13. When you click the QUIT button, the chirp signal sound will be played, and the image shown in Figure 3-14 will pop up. Where the exported App Designer M-file (SINCfun_export.m) is stored, SINCfun_export.mat file will also be stored.

Example 3: Solving a Quadratic Equation

Let's take another example of creating a GUI model that solves a quadratic equation for any given coefficients for *a, b,* and *c* and displays a plot for the user's specified value ranges of the variable *x*. The quadratic equation is formulated by $ax^2 + bx + c = 0$,

275

and if the numerical values of a, b, and c are given, you can compute the roots of the given equation numerically. This exercise is completed in two stages, building the GUI interface and editing the callback functions.

Building the GUI

You start again with a blank GUI window and choose a Panel block, onto which you drag and drop several blocks, such as Static Text, Edit Text, Push Button, and Axes. There is a good reason for employing the Panel block when building a GUI model. It helps you manipulate all the GUI blocks within it and around the Design Area with respect to each other.

We place the blocks in a, b, c order and then compute the results for D (discriminant), the root for $x1$ and $x2$, and the Push Button block to make the GUI model compute the discriminant, root $x1$, and root $x2$ values. After that, we edit the block properties by changing the font size, type, string, tag, and background color of each block. Moreover, we edit the user's entry blocks for x_{min} and x_{max} and use a Push Button block to plot the computation results in 2D. In addition, we change the name of the panel window to "Quadratic Equation Solver" by double-clicking the Design Area and changing the Title in the Property Inspector. Figure 3-16 shows the completed GUI model window.

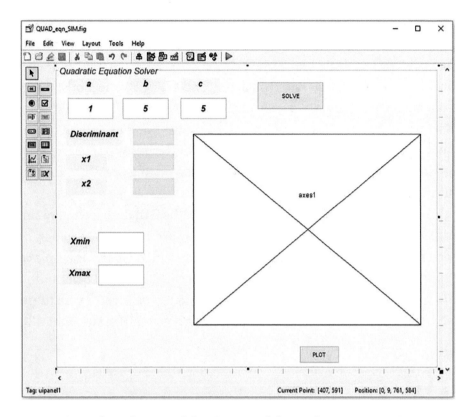

Figure 3-16. *Complete design of the GUI model window*

Note   In this GUI model (shown in Figure 3-16), the component names are not shown in order to save space in the palette window. You can turn them on/off by choosing File ➤ Preferences ➤ GUIDE ➤ Show Names in the Component palette.

Note that in the GUI model shown in Figure 3-16, the following properties of the blocks are altered: in Static Text blocks, such as a, b, c, x1, x2, Xmin, and Xmax, the string, font size and type, and background color are changed. The properties of the three Static Text blocks used to display the computation results for Discriminant, x1, and x2 are altered. For instance, for the Static Text block for the Discriminant value display, background color, font type, size, string, and tag are also changed. Its string is changed to be an empty space, and its tag is renamed to D.

The background color, font type, size, string, and tag of the Static Text blocks used to display the values of x1 and x2 are altered. The strings are set to be empty and the tags are renamed to x1 and x2, for *x1* and *x2*, respectively. There are two Push Button blocks used to solve the given equation (based on a, b, c) and to display the plot of the given quadratic equation with respect to the user entries for Xmin and Xmax.

The first Push Button block's string is changed to SOLVE and its tag is SOLVE_eqn, and the second Push Button string is called PLOT and its tag is called PLOT_eqn. As stated earlier, tags are vital and require careful attention when editing or rewriting subfunctions in the M-file of the GUI model.

Up to this point, all we have worked with is within the GUI design window and the properties of our chosen blocks within the Property Inspector, which we did by double-clicking each block individually. Another change that we made in this example was altering the panel window; we altered its background color, title, font type, and size. Finally, after completing all the changes in our GUI model, we saved it with the file name QUAD_eqn_SIM.fig (see Figure 3-17), which saved QUAD_eqn_SIM.m automatically as well.

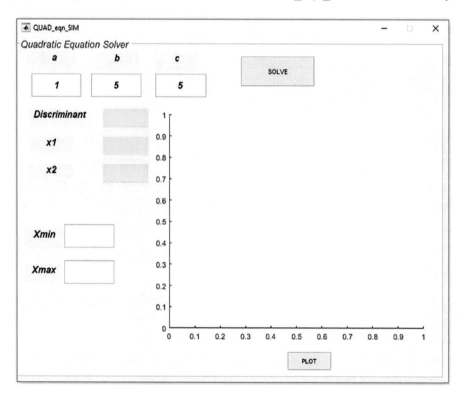

Figure 3-17. *Completed GUI model to compute roots and display a 2D plot of the quadratic equation*

Editing the Callback Functions

The model shown in Figure 3-17 does not reveal any results when you click the SOLVE or PLOT button. You can edit and rewrite two callback functions of the M-file— QUAD_eqn_SIM.m. Note that all of the automatically generated comments are removed from the script.

```
function varargout = QUAD_eqn_SIM(varargin)
% QUAD_EQN_SIM MATLAB code for QUAD_eqn_SIM.fig
...
gui_Singleton = 1;
gui_State = struct('gui_Name', mfilename, ...
'gui_Singleton', gui_Singleton, ...
'gui_OpeningFcn', @QUAD_eqn_SIM_OpeningFcn, ...
'gui_OutputFcn', @QUAD_eqn_SIM_OutputFcn, ...
'gui_LayoutFcn', [] , ...
'gui_Callback', []);
if nargin && ischar(varargin{1}) gui_State.gui_Callback =
str2func(varargin{1});
end
...
handles.output = hObject;
guidata(hObject, handles);

function varargout = QUAD_eqn_SIM_OutputFcn(hObject, eventdata, handles)
...
varargout{1} = handles.output;

function a_Callback(hObject, eventdata, handles)
...
function a_CreateFcn(hObject, eventdata, handles)
...
function b_Callback(hObject, eventdata, handles)
...
function b_CreateFcn(hObject, eventdata, handles)
...
function edit3_Callback(hObject, eventdata, handles)
```

```
...
function edit3_CreateFcn(hObject, eventdata, handles)
...
if ispc && isequal(get(hObject,'BackgroundColor'),
get(0,'defaultUicontrol BackgroundColor'))
set(hObject,'BackgroundColor','white');
end

function Xmin_Callback(hObject, eventdata, handles)
...
function Xmin_CreateFcn(hObject, eventdata, handles)
...
function Xmax_Callback(hObject, eventdata, handles)
...
function Xmax_CreateFcn(hObject, eventdata, handles)
...
if ispc && isequal(get(hObject,'BackgroundColor'),
get(0,'defaultUicontrol BackgroundColor'))
set(hObject,'BackgroundColor','white');
end

function SOLVE_eqn_Callback(hObject, eventdata, handles)
...
function PLOT_eqn_Callback(hObject, eventdata, handles)
...
```

The callback functions that we will edit are PLOT_eqn_Callback(hObject, eventdata, handles) and SOLVE_eqn_Callback(hObject, eventdata, handles). These callback functions make the GUI model compute and display discriminant, two real-valued roots in pre-defined blocks and display a 2D plot according to our entries for Xmin and Xmax. Here are the edited callback functions for the two Push Buttons called SOLVE and PLOT:

```
...
function SOLVE_eqn_Callback(hObject, eventdata, handles)
a=str2double(get(handles.a, 'string'));
b=str2double(get(handles.b, 'string'));
c=str2double(get(handles.c, 'string'));
```

280

```
D=b^2-4*a*c;
if D>0 % There are two real valued roots;
x1=(-b+sqrt(D))/(2*a);
x2=(-b-sqrt(D))/(2*a);
elseif D==0 % There is a unique root;
x1=-b/(2*a);
 x2=x1;
else % No real valued roots exist; x1='No Root';
x2='No Root';
end

D=num2str(D); set(handles.D, 'string', D); x1=num2str(x1); set(handles.x1,
'string', x1); x2=num2str(x2); set(handles.x2, 'string', x2);
function PLOT_eqn_Callback(hObject, eventdata, handles)
a=str2double(get(handles.a, 'string')); b=str2double(get(handles.b,
'string')); c=str2double(get(handles.c, 'string'));
D=str2double(get(handles.D, 'string')); x1=str2double(get(handles.x1,
'string'));
x2=str2double(get(handles.x2, 'string'));Xmin=str2double(get(handles.Xmin,
'string'));
Xmax=str2double(get(handles.Xmax, 'string'));
x=linspace(Xmin, Xmax, 200); y=a*x.^2+b*x+c;
plot(x, y, 'r-', 'linewidth', 1.5);
xlabel('x'), ylabel('y = ax^2+bx+c')
if D>=0 % Roots x1 and x2 will be plotted if the equation has real roots
hold on    % Plot is held if D>=0
plot(x1, 0, 'rs', x2, 0, 'gd', 'markersize', 7, 'markerfacecolor', 'y')
legend('plot of quad. eqn', 'root: x_1', 'root: x_2')
end
title(['Plot of: ' num2str(a) 'x^2 + ' num2str(b) 'x + ' num2str(c) ' =
0']); grid on
```

Note that in the subfunction SOLVE_eqn_Callback, you use the get() command to obtain string values of a, b, c and convert them into numerical data with str2double(). Using the set() command, the computed values of D, x1, and x2 are assigned to D, x1, and x2 in order to display them in their respective Static Text blocks called D, x1, and x2.

In the latter subfunction, you obtain the numerical values of a, b, and c with the get() command. You run the whole M-file called QUAD_eqn_SIM.m by clicking the Run button or calling the file from the Command window (for example, $3x^2 + 5x + 1 = 0$) and then plot the given equation for $x = -1...3$.

Note that in the SOLVE_eqn_Callback callback function, if the discriminant is $D \geq 0$, then the computed roots will be also plotted. Moreover, there are several plot tools used here (hold on, legend, markersize, and markerfacecolor) that are explained in detail in Chapter 6.

Figure 3-18 shows the final GUI model with its results for the given example $-3x^2 + 5x + 1 = 0$ and for $x = -1...3$.

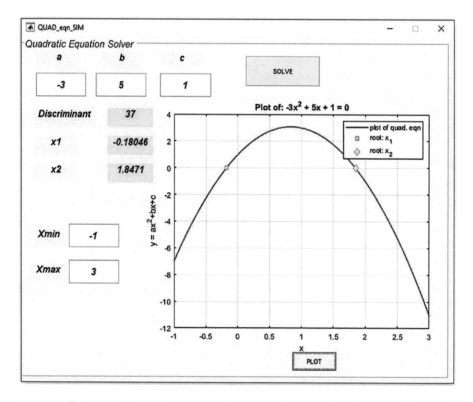

Figure 3-18. *The GUI model to compute roots of a quadratic equation and plot its values with user entries for Xmin and Xmax*

Let's review some of the most common errors made while editing and rewriting subfunctions. They are:

- Misspelled tag names of blocks that are recalled within subfunctions or set values to display in a GUI model window.

- Necessary conversion operators (such the get() and set() operators) are not employed appropriately within subfunctions for numerical calculations.

- Editing the wrong subfunctions (callback functions) of buttons to make a GUI model perform its aimed operations.

To avoid these common mistakes, it is recommended to choose tag names carefully and double-check their names while writing and editing nested callback functions. Another recommendation is to write down all the tag names and type them in with care. When converting variable values from one type to another with the get() and set() operators, look at the task specifics. For instance, what type of data is needed for processing, and by which handle names (such as handles.WHAT) are values of variables or data obtained. There are several ways to avoid editing the wrong callback functions. One is to look up an assigned name tag for each button that makes the GUI model perform its anticipated operations. MATLAB automatically assigns a subfunction name to every operational button and entry block with its given name tag (a given name tag by a user) with an underscore sign and callback(). For instance, if an operational button's name tag is PLOT_all, it will have an assigned subfunction called PLOT_all_Callback(hObject, eventdata, handles). Another approach, after saving the GUI model and the automatically generated M-file, is to click an operational button (e.g., a Push Button) in a GUI model design window. Then right-click and choose View Callbacks ➤ Callback to directly reach the right subfunction subject to edit.

GUI Dialogs and Message Boxes

MATLAB has a few commands that create/call ready-to-use pop-up dialog boxes, including Error, Warning, Help, Information, User Entry Input dialog, and so forth. They can be created without writing scripts and callback functions or building GUIs. One of the most common purposes of these message boxes is to deliver messages to the users. In general, creating these pop-up message boxes is straightforward and requires only one command. The dialog and message boxes (except for the user input box) can all be called and generated with one general command: errordlg(), warndlg(), helpdlg(), and msgbox(). On the other hand, a user input dialog box requires several commands, such as the name of the input dialog, the input dialog window title/message, and the input dialog window. Let's look at a few examples to see how to create these message boxes, a user entry dialog window, and an entry status box.

Error Dialog

General command syntax: `errordlg('Add Notes')`

    Example:

```
errordlg(['Time is gone!' 'Today is ' date])
```

Warning Message

General command syntax: `warndlg('Add Notes')`

    Example:

```
warndlg(['Outside temperature is ' num2str(-20) ' C; thus, be dressed up
accordingly!!!'])
```

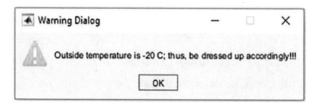

F1 Help/Message Box

General command syntax:

```
helpdlg('Add Notes')
```

Example:

```
helpdlg('What is your problem?', 'This is a help line');
```

A second example:

```
CreateStruct.WindowStyle='replace'; CreateStruct.Interpreter='tex';
msgbox('f(\alpha) = e^{cos(\omega*\alpha)}','Function: ',CreateStruct);
```

Note that CreateStruct, used in this example, is a variable type of struct that creates a structure type of variable with two fields of a character type—called WindowStyle with the value of replace and Interpreter with the value of tex. They are recognized automatically by MATLAB. Without creating the CreateStruct variable with its two fields, WindowStyle and Interpreter, the equation would be displayed as shown in the message box: $f(\alpha) = e^{\cos(\omega\alpha)}$.

Finally:

```
msgbox('\copyright by NDSU', 'Copyright',CreateStruct)
```

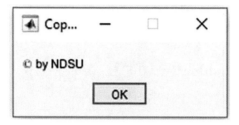

General Syntax

There are alternative ways to create help/message/error/warning messages. The general command syntax is as follows:

```
HM = msgbox('Message', 'Box Title', 'Icon')
```

Here is just a message box:

```
msgbox('Hello World', 'My Message')
```

Here's a warning message:

```
msgbox('Hello World', 'My Message', 'warn')
```

Here's a message box with an error sign:

```
msgbox('Hello Students', 'My Message', 'error')
```

Here's a message box with an information sign:

```
msgbox('Hello Students', 'My Message', 'help')
```

Input Dialog

General command syntax:

```
ANSWER = inputdlg(PROMPT,NAME)
```

Here is a short script:

```
Enter1={'Enter # of FS terms to evolve & Hit [ok] & Wait: '}; % Message
Name1='Input for TERMS of Fourier Series'; % Dialog box name
Numlines = 1; % Number of lines for Input dialog
ANSWER=inputdlg(Enter1, Name1, Numlines); % Generate and Read Input Dialog
[n status] = str2num(ANSWER{1}); % Picks up the user entry under variable
name of n
```

Question Dialog

General command syntax:

```
ButtonName = questdlg('Your Question ...','Question?', 'Option1','Option2')
```

An example:

```
YA= questdlg('Select your answer to the question: "How do you feel?"
','Question', 'Superb', 'I Can Work for a while', 'Tired a bit', ' ')
```

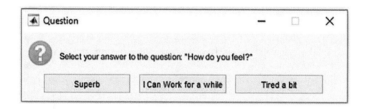

Let's look at a more involved example that has several choices.

Making a Choice

You are in a café and would like to make a purchase. You would like to choose from the menu coffee, tea, or some sweets. The café has some shortages that you are not aware of. The message and dialog boxes such as warning, help, input, or error messages need to be displayed according to your order and the café shortages. Here are the steps to be implemented in this script:

Step 0. You are in a café, and a waiter comes up to take your order.

Step 1. You decide between coffee, tea, and chocolate cake.

Step 2. The waiter responds that they don't actually carry tea and are out of chocolate cake.

Step 3. You then choose your coffee type: Normal, Strong, or Special (your taste).

Step 4. You taste and grade the coffee.

Step 5. You pay the bill and leave a tip.

Here is the answer script:

```
clearvars
% It is time for coffee, let's go to a cafe ...
% Step 0
H1 = msgbox('You are in a cafe and ordering coffee of your taste', 'Cafe');

pause(2), delete(H1)
% Step 1. Ordering your coffee
H2 = msgbox('What can I do for you today? Do you want coffee or sweets',...
'Waiter has arrived');
pause(3), delete(H2)
% Answer what you like to have
YA= questdlg('Your answer: "Do you want coffee or sweets?" ','Question',
... 'Coffee', 'Tea', 'Chocolate Cake', ' ');
  switch YA
  case 'Coffee'
  CA = questdlg('Your answer: "OK. What kind of coffee?" ','Question',
  ... 'Normal', 'Strong', 'Special', ' ');
  case 'Tea'
  H3=warndlg(['Sorry! We do not serve' ' Tea. Maybe something else?']);
  pause(2) delete(H3)
  case 'Chocolate Cake'
  H4=errordlg(['Very Sorry!' 'No cakes Left ']); pause(2) delete(H4)
end
Response=exist('CA', 'var'); if Response ==1
  switch CA case 'Normal'
  HN=warndlg(['Normal coffee is', ...
```

```
'3 tea-spoons of coffee+3 Table-spoons of Milk+1 tea-spoon of sugar']);
pause(3), delete(HN)
case 'Strong'
HS=warndlg(['Strong coffee is', ...
'4 tea-spoons of coffee+2 Table-spoons of Milk+2 tea-spoon of sugar']);
pause(3), delete(HS)
case 'Special'
HS = msgbox('Select: how much coffee, milk and sugar to add?',
'Selection'); pause(3), delete(HS)
Call = 'Order your coffee';
Selection = {'how many tea-spoons of coffee: ',...
'how many tablespoons of milk: ', 'how many teaspoons of sugar: '};
n_lines=[1, 13];
ANS=inputdlg(Selection, Call, 1, {' ', ' ', ' '}); end
OK = msgbox('You have had your coffee','You are about to leave ');
pause(5), delete(OK)
PB= questdlg('Waiter asks: "How was the coffee?" ','Question', ... 'Just
Superb', 'Nice', 'Umm...but OK', ' ');
switch PB
case 'Just Superb'
helpdlg('Thank you so much! 4$+1$ (tips)', 'Here is the payment');
case 'Nice'
helpdlg('Thanks! 4$+0.50$ (tips)', 'Here is the payment'); case
'Umm...but OK'
helpdlg('Thank you! 4$ (No-tips)', 'Here is the payment'); end
else
H5 = msgbox('Bye - Bye! See you next time!', 'Waiter leaves'); pause(4),
delete(H5)
end
```

Note that in this script, we used MATLAB's built-in function, which verifies whether a specific variable exists in the workspace. It uses the command syntax of exist('N', 'var') to determine whether the variable N is available in the workspace. It can also be used to verify whether any specific file exists in the current directory.

For example, `exist('MY_fun.c', 'file')` checks whether the file called `MY_fun.c` is present in the current directory of MATLAB. The output of the built-in function `exist()` can be 0, 1, or 2. The 0 means the variable or file does not exist, 1 means the variable exists, and 2 means the file exists in the current directory.

When the script is executed, two message boxes (H1 and H2) are displayed for two to three seconds, and then closed automatically. Then the next question dialog box pops up, and you need to choose Coffee, Tea, or Chocolate Cake.

Depending on your selection, other options and responses will appear. If you select Coffee, the question dialog (CA) will pop up and ask you to select your coffee type: Normal, Strong, or Special.

If you select Normal or Strong in the dialog box (CA), there will be a pop-up box displaying the contents of your selected coffee in terms of coffee, milk, and sugar. If you select the Special type of coffee, you need to input (input dialog ANS) the amount of coffee, milk, and sugar you want in your coffee.

The message box (OK) pops up and closes down automatically after five seconds. Finally, the waiter asks you how your coffee tasted.

Based on your answer, you get the help dialog box indicating how much to pay, including the tip.

If you select Tea, the warning dialog box (H3) will pop up and close down automatically after two seconds. If you choose Chocolate Cake, the error dialog (H4) will pop up and close down automatically in two seconds.

The last message box (H5) is displayed if you select Tea or Chocolate Cake in the first step.

These message and dialog boxes can also be employed within function files in a very similar manner.

Providing Input to an Equation

This simple numerical simulation exercise computes the values of a mathematical equation by writing a number to the function file: $F = e^{cos(\omega t)}$. The default values of input arguments for simulation are $\omega = 13; t = 0 : \dfrac{\pi}{100} : \pi$.

The varying inputs are as follows:

- No input; use default values

- One input, ω or t, whereby the missing argument takes the default values

- Two inputs, ω and t, whereby an error dialog pops up.

Here is the complete answer script of this exercise:

```
function F = ERR(varargin)
% HELP. ERR.m simulates how to employ ERROR dialog box
warndlg('Note that this is a varargin function file ');
if nargin == 0
  omega = 13; t = 0:pi/100:pi;F = exp(cos(omega*t)); plot(t, F, 'bo--'),
  shg elseif nargin == 1 && numel(varargin{1})==1
omega=varargin{1}; t = 0:pi/100:pi;F = exp(cos(omega*t)); plot(t, F,
'bo--') elseif nargin == 1 && numel(varargin{1})>1
  t = varargin{1}; omega=13;F = exp(cos(omega*t));
  plot(t, F, 'bo--')
  elseif nargin ==2 && numel(varargin{1})==1
omega=varargin{1}; t =varargin{2};
F = exp(cos(omega*t));
plot(t, F, 'bo--')
elseif nargin == 2 && numel(varargin{1})>1
omega=varargin{2}; t = varargin{1};
F = exp(cos(omega*t)); plot(t, F, 'bo--')
```

```
else
  errordlg(['Wrong Number of Inputs!' num2str(nargin) ' are too many
  INPUTs'])
disp('No OUPUTS')
F=warndlg(['No OUTPUTS because ' num2str(nargin) ' are too many INPUTS']);
end
end
```

Let's simulate the previous function file (ERR.m) from the Command window:

```
>> omega = 13; t = 0:pi/100:pi; OmegaN=13;
>> F = ERR(omega, OmegaN, t); No OUPUTS
```

These warning and error dialog boxes are displayed:

This concludes our brief discussion of GUI tools and functions. These tools are handy and can be associated with M/MLX-files and function files to make scripts more user-friendly with GUI tools.

Summary

In this chapter, we briefly covered how to employ GUI tools and functions via GUIDE and the App Designer to develop GUI models via a few examples. In addition, we demonstrated via a few examples how to design error, warning, and help message dialog boxes and user entry interfaces in association with M/MLX-files and function files.

Exercises for Self-Testing

Exercise 1

Design/model and rewrite an M-file for the next GUI model to compute a torus function and plot its 3D plot (shown in the next GUI model figure called PLOTall) by taking user-entered input. (See Chapter 6 for how to build 3D plots.)

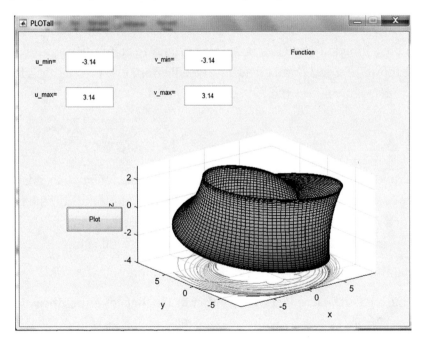

Use the umbilic torus function defined by the following:

$$x = \sin u \left(7 + \cos\left(\frac{u}{3} - 2v\right) + 2\cos\left(\frac{u}{3} + v\right) \right),$$

$$y = \cos u \left(7 + \cos\left(\frac{u}{3} - 2v\right) + 2\cos\left(\frac{u}{3} + v\right) \right),$$

$$z = \sin\left(\frac{u}{3} - 2v\right) + 2\sin\left(\frac{u}{3} + v\right)$$

For $-\pi \le u \le \pi$, $-\pi \le v \le \pi$.

– Create a stand-alone application of the created GUI model to plot an umbilic torus function.

– Associate the created GUI with another M-file to display the existing (external) image in *.jpg format, for instance.

– Create a function file called umbilic_torus.m to compute x, y, z for the inputs of u and v and create an .mex file from it. (See Chapter 4 for how to create an .mex file.)

Exercise 2

Create a GUI model to compute the volume of a cylinder and rectangular prism with the necessary user entries. For example, for a cylinder, a user should add the radius and height, and for a rectangular prism, they should add the length, width, and height. All units should be in the SI Unit System only. Also, perform the following tasks:

– Create a stand-alone application of the created GUI model.

– Associate the created GUI model with another M-file that computes the values of $f(t)$ $= e^{\sin(250t) + 2.5 \cos(750t)}$ for $t = -\pi...\pi$ with $\Delta t = 1/10000$ and plays a sound of $f(t)$ with a sampling frequency of $f_s = 10000$ Hz by using sound().

Exercise 3

Build a new GUI model. The new GUI model should have one drop-down option

(a Pop-up Menu button) to select a function type ($sin\left(\dfrac{2\theta}{3}\right)$, $cos\left(\dfrac{5\varphi}{3}\right)$, and quadratic

polynomial $x^2 + 6x + 3$) that is computed by separate M-files to plot it. Also, include
an Edit Text block so a user can specify value ranges for arguments: θ, φ, x. Also, add a
Push Button to close the GUI model with a message dialog box and the message ALL
Completed! and a sound of rectangular pulses using MATLAB's built-in rectpuls()
function.

CHAPTER 4

MEX Files, C/C++, and Stand-Alone Applications

MEX stands for *MATLAB executable*, and it's a way to call custom C, C++, or FORTRAN routines/code directly from MATLAB, thereby treating them as if they were built-in MATLAB functions. Since MATLAB is compatible with several programming languages, including C/C++, FORTRAN, C#, and Java, the MEX files can be called exactly like M-files or M-functions within the MATLAB environment. It is also possible to create stand-alone applications in MATLAB as self-executable files.

MATLAB's powerful Coder toolbox is capable of creating stand-alone and self-executable C code. It is beneficial to know how to create C code from MATLAB's M-files since the C programming language is free and does not have any licensing requirements. It can be employed to create executable files, dynamic link libraries, and libraries of large applications. In fact, C can be used to solve a wide range of tasks with myriad types of applications.

There are several reasons that we would create or employ MEX files. One of them is that they are capable of calling large existing C, C++, or FORTRAN programs directly from MATLAB without rewriting them as M-files. The other reason is to speed up computation processes where M-files have bottlenecks. For instance, MEX files can be employed on large loop-based computations to avoid bottlenecks with time-consuming computation processes. The MEX files are better than M-files for several reasons. They are called and executed as executable files without having to be compiled, unlike M-files. They can call large existing C or FORTRAN routines directly from MATLAB without having to be rewritten as M-files. Moreover, the MEX files allow parallelism, so we can write multithreaded C code for various costly (in terms of machine time) computations and simulations that cannot be vectorized.

© Sulaymon Eshkabilov 2022
S. Eshkabilov, *Beginning MATLAB and Simulink*, https://doi.org/10.1007/978-1-4842-8748-4_4

Verifying Compilers

Before you start generating MEX files, it's a good idea to verify which compilers are installed in the current MATLAB package. You can do this by typing the following command in the command window prompt:

```
>> mex –setup
```

Then you follow the instructions to select the appropriate compiler from the list of installed compilers. If there is no compiler installed, you need to install the one that is compatible with MATLAB to generate MEX files (a list of compatible compilers can be found on the MathWorks website: `https://www.mathworks.com/support/requirements/previous-releases.html`). Figure 4-1 shows a general flowchart of C/C++ code generation in MATLAB.

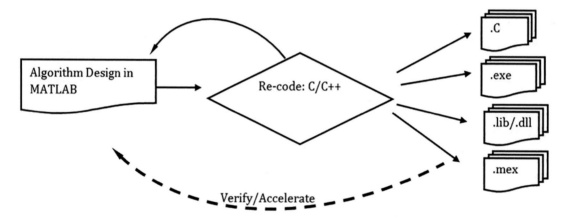

Figure 4-1. *Flowchart of code generation in C/C++*

Hereafter, we demonstrate several examples of converting M-files into C/C++, MEX, and executable stand-alone applications, compiling C code, and testing it in the MATLAB environment.

Generating C Code

This simple example will generate C code in MATLAB to compute the sum of odd numbers based on the user's specified integer number. We first write an M-file and then convert it into an MEX file in MATLAB using a code generator application tool called

MATLAB Coder.[1] This tool is used to generate C and C++ code from MATLAB code for a variety of hardware platforms, embedded systems, and desktop platforms. The generated code can be integrated in existing programs and projects as source code, libraries, and/or dynamic libraries.

Note   MATLAB Coder is a separate application that does not come with the base MATLAB package and has to be installed separately.

Let's take this SUModd.m M-file function, which computes the sum of odd integers up to the user-specified integer, N.

```
function SUModd(N)
%#codegen Sum=0; ii=1; while ne(ii,N)
    if ne(mod(ii,2),0) Sum=Sum+ii;
else
    Sum=Sum;
end
ii=ii+1;
end
Sum
```

Note that the second line of the SUModd.m function is a command (comment), %#codegen. It's recommended that you include this comment in M-files that are subject to conversion into MEX files. The conversion of the M-file can be achieved in three ways. First, you can use this command:

```
>> mex SUModd.m
```

Second, you can type >> coder in the Command window. Finally, you can use the MATLAB Coder[2] option ⊞ MATLAB Coder, which is on the APPS tab in the main menu. When you click the MATLAB Coder icon, the GUI window of the MATLAB Coder pops up, as shown in Figure 4-2. After this step, the input variable type is defined by clicking the hypertext "Let me enter input or global types directly," as shown in Figure 4-3.

---

[1] MATLAB Coder is a registered trademark of The MathWorks Inc.
[2] MATLAB Coder is a registered trademark of The MathWorks Inc.

In the Define Input Types box, as shown in Figure 4-4, you can input the variable type N (single, double, int8, uint8, int16, uint16, ... structure, complex number, etc.). For this exercise, select a double data format from all values. Note that the selected data format should be followed while using the generated *.mex file. Then you select the size (1x1, ... mxn) for N and click Next.

You then specify the name of the M-file to be converted into C/C++ and MEX in the Generate Code for Function box.

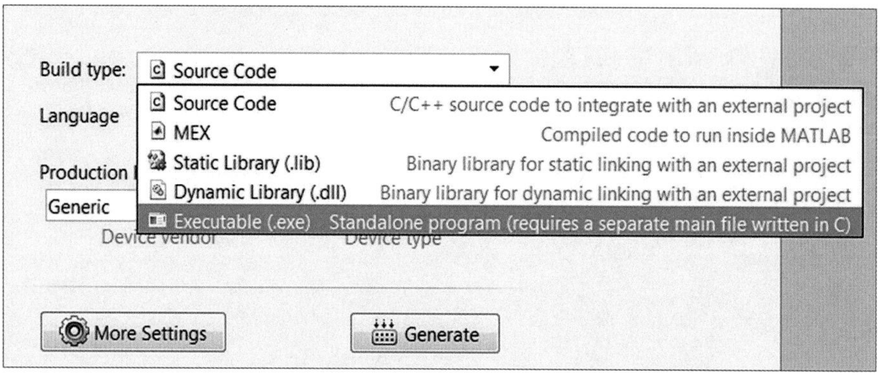

Figure 4-2. *Select None and click Next in the bottom-right corner*

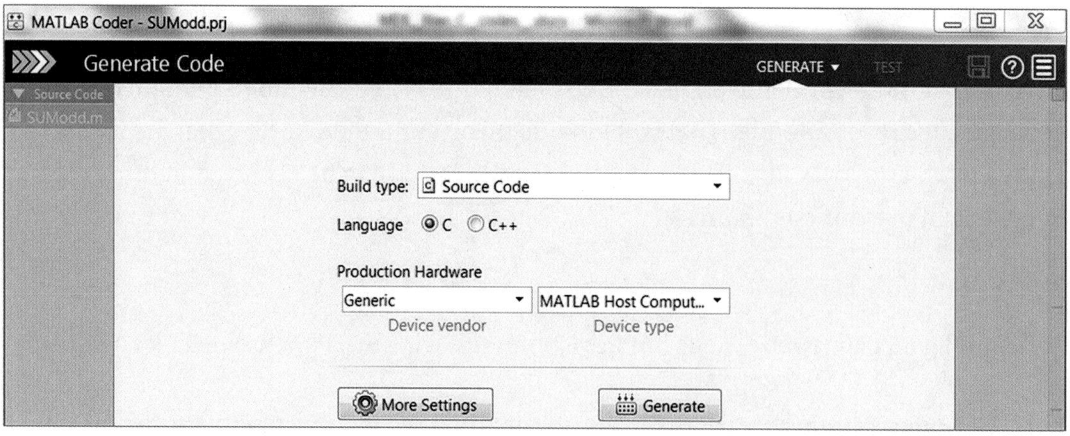

Figure 4-3. *Click the hypertext "Let me enter input or global types directly"*

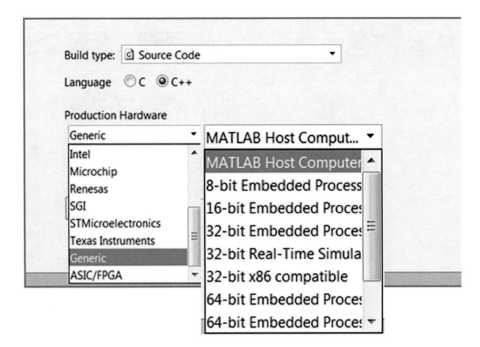

Figure 4-4. *Define Input Types box*

Note   that the selected data format for the input variable should be respected while executing the created *.mex file. Otherwise, there will be data format errors.

After you press Enter, the MATLAB Coder will automatically generate the MEX file with all the source code in *.c and *.h files if there are no issues converting the MATLAB functions. You start the code generation process by selecting from the drop-down options which source code type you want to generate. In this case, you can choose from C/C++, MEX, Static Library (.lib), Dynamic Library (.dll), and Executable (.exe), as shown in Figure 4-5. Next, you need to choose in which language—C or C++—you want to create the MEX file (see Figure 4-5).

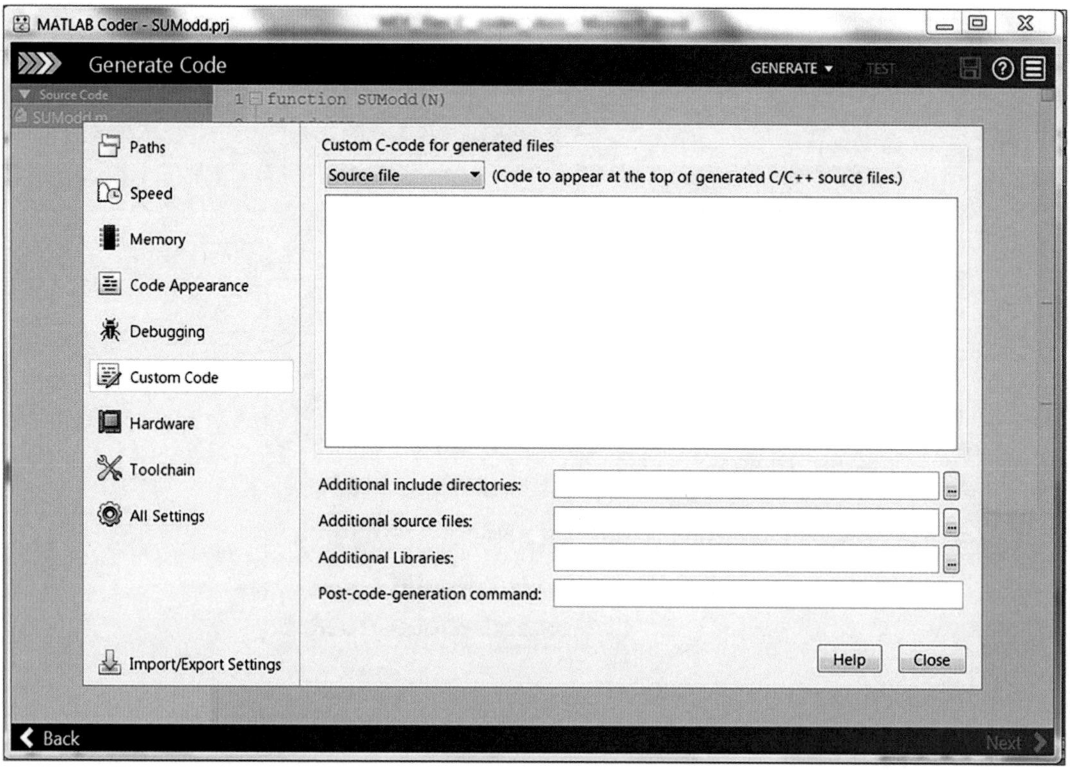

Figure 4-5. *Source code type selection generated in C/C++*

Note that these options are operating system and platform dependent. In our case, we are working in the Windows OS environment, so these libraries are applicable to the Windows OS. Then you select the hardware board type and device vendor—MATLAB Host Computer and Production Hardware (see Figure 4-6).

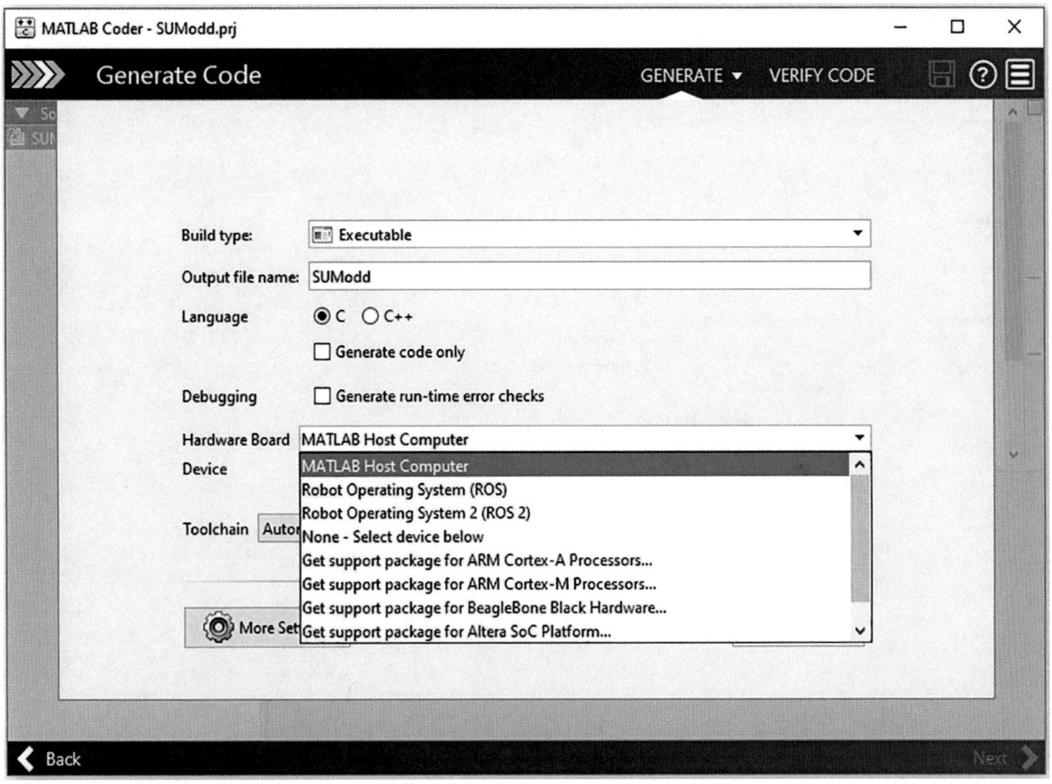

Figure 4-6. *Hardware board type selection options*

You can make a few more adjustments in the More Settings area, as shown in Figure 4-7.

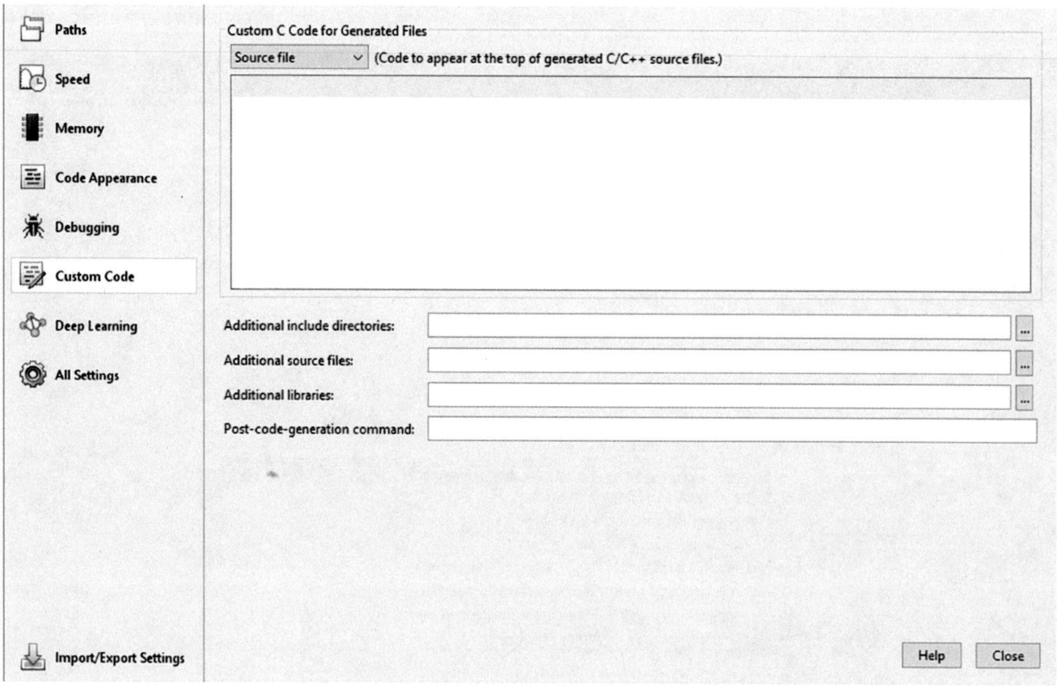

Figure 4-7. *More Settings option of the Coder's GUI window tools*

If there are no errors or warnings, the C-generated code with the target source code will appear (see Figure 4-8). The MEX file can be opened in the MATLAB workspace or within any M-file, like any MATLAB function file or M-file would be.

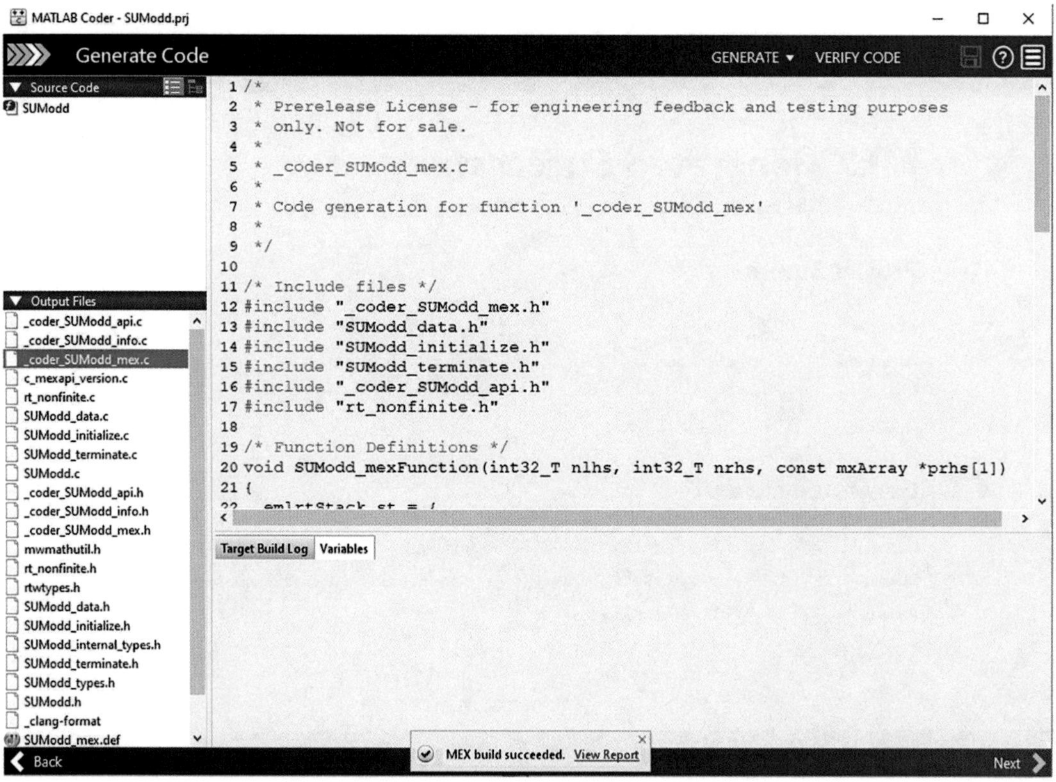

Figure 4-8. *Generated MEX file contents and libraries from the source code called SUModd.m*

Figure 4-8 shows the successfully generated MEX file called SUModd_mex.c. It is saved in the current (working) directory of MATLAB. The successfully generated MEX file, as shown in Figure 4-9, is obtained after clicking Next button in the bottom-right corner.

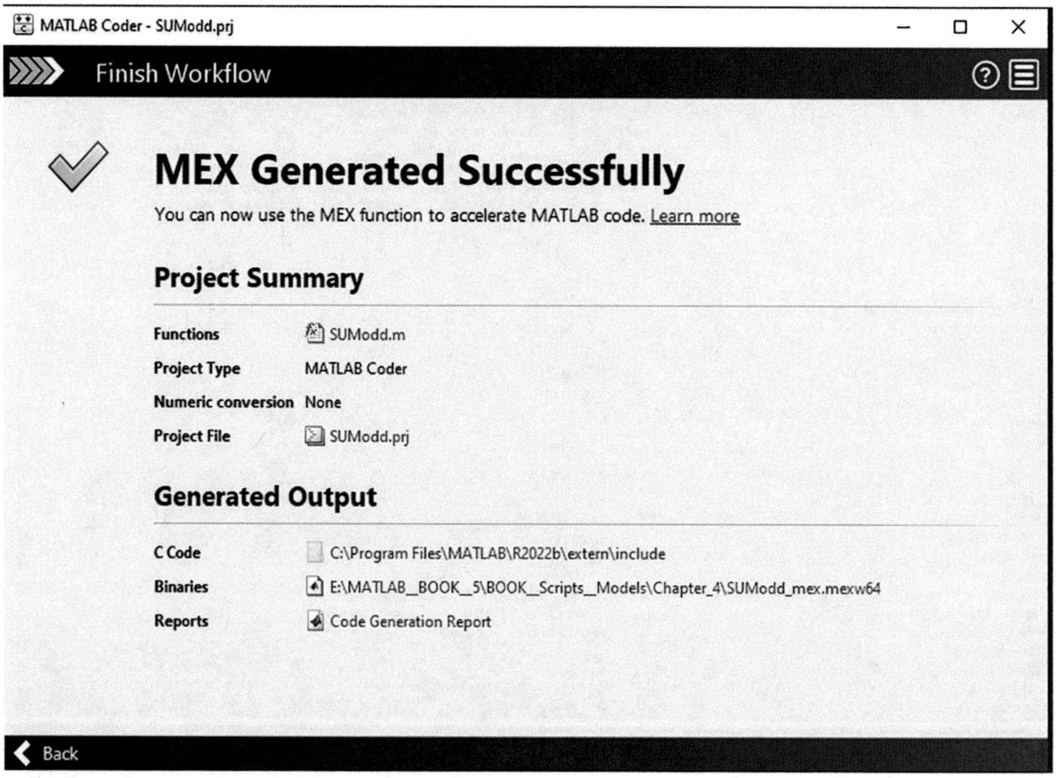

Figure 4-9. *Successfully generated MEX file from the source code called SUModd.m*

There might be issues linked to the conversion and encoding processes of M-files to MEX files, as a number of MATLAB's functions are incompatible with C. As an example, let's modify the M-file (in our initial M-file called SUModd.m) by including the tic and toc MATLAB commands, which compute elapsed time, and then try to re-create the MEX file. With these modifications in the M-file, the Coder prompts you with the two compatibility issues illustrated in Figure 4-10. After removing the two commands—tic and toc—from our M-file and saving the updated version, we re-encode and then obtain no warnings, as shown in Figure 4-8.

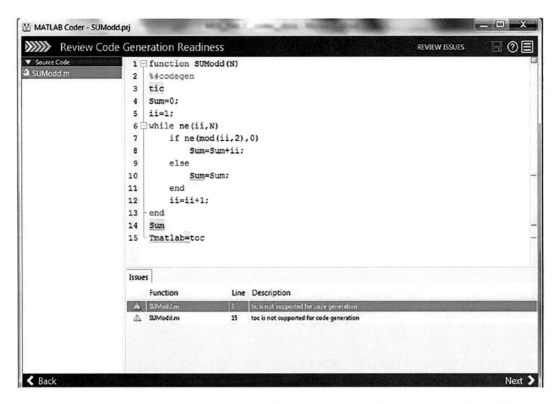

Figure 4-10. *Two errors/issues occurred with the tic and toc commands while converting the M-file to the MEX file*

The process of recalling and using MEX files is similar to any M-file. For this example, the created MEX file can be recalled as follows:

```
>> SUModd_mex(N)
```

In addition, while executing this MEX file, you can also use tic and toc commands to compute the elapsed time, e.g.:

```
>> tic; SUModd_mex(N), Tcalc = toc;
```

It also can be recalled within any M-file in a similar way. Note that N is predefined in the MATLAB workspace according to the variable type and size used in the MEX file during the initial step. One of the most important advantages of MEX files over M-files when dealing with for and while loops is their speed, in other words, computation efficiency. For instance, from the previous example, we can demonstrate the efficiency of the MEX file over its parent M-file.

```
% Compare computation efficiency of MATLAB and MEX files clearvars; N=1e+8;
Tmatlab=cputime; SUModd(N), TM=cputime-Tmatlab; fprintf('M-file Comp. Time
is %5.5f sec \n', TM); clearvars; N=1e+8;
Tmex=cputime; SUModd_mex(N), TMEX=cputime-Tmex; fprintf('MEX-file Comp.
Time is %5.5f sec \n', TMEX);
```

The computation (elapsed) time for each file is as follows:

M-file comp. time is 9.26115 sec

MEX file comp. time is 1.06168 sec

So, the computation time of the MEX file is about nine times shorter than the original M-file. Also, it must be noted that MEX files take precedence over M-files when the same file names exist in both M-files and MEX functions.

Note   MEX files take precedence over M-files when the same file names exist in both M-files and MEX functions.

Let's look at another example. Compute the Leibnitz series, an approximation of $\pi/4$ using the sum of series with the following formulation:

$$Error \cong \frac{\pi}{4} - \sum_{k=0}^{N} \frac{(-1)^k}{2k+1}$$

The only user entry will be Error margin based on the entry. Write a function file to compute the N term value using the previous formulation and create its MEX file.

Here is the function file called Leibnitz.m with a single user entry and two outputs, such as the computed Error evolutions and the final N term.

```
function N = Leibnitz(E0)
S=0;          % Initial value of summation
Error=pi/4;   % Initial Error value
k=0;
while abs(Error)>E0
    p=(-1)^k;
    S=S+p/(2*k+1);
```

```
Error=pi/4-S; % Accumulates all of the values of the Error from all
iterations
k=k+1;
```
end
N=k;

We follow the previously explained steps that were shown in Figures 4-2 to 4-9. While defining input types (Figure 4-4), you can select double as a data format and 1x1 for the size of E0 (input argument). After successful implementation of the steps given in Figures 4-2 to 4-9, Leibnitz_mex.mex will be created, as shown in Figure 4-11. That can be verified by clicking VERIFY CODE in the upper-right corner of the created MATLAB Coder window (see Figure 4-11).

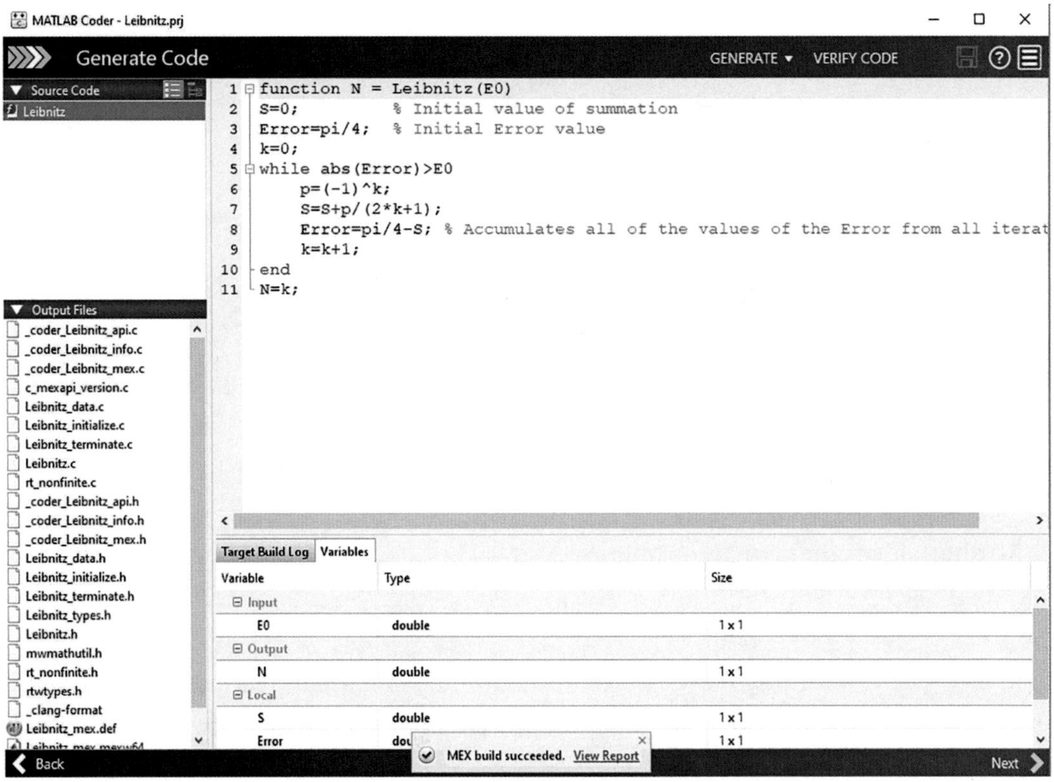

Figure 4-11. *Successfully generated MEX file contents and libraries from the source code called Leibnitz.m*

Now, we test the created *.mex file from the MATLAB Coder window; see Figure 4-12 using the input entry for an error variable of 0.001.

Figure 4-12. *Verifying the successfully generated MEX file*

This verification can be also done from the Command window using the same command as shown in Figure 4-12.

```
>> Leibnitz_mex(0.0001)
ans =
      2500
```

Creating MEX Files from Other Languages

MEX files can be created from C/C++ or FORTRAN source code as well. That can be done with a single command, as shown here:

```
>> mex HELLO.c
```

An alternative command is as follows:

```
>> mex -v HELLO.c
```

In the latter case, the -v flag displays the compiling and linking process. Note that the source code HELLO.c in C (given here) is already created and put in the working directory of MATLAB.

```
#include "mex.h" /* Always include this */
void mexFunction(int nlhs, mxArray *plhs[], /* Output variables */ int
nrhs, const mxArray *prhs[]) /* Input variables */
```

```
{
    /* Do something interesting or fun */
mexPrintf("Hello WORLD:)...:), This is SALOMON ... world!\n"); return;
}
```

The created MEX file, called HELLO_mex.mexw64 from the source code HELLO.c, can be executed by recalling it in the Command window:

```
>> HELLO_mex
```

That command outputs the following:

```
>> Hello WORLD:)...:), This is SALOMON ... world!
```

Note that the file extension of the MEX files, either mexw32 or mexw64, indicates which processor type (OS system) is running on the user computer.

Here, we have covered the generic procedures for creating MEX files; however, the exact procedures are system dependent. In other words, they depend on which operating system is installed, which MATLAB version and compilers are installed and used, and so forth. Thus, it is necessary to recompile MEX files for every platform. One of the most essential challenges in developing MEX files is the problem statement that should be in vector form. If it's not, then the created MEX application will be rather slow.

Building Stand-Alone Applications

Stand-alone applications can be installed and run on machines/computers that don't have the MATLAB package. To compile/create a stand-alone application from a MATLAB code/script, the MATLAB Application Compiler is required along with the MATLAB package. Generating stand-alone applications from existing MATLAB scripts is a straightforward process. It can be accomplished in several ways. You can run this command:

```
>> mcc
```

Or you can use the MATLAB Application Compiler with GUI tools, which can be accessed by choosing the APPS tab and selecting APPLICATION DEPLOYMENT ➤

MATLAB Application Compiler .

When you click the icon, the MATLAB Application Compiler's GUI window will open, as shown in Figure 4-13. You can perform the following actions shown in Figure 4-13: (1) You can add a main filename (M-file) by clicking the add + button. (2) You may also need to add information, such as the application's name, author's name, email, company, summary, description, and so forth, to the application. (3) You can add/change the custom splash screen by clicking the placeholder and adding a photo/screenshot.

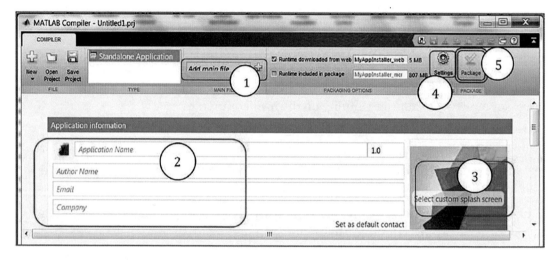

Figure 4-13. *MATLAB Compiler's stand-alone application GUI developer*

(4) You can also adjust settings, such as any additional parameters passed to the MATLAB Compiler, the location of the output folders for test and end user files, and the package installers via the Settings icon. (5) After you add a filename in step 1 to create the executable stand-alone application, the Package icon will become active (see Figure 4-14).

Note that in this example, you are using the function file called QUAD_eqn_SIM.m (the GUI model created in Chapter 3) to generate a stand-alone application. Moreover, you have added some information about the project and added a photo for the custom

splash screen. After completing all of these steps, you click the icon and indicate where to save the stand-alone application project.

The process of creating the stand-alone application then starts, as shown in Figure 4-14.

This may take a few minutes depending on the complexity of the original M-file and whether it also calls/uses other M-files and data (.mat files), uses GUI tools, and so forth. If no problematic issues or errors are encountered during the packaging process of a stand-alone application, the new application will be created and packaged in three folders—for_redistribution, for_redistribution_files_only, and for_testing— and in one log file called PackagingLog.txt. These are generated automatically by the compiler. To verify the generated stand-alone application, we open the folder called for_testing (or click the hyperlinked Open Output folder), and within this folder, we run QUAD_eqn_SIM.prj 	█ QUAD_eqn_SIM by clicking it.

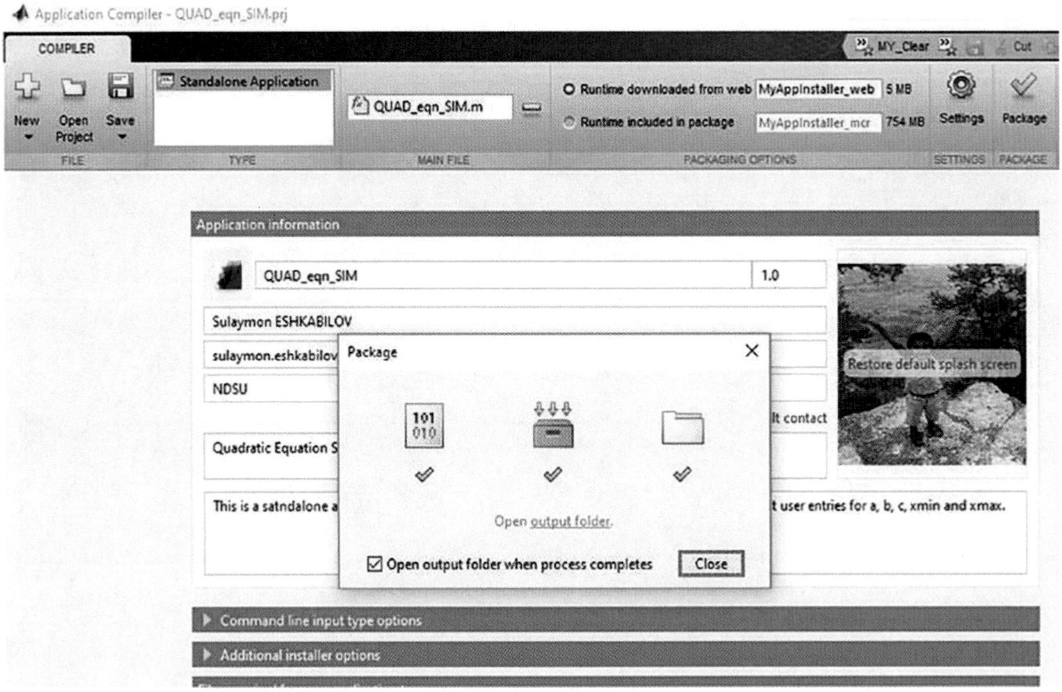

Figure 4-14. *Note indicating successful completion of the compiled stand-alone application project QUAD_eqn_SIM.prj*

Note   When the packaging/compiling process of the stand-alone application is completed, there will be three folders—for_redistribution, for_ redistribution_files_only, and for_testing—and one log file called PackagingLog.txt. These are generated automatically by the MATLIB application compiler.

The created stand-alone application opens, as shown in Figure 4-15. Now it can be tested with some entries.

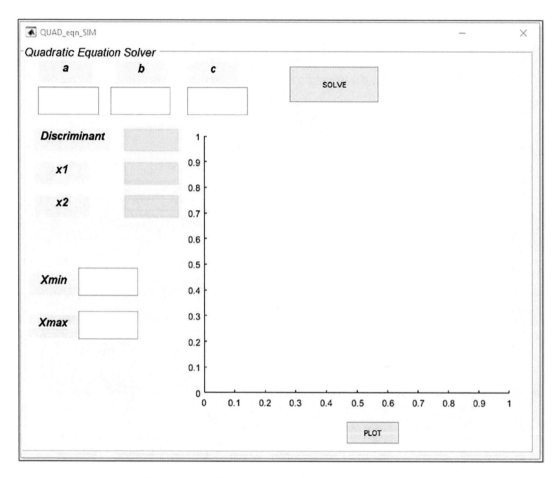

Figure 4-15. *Complete stand-alone application*

You can test your newly created application with the entries—a = 2, b = 7, and c = 3— and plot for Xmin = -5 and Xmax = 5. Figure 4-16 shows the results. So now this stand-alone application can be installed on any computer, and it works without the MATLAB package. The installation package of the stand-alone application is one executable file, called MyAppinstaller_web.exe. This *.exe file resides inside the folder called for_redistribution. This application can be installed by double-clicking this *.exe file on any computer. It runs without the MATLAB packages seamlessly, as shown in Figure 4-17. You simply select the folder where you want to install the application and follow several other standard steps for installation of software packages. Note that this installation process may take a few minutes.

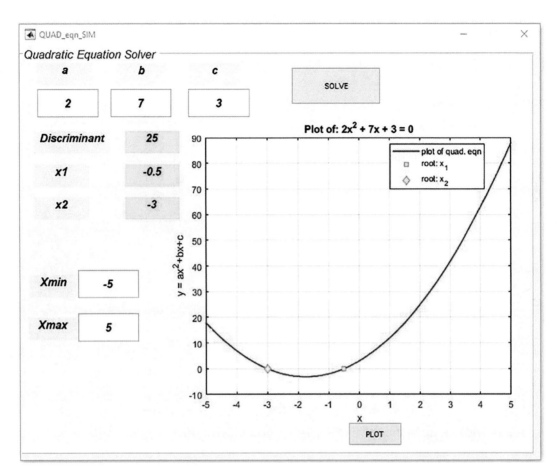

Figure 4-16. *Complete stand-alone application with simulation results*

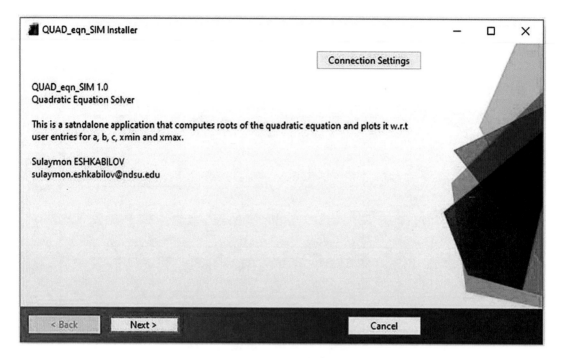

Figure 4-17. *QUAD_eqn_SIM stand-alone application installation process*

Once the stand-alone application is installed, it appears in the program's list of computers and can be launched like any other program, by clicking it or using its shortcut icon.

Note   The only installation *.exe (called MyAppinstaller_web.exe) resides inside the folder called for_redistribution. It's the only file required to install the developed application on any machine.

Summary

Note that, like MEX files, stand-alone applications are also system dependent. In other words, if we create a stand-alone application on a 64-bit processor OS computer in MATLAB, it can be run only on 64-bit processor OS system computers. Thus, for each system type, we have to recompile the stand-alone application. In addition, it must be noted that in MATLAB, there are many functions and tools within various toolboxes, and if we employ specific functions or tools from various toolboxes of MATLAB and/ or packages and intend to create a stand-alone application, the MATLAB compiler may fail to package these applications for obvious reasons. On the other hand, it is viable to create stand-alone applications from GUI models with some limitations based on MATLAB's incompatibility with those stand-alone applications.

Exercises for Self-Testing

Exercise 1

Use the umbilic torus function defined here:

$$x = sin u \left(7 + cos\left(\frac{u}{3} - 2v\right) + 2cos\left(\frac{u}{3} + v\right) \right),$$

$$y = cos u \left(7 + cos\left(\frac{u}{3} - 2v\right) + 2cos\left(\frac{u}{3} + v\right) \right),$$

$$z = sin\left(\frac{u}{3} - 2v\right) + 2sin\left(\frac{u}{3} + v\right)$$

for $-\pi \le u \le \pi$, $-\pi \le v \le \pi$.

Create a function file called `umbilic_torus.m` to compute x, y, z for inputs of u and v and create an `.mex` file from it.

Exercise 2

Create a function file called `My_fun.m` to compute the values of $f(t) = e^{sin(250t) + 2.5\,cos\,(750t)}$ for the user-specified values of $t = t_{min} : dt : t_{max}$, and create an `.mex` file from it.

Exercise 3

Write some C code called `Say_HELLO.c` to display the next words:

```
Hello World!
This is Great DAY!
Live and Learn...
```

Create an .mex file from the C code called Say_HELLO.c and test it in the MATLAB workspace.

Exercise 4

Create a stand-alone application from SINCfun.m created in Chapter 3 to compute and plot the values of $f(t) = \dfrac{sin(\pi t)}{\pi t}$ for the user-specified values of $t = t_{min} : dt : t_{max}$.

Exercise 5

Create a stand-alone application to compute the Fibonacci sequence of numbers according to a user-entered number of elements up to 50 terms and display the found results.

CHAPTER 5

Simulink Modeling Essentials

Simulink[1] is the graphical programming package that works in association with MATLAB and interacts with it as one combined package. It is employed for modeling, simulating, and analyzing dynamic systems, control algorithm development, and so forth. It supports linear and nonlinear systems, continuous and discrete systems, and multirate systems. With Simulink, you can model myriad types of systems, processes, and problems, and you can use top-down and bottom-up approaches.

The Simulink package, like MATLAB, is expandable. By using its standard blocks, you can develop your own library of blocks and subsystems and add to and expand existing Simulink libraries. In your Simulink models, you can combine continuous systems with discrete ones. One of the main advantages of using the package from a user's perspective is that it is much easier to model systems via block diagrams because it doesn't require any preliminary programming skills or experience from users. In Simulink, you simply drag, drop, and connect blocks, and of course, adjust parameters, solvers, and other components in the model. Another advantage of the Simulink package is that its models can work interactively with MATLAB and can be manipulated and executed from the MATLAB Command window and the M/MLX-files and function files.

Simulink Modeling

To launch the package from MATLAB, you just type >> `simulink` in the Command window, click the Simulink icon ⬚ in the menu panel, or click ⬚ and ⬚ Simulink Model icons. The window in Figure 5-1 will pop up. From the Simulink startup window, you can

---

[1] Simulink is a registered trademark of MathWorks Inc.

© Sulaymon Eshkabilov 2022
S. Eshkabilov, *Beginning MATLAB and Simulink*, https://doi.org/10.1007/978-1-4842-8748-4_5

open existing models (recently worked on ones) from the right-side pane (Open) or create a new model by clicking Blank Model or the other options there. Also, from the Examples tab, users can open, study, and change existing examples. There are dozens of examples from different areas of engineering, physics, computing, image processing, code generation, and so forth.

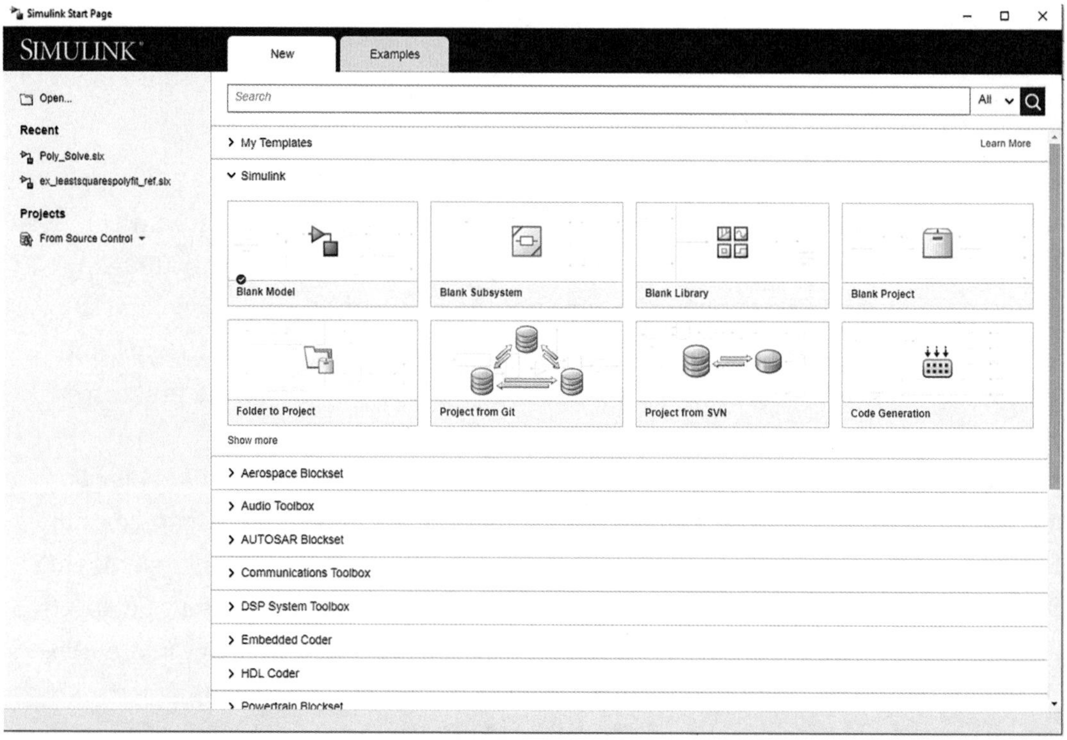

Figure 5-1. *Simulink's startup window*

Figure 5-1 shows the default startup window of the Simulink package. Note that Simulink is a stand-alone package, and there are a few toolboxes and add-ons (see Figure 5-2) that can be installed. All of the blocks of the additional libraries and add-ons will be accessible once they're installed from the Simulink Library browser.

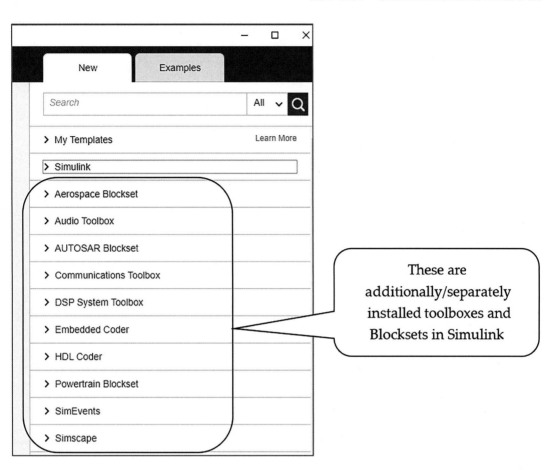

Figure 5-2. *Simulink's startup window and additional toolboxes installed in it*

Example: Arithmetic Calculations

Let's look at a simple example to demonstrate how to build a Simulink model that computes the addition, multiplication, and subtraction of four different scalar numbers. The complete model is shown in Figure 5-3. It's composed of Constant, Gain, Sum, and Display blocks. The scalars entered in constant blocks are 13, 22, 3i/2, and $\sqrt{5}$. When this model is executed, it displays its computed results in the Display block. Note that all of the blocks employed in this model are available in the Simulink Library.

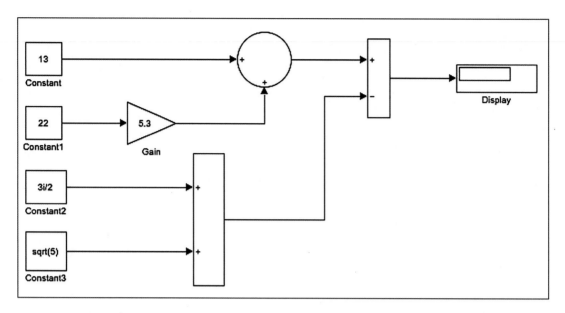

Figure 5-3. *Complete Simulink model example with arithmetic calculations*

To open a blank model, click Blank Model, as shown in Figure 5-1. Note that a blank model can also be opened from the already opened Simulink window by clicking or by pressing Ctrl+N on the keyboard. After clicking the Blank Model icon (see Figure 5-1), the model shown in Figure 5-4 opens. This is called untitled.slx by default, just like MATLAB's M/MLX-files.

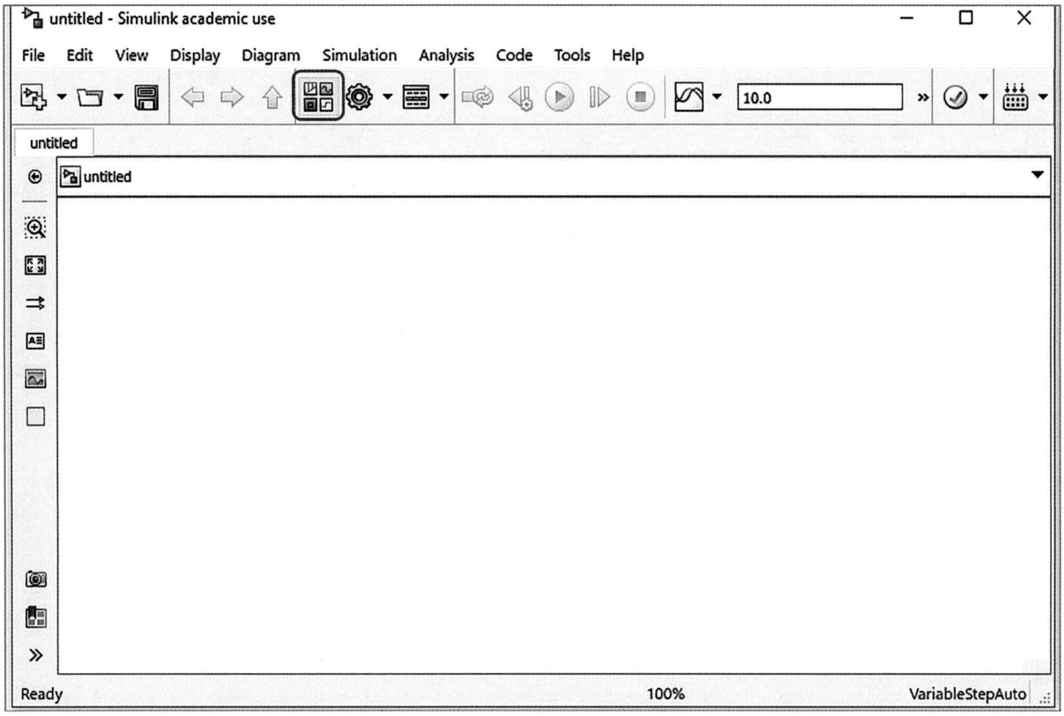

Figure 5-4. *New Simulink model window (untitled by default)*

From the opened blank model window, click  (the Simulink Library icon). That will open a Simulink Library window, as shown in Figure 5-5. It must be noted that the toolboxes that are available in the library are defined by which toolboxes are installed and which user developed/created custom libraries are installed. If necessary, you can also create a new blank model by clicking the New Model icon  in the Library Browser (see Figure 5-5). This also creates a blank model window.

The Simulink Library (see Figure 5-5) looks different in different versions, but most of its general blocks function using the same principles. In this regard, it is worth pointing out that the package has been developed and subject to constant improvement with novel add-ons/blocks, and therefore, the models created in recent versions of Simulink are not fully compatible with its older versions. All of models are forward compatible; in other words, models created in older versions of Simulink work in later versions of the package.

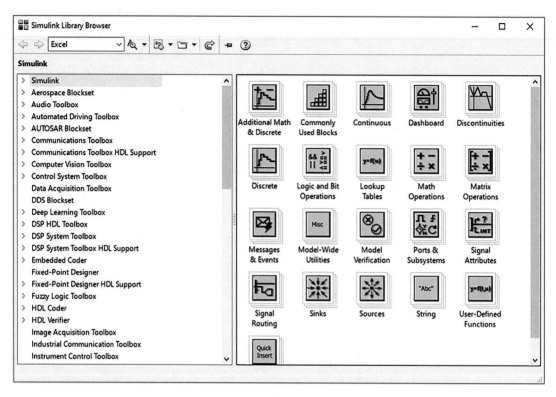

Figure 5-5. *Simulink Library browser*

Let's build a new model where you drag and drop all the necessary blocks from the library. Dragging and dropping a block into a new model area is the most common practice in Simulink-based modeling or programming; however, in recent versions of Simulink (starting from MATLAB/Simulink 2018a), there is an alternative way of obtaining blocks within a model area. It is also possible to obtain any block by double-clicking a desired spot of the model area. For example, if you want to use an Input block, you can double-click and type in the Create Annotation search box. The prompt drop-down options will appear, as shown in Figure 5-6. From the drop-down option, you can select the desired block name. There is another optional step to specify the port number in this case. For other blocks, this option differs.

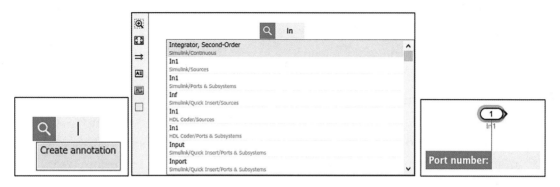

Figure 5-6. *Creating/obtaining blocks within the model area by using the annotation option*

Once all of the necessary blocks are placed in the model area, they need to be connected. There are two ways to connect blocks in the Simulink model window. You can connect a block in the blank model window by clicking the left mouse button and holding the Ctrl key on the keyboard. You then click another block to connect the two. An alternative way to link blocks or connect signals in between blocks is to drag a signal arrow (see Figure 5-7) to the block you want to connect.

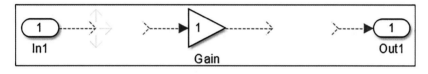

Figure 5-7. *Connecting blocks*

Some blocks have input and output ports, some have only input ports, and others have only output ports. You can only connect signals from their output port to another block's input port. In other words, it is not possible to connect signals from output port to output port or input to input.

Note   Blocks can be resized easily. You click the block to be resized and drag it from one of the four corners of the rectangular boundary around the block.

Before you start working with Simulink modeling, you must adjust one tool to ease the process of working with the library of blocks. To keep the Simulink Library on top of all the model windows, the 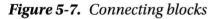 icon must be clicked to a position of "on top," which looks like this: ▣. This is done with a single click.

Note   To keep the Simulink Library on top of all the windows, including the model window, click the "stay on top" 🔲 icon.

Moreover, at the beginning you need to work within the Simulink Library. That can be accessed by clicking the + before Simulink. The Simulink Library contains a number of block sets grouped into Commonly Used Blocks, Continuous, Dashboard, Discontinuities, Discrete, Logic, and Bit Operations. Let's look at several examples to explore the modeling tools and aspects in the Simulink environment.

Example: Modeling Simple Arithmetic Operations

Let's compute simple arithmetic operations using these +, -, /, *, $\sqrt{\ }$, etc., operators to compute this $13 + 22 * 5.3 - (3i / 2 + \sqrt{5})$ and display the result.

To model these arithmetic computations of the exercise, you need the following Simulink/Commonly Used Blocks from the Library: four Constant blocks, three Sum blocks, one Gain block, and one Display block from Simulink/Sinks to output the computation results.

1. Drag and drop all of these blocks in the model area.

2. All of the Constant blocks' constant values must be changed according to the given task (13, 22, 3i/2, $\sqrt{5}$)—see Figure 5-8. Similarly, add one Gain block (for 5.3) by double-clicking each block, one after another (see Figure 5-9).

3. Link a Constant block called Constant (13) to a Sum block with the + sign. You do this by clicking the block and holding the Ctrl key on the keyboard. Then you click a Sum block.

4. Link a Constant block called Constant1 (22) to a Gain block, which is linked to the Sum block with the + sign, as shown in Figure 5-6.

5. Link a Constant block called Constant2 (3i/2) to a second Sum
    block. This rectangular shape can be changed by double-clicking
    a Sum block with the + sign. Similarly, to this Sum block, the
    Constant block called Constant3 ($\sqrt{5}$) is linked using a + sign.

6. Connect two Sum blocks to a third Sum block with the +
    and –signs. Subsequently, link the third Sum block to the
    Display block.

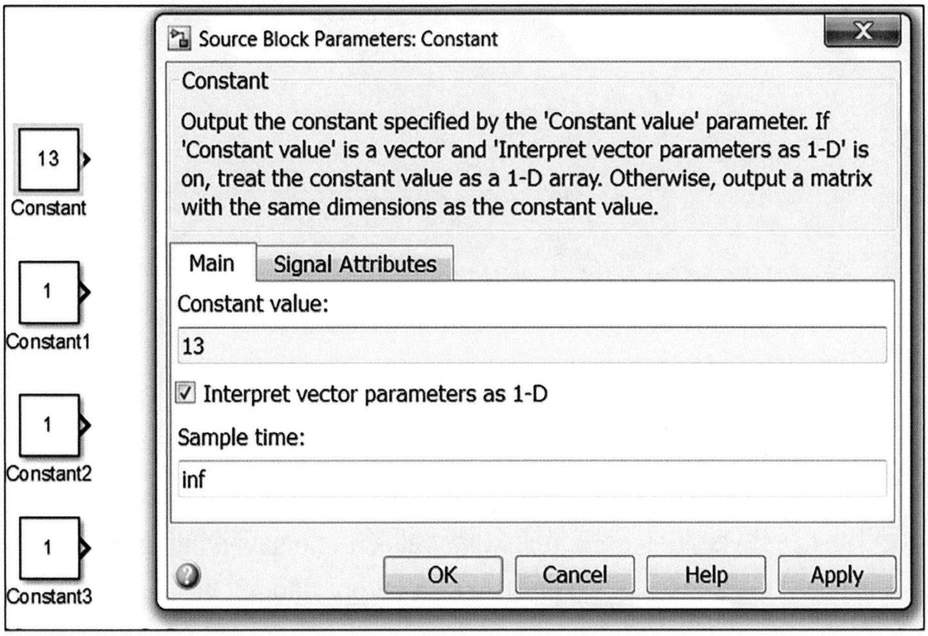

Figure 5-8. *Blocks*

Note   Any block can be copied numerous times by holding Ctrl and clicking the
block and then dragging the block over any spot in a model space. This method is
faster and more efficient than dragging the same block from the library.

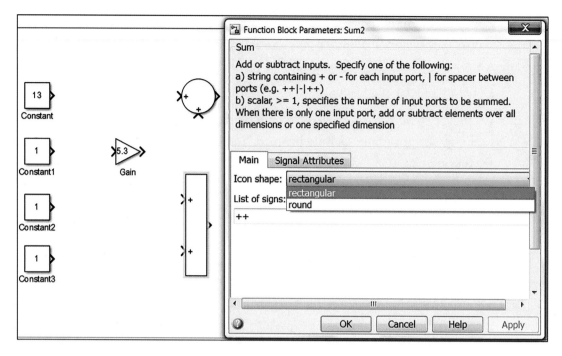

Figure 5-9. *Altering the Sum block's signs and icon shape*

That completes our simple computation Simulink model.

This model is saved with the filename of Ex1_Arithmetic_operations.mdl.

Note   In the latest versions of Simulink, models can be saved either in *.mdl (backward-compatible file format) or in *.slx (supported in later versions). Any Simulink model can be exported to previous versions via File ➤ export model to ➤ previous version. For the "Save as type" option, you select the appropriate version from the drop-down.

To see the computation results, click the Run button 	⏵ or press the Ctrl+T keys on the keyboard. The results are displayed in the Display block. Figure 5-10 shows the complete model.

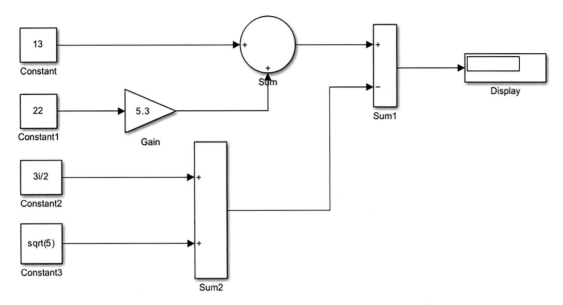

Figure 5-10. *Completed Simulink model called Ex1_Arithmetic_operations.mdl*

Note   To view a full-screen model, just press the spacebar on the keyboard. To zoom in and out, press the Ctrl++ and Ctrl+- keys (in later versions) on the keyboard, or click ⬚ to get a Fit view. You can also use ⊕ to zoom in.

Note that you type in $\sqrt{5}$ as a value of the Constant block called Constant3 by sqrt(5) and 3i/2 as 3i/2 since Simulink (like MATLAB) recognizes imaginary numbers via the letters i and j automatically. Any Simulink model can be simulated/executed from MATLAB via the workspace or within any M-file using the sim() command. The simulated model must reside in the current working directory. For instance, you can run the previous model with this command:

```
>> sim('Ex1_Arithmetic_operations.mdl')
```

Note   You can alter the name of any block by clicking its name tag. You can alter the properties and parameters of any block by opening the property window (by double-clicking it) and inserting the necessary changes or selecting necessary options. For instance, in a display block, you can format the data as short, long, short_e, long_e, hex, etc.

Performing Matrix Operations

Let's build a Simulink model that performs the following matrix operations:

$$\begin{bmatrix} 3 & 2.5i \\ -2i & 3.2 \end{bmatrix}^T \frac{1}{5} + \begin{bmatrix} \sqrt{-3i} & \sqrt{2.2} \\ \sqrt{2.5} & \sqrt{5.4} \end{bmatrix}^{-1} \sqrt{3.3i}$$

Figure 5-11 shows the complete model. The model performs matrix operations, such as inverse, square root, transpose, sum, and division by a scalar (real and imaginary numbers). The following matrix operations (transpose and inverse) are performed by employing the Interpreted MATLAB Function block:

$$\begin{bmatrix} 3 & 2.5i \\ -2i & 3.2 \end{bmatrix}^T, \begin{bmatrix} \sqrt{-3i} & \sqrt{2.2} \\ \sqrt{2.5} & \sqrt{5.4} \end{bmatrix}^{-1}$$

Math Function and Product of Elements blocks are compared with the computed results and shown via Display blocks.

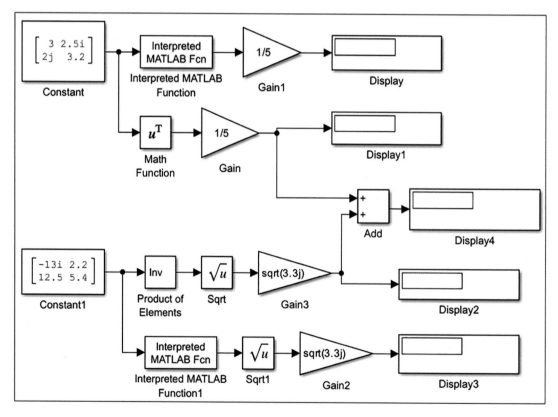

Figure 5-11. *The complete model: example 2, matrix operations*

To model this exercise, follow these steps:

1. For this model, you need two Constant blocks, four Gain blocks, five Display blocks, an Add block from the Simulink/Commonly Used Blocks library, two Sqrt blocks, one Math Function block, one Product of Elements block from the Simulink/Math Operations library, and two MATLAB Interpreted Function from the Simulink/User-Defined Functions (see Figure 5-12).

2. Enter the elements of two matrices into two Constant blocks as [3 2.5j; 2j 3.2] (see Figure 5-13) and [-3i 2.2; 2.5 5.4]. These will be connected with two separate Math Function blocks, one of which is Math Function (transpose of a matrix) and the other is Product of Elements (inverse of a matrix). These two blocks are edited accordingly; for instance, to obtain a transpose operator of the Math Function block, a Function type is chosen to be a transpose from drop-down options. In the Product of Elements block, the "number of inputs" option is changed to be a division (/), and the multiplication option is selected to be a matrix(*) multiplication.

3. Link the Math Function (transpose) block to the Gain block (1/5), which is subsequently linked to the Sum block. The Product of Elements block is connected to the Sqrt block, which is linked to the Gain block ($\sqrt{3.3i}$). Finally, signals from the Gain block (1/5) and the Gain block ($\sqrt{3.3i}$) are connected with the Sum block (see Figure 5-13), which is linked to the Display block.

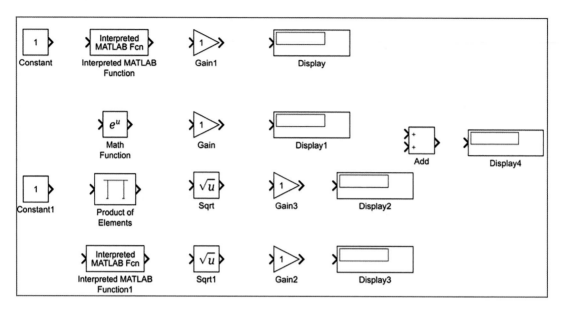

Figure 5-12. *All blocks necessary for this model*

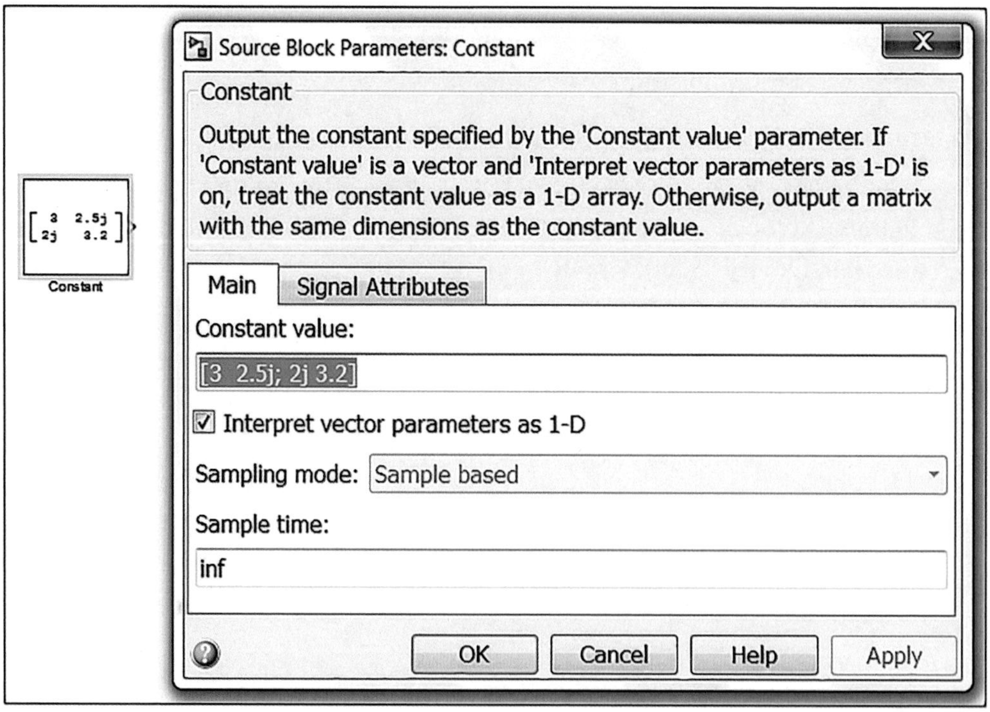

Figure 5-13. *Inputting matrix elements into the Constant block's constant value*

Note   In Simulink modeling, you can attain the same results by altering the properties and parameters of blocks. For instance, matrix inverse can be obtained by inserting the MATLAB function called inv(u) in the interpreted MATLAB Function block.

Matrix transpose is obtained from the Math Function block's options, as shown in Figure 5-14.

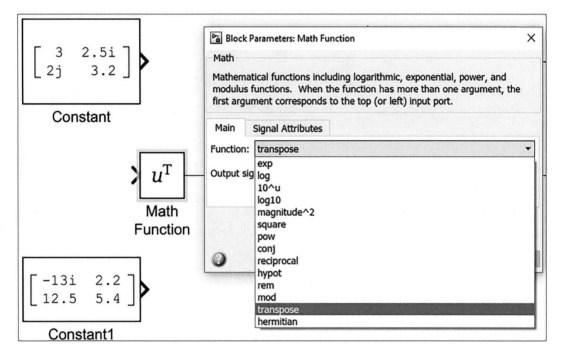

Figure 5-14. *Selecting the Function type in the Math Function block*

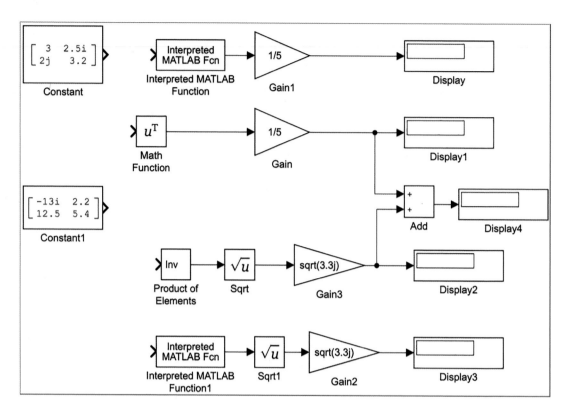

Figure 5-15. *All blocks with adjusted options and entered values/elements*

Note that there are two ways to obtain the inverse of matrices demonstrated in the completed model, as shown in Figure 5-16. By running the completed model, you can see that the results in both matrix inverse operations are the same. Note that the Display block is resized to show all calculation results.

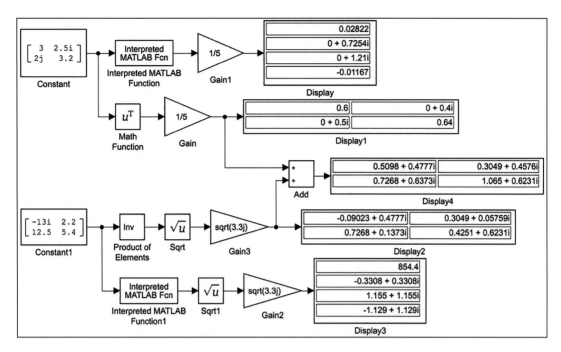

Figure 5-16. *Complete matrix operations and Simulink model called Ex2 MATRIX_operations.slx*

Note   There are two file formats used to save Simulink models—*.mdl and *.slx. The latter model type is supported in later versions (starting with MATLAB 2010) of the Simulink package, and the former format can be opened and simulated by most versions, depending on the blocks used. Via a model export option, you can save models in the previous versions of Simulink.

There are a few new blocks with different properties included in later versions of the Simulink package. Therefore, the models developed by employing such new blocks cannot be simulated by earlier versions of the package.

Computing Values of Functions

In this section, you learn how to compute values of the following math functions, save the computation results in a separate *.mat file and MATLAB workspace simultaneously, and display them in a plot figure. Given $H(t) = e^{sinc(t)} + e^{sin(250t)}$, $t = -3\pi...3\pi$, $\Delta t = \pi/3000$.

There are several ways to build a computation model of the given example. Let's start with a simple and straightforward way. You first take the necessary blocks from the Simulink Library, like the previous two examples. For that, you need the following blocks: Clock, Scope, Math Function, Gain, Trigonometric Function, To File, and Add To Workspace. These are from Simulink/Sources, Simulink/Math Operations, Simulink/Sinks, and Simulink/User-Defined Functions, respectively. The modeling process starts with dragging all of the blocks from the Simulink Library and connecting them in the order of Clock+Interpreted MATLAB Function+Math Function1+Add+Scope2+To File and Clock+Trigonometric Function+Math Function+Scope1+Add+To Workspace. Figure 5-17 shows the completed model that is saved under the file name Ex3_Function_ Compute.slx.

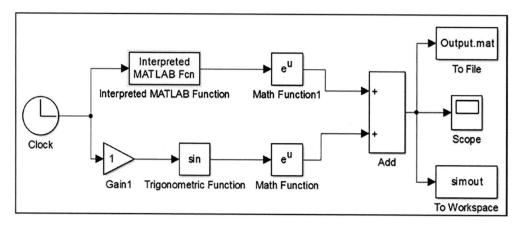

Figure 5-17.   *The complete simulation model called Ex3_Function_Compute.slx*

This completed model, as it is, cannot be executed because the properties and parameters of several blocks used in the model need to be fixed according to the given tasks of this exercise.

For instance, the simulation period should be in the range of $t = -3\pi...3\pi$ with the time interval of $\Delta t = \pi/300$. The MATLAB function of the block Interpreted MATLAB Function must be altered to sinc(). Two optional editing points—the To File and To Workspace blocks—should be renamed as external *.mat files, and a variable should be saved in the workspace in structure format.

You can start adjusting the simulation time interval and the time step via the Model

Configuration Parameters (see Figure 5-18). They can be accessed by clicking the icon or by pressing the Ctrl+E keys on the keyboard. Insert the start time of -3*pi and the stop time of 3*pi, and change the solver type to a fixed-step from the variable step that is chosen by default. In addition, in the Solver options, you should set Type to Fixed-step and, in the Additional options, a fixed-step size (fundamental sample time) of pi/3000. See Chapter 8 for a more detailed explanation on solvers and how to choose their types and parameters, including solver algorithm, step size, relative error, and absolute error tolerances.

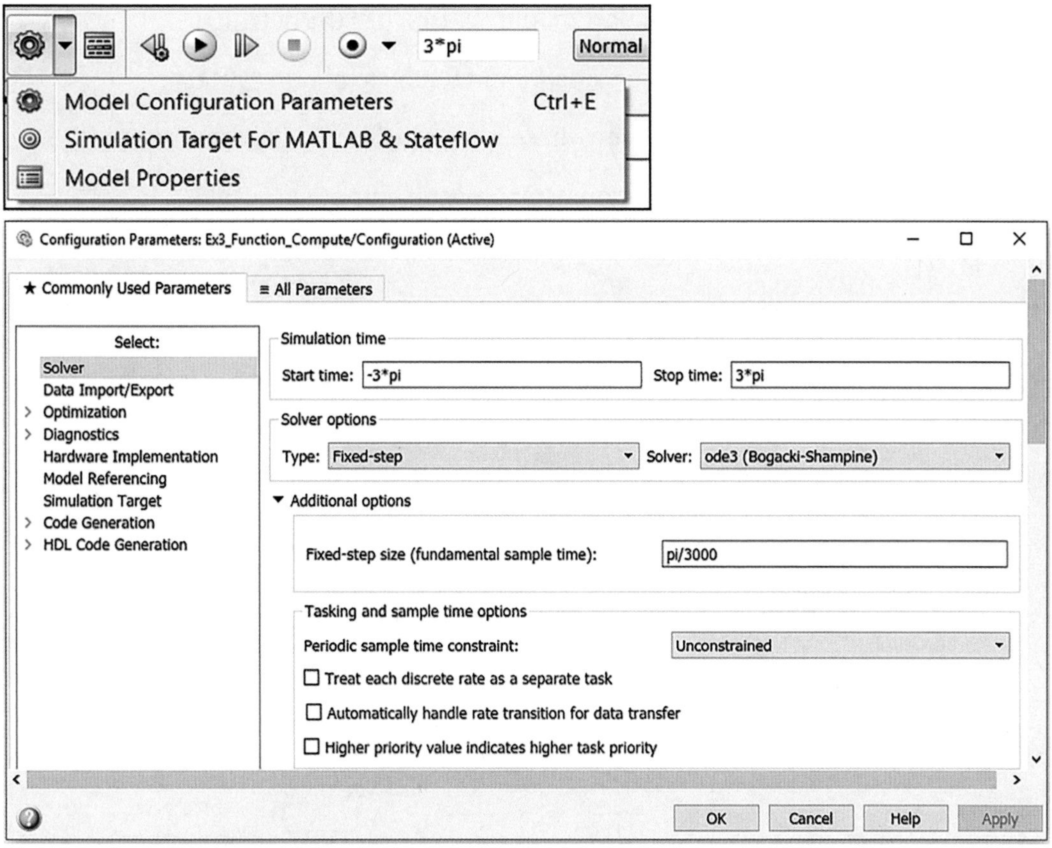

Figure 5-18. *Adjusting the configuration parameters*

Note   For many models, the accuracy of the simulation results depends on the chosen solver type and step size.

Also, before proceeding with the simulations, you need to make the following adjustments to the model:

1. The MATLAB function of the Interpreted MATLAB Function block should be sinc(u). Note that the input variable name u is defined by default.

2. In both MATH Function blocks, a Function type (from the drop-down options) of exp function should be used (it's the default).

3. In the Gain block, the Gain value should be 250 to obtain 250*t.

4. In the Trigonometric function, make it the sin function by default.

5. Even though this step is optional, rename the output file Output_Data.mat and the output variable by output saved in the MATLAB workspace.

Note   When you're saving the computation results via an output file in *.mat format and an output variable, there are several options (formats) to save data—time-Series, array, Structure, and Structure with time series.

After making all these adjustments, the model should look like Figure 5-19.

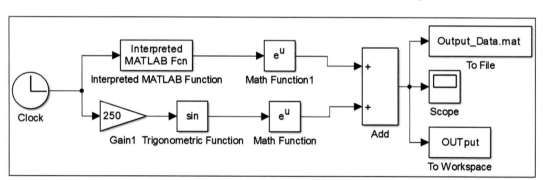

Figure 5-19. *Finalized model with all necessary adjustments and tunings*

Finally, you can simulate the Ex3_Function_Compute.slx model by pressing Ctrl+T on the keyboard or clicking the Run button. You can view the simulation results by double-clicking the Scope block. Figure 5-20 shows the simulation results, plotted in the Scope block.

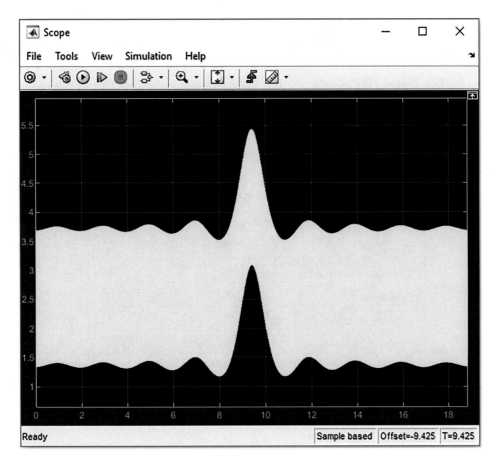

Figure 5-20. *Simulation results*

Note   in the Scope display, the simulation results can be zoomed in or out proportionally, horizontally, and vertically. They can also be auto-zoomed in or out, all by using the ▣, ▣, ▣, ▣ tools.

The simulation results, shown in Figure 5-20, are not coherent with the set simulation period that is set to be within -3*pi ... 3*pi. The problem is in the Scope block's Time display offset, which is set to 0 by default. The time offset needs to be set to -3*pi, which can be fixed via the Configuration properties of Scope. The Configuration properties of Scope can be accessed by clicking the ▣ icon from its drop-down options of ▣ Configuration Properties. After opening the Configuration Properties, click the Time tab and enter -3*pi for "Time display offset." Subsequently, click Apply and OK (see Figure 5-21) and you'll obtain a correct display of the results in Scope, as shown in Figure 5-22. In addition, the simulation results displayed in the scope can be saved as a variable in the format of structure with time series, array, or structure. You do this by selecting the Log Data to Workspace option on the Logging tab (see Figure 5-21).

Moreover, the Scope block has many graphical display options that can be accessed via the ⊙ drop-down options, by clicking Style ▧. Via the Style ▧ options, you can adjust the figure color, axes colors, lines, and markers. See Figure 5-22.

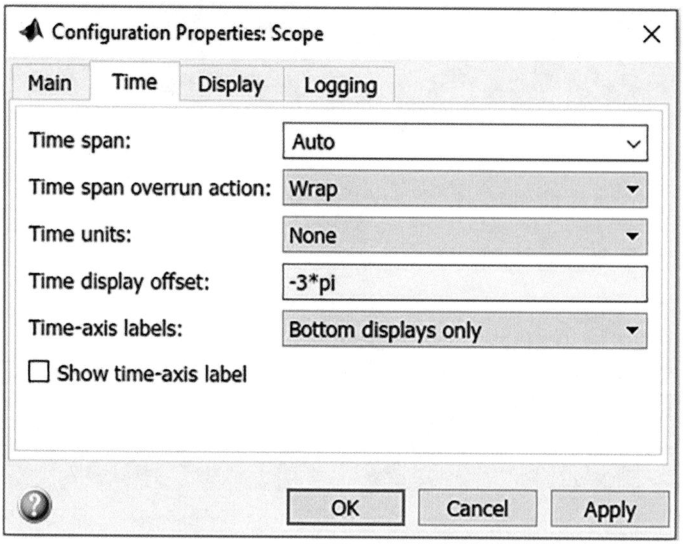

Figure 5-21. *Setting up the Time display offset at -3*pi in Configuration Properties: Scope*

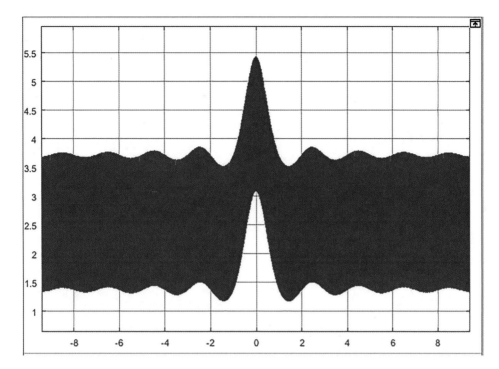

Figure 5-22. *A complete view of the simulation results*

To improve the readability and manageability of large and complex Simulink models, a whole model or part of a model can be associated in one Subsystem block.

It is easy and straightforward to create subsystems from existing models. The easiest way to create subsystems from existing models is to select the blocks (interlinked ones) by holding the left mouse button and dragging the cursor over the desired blocks and then pressing Ctrl+G on the keyboard. To create one subsystem out of a whole model, press Ctrl+A first and then Ctrl+G on the keyboard.

Let's look at the complete model (see Figure 5-19) and create a subsystem out of the whole model, excluding the Input signal to Clock and the Output signal to blocks—To File, Scope, and To Workspace. You first select the model blocks except for the Input block and all the Output blocks, by using the left mouse button, and then press Ctrl+G on the keyboard. The subsystem is created from the selected blocks of the model, as shown in Figure 5-23. To have access to what is under the created subsystem, you need to double-click it. See Figure 5-24.

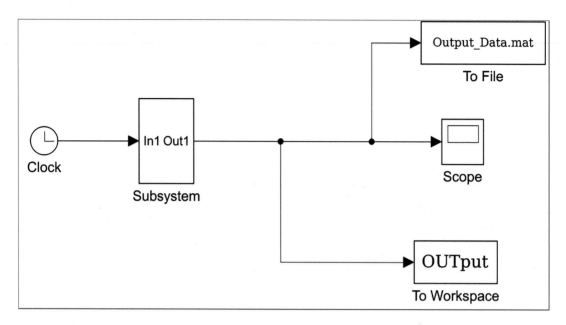

Figure 5-23. *Subsystem created out of the completed model (Figure 5-19), excluding Input and Output blocks*

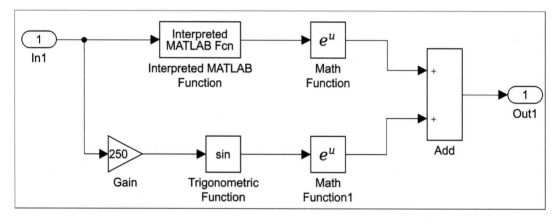

Figure 5-24. *The created subsystem components*

Note that this subsystem in Figure 5-24 has one input block called In1 and one output block called Out1. They are linked with the Input block, Clock, and Output blocks—To File, Scope, To Workspace—to receive and send signals, respectively.

Note that if you create a subsystem from the whole model (see Figure 5-19) excluding only the Input block, then the subsystem will contain only one input block, called In1. If you create a subsystem from the whole model (see Figure 5-19) excluding the output blocks—To File, Scope, To Workspace—then it will contain one output block Out1. If you create a subsystem from the whole model including all blocks, as well as the Input and Output blocks, the created subsystem does not contain any input and output blocks.

The completed model can be simplified. In other words, the number of blocks used in this model can be reduced by employing the Interpreted MATLAB Function block with appropriately edited function formulation expressed by exp(sinc(u))+exp(sin(250*u)) and adjusting Scope parameters that save simulation results in the MATLAB workspace as a structure with time series. Figure 5-25, together with the subsystem created from the model in Figure 5-23, produces the same results as the subsystem in Figure 5-19.

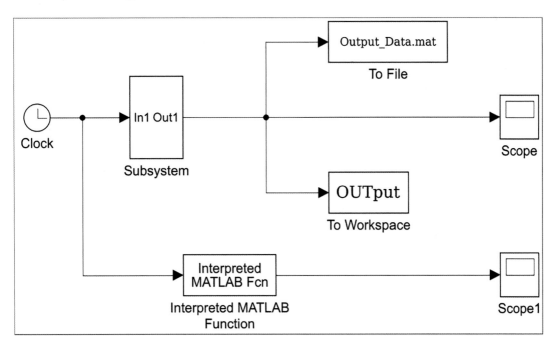

Figure 5-25. *Simplified model Ex3_Function_Compute_Simple.slx*

Input/Output Signals from/to the MATLAB Workspace

As stated, the Simulink model works interactively and flawlessly with the MATLAB workspace. For model development and simulation purposes, input signals can be generated in the MATLAB workspace or within an M-file and then transferred to the Simulink model environment. Similarly, all final simulation results of Simulink models can be sent to the MATLAB workspace. To have an input signal loaded from the MATLAB workspace and an output signal (simulation results) sent back to the MATLAB workspace at the same time, you use input and output blocks and adjust the model configuration parameters ⚙▾, which can be accessed by clicking the ⚙▾ icon from the menu panel or pressing Ctrl+E on the keyboard. In the previous example shown in Figure 5-25 (simplified part is considered), we substitute the input signal Clock block with In1 (the Input block) and the Scope block with Out1 (the Output block); see Figure 5-26.

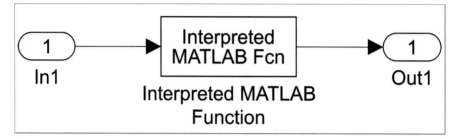

Figure 5-26. *Ex3_Function_Compute_In_Out.slx model with Input/Output from/ to MATLAB workspace*

In addition, the Input and Output signal options in the model configuration parameters ⚙▾ are checked (see Figure 5-27) for input as t, u, and for output as tout and yout, respectively. Note that before starting the simulation, you must define the input signal in the form of t, u in the MATLAB workspace as two column vectors. That can be done for this example as follows:

```
>> t=[-3*pi:pi/3000:3*pi]'; u=t;
```

In this example, the input signal is defined by time only, and thus, t = u. Moreover, we make several adjustments (Start time: -3*pi; End time: 3*pi; Type: Fixed-step; Fixed-step size (fundamental sample time): pi/3000) in the Solver options of this model via the Configuration Parameters window, as shown in the example in Figure 5-18.

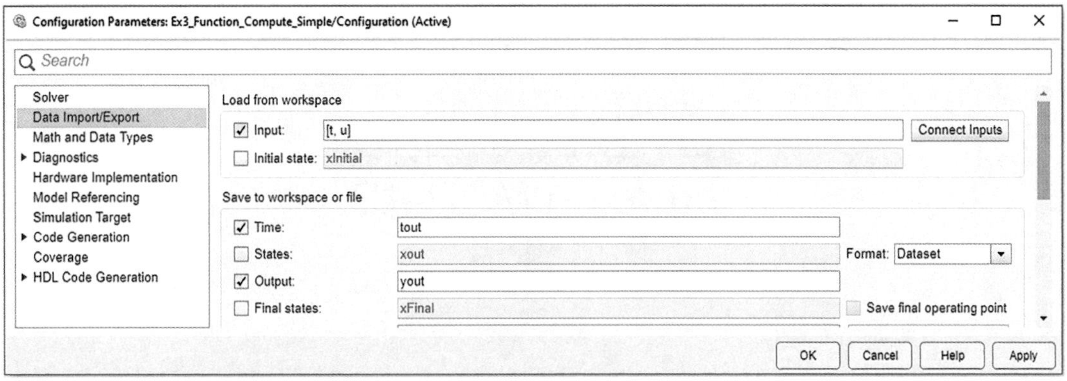

Figure 5-27. *Configuration parameters changed to load an input signal from the workspace and export an output signal back to the workspace*

In addition, we can add the Scope block to the model by clicking the signal going to the Out1 block and using the right-click options (Create & Connect Viewer ➤ Simulink ➤ Scope), as shown in Figure 5-28 (top). After selecting Scope, the scope sign shows up on top of the signal going to the Out1 block, as in Figure 5-28 (bottom).

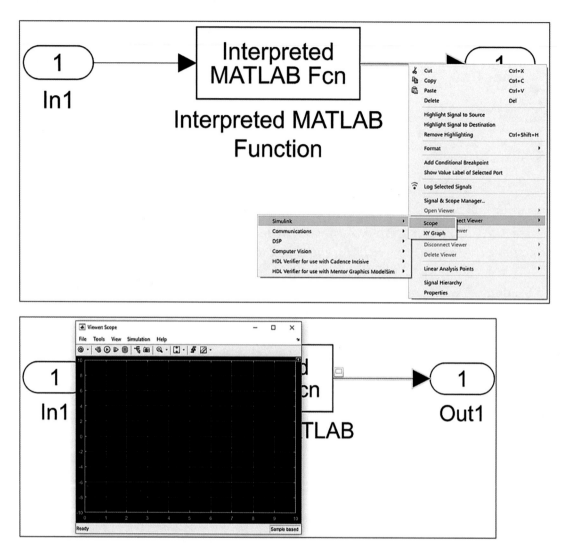

Figure 5-28. *Adding a Scope block to a signal*

In addition, make one important adjustment (Time display offset: -3*pi) in the Scope block to make it display the whole simulation results completely, as demonstrated in the example and shown in Figure 5-21. Now, you save the model (Ex3 Function_ compute_In_Out.slx) and simulate it after entering this in the MATLAB workspace:

```
>> t=[-3*pi:pi/3000:3*pi]; u=t;
```

After simulating the three alternative models, Ex3 Function_compute.slx, Ex3_Function_Compute_Simple.slx, and Ex3_Function_Compute_In_Out.slx, the simulation results are identical. Via these examples, you have seen how easily one block can be substituted for another, how easily subsystems can be created from the existing models by associating parts of interconnected blocks, and how models can be simplified by reducing the number of blocks used in them.

Simulating a Mechanical System

Let's consider a mechanical spring-mass-damper system with Newtonian friction that is formulated by the following differential equation:

$$m\,\ddot{x}(t) + b * sign(\dot{x})\dot{x}^2(t) + kx(t) = f(t)$$

Let's treat the given model of the system as continuous and discrete systems in order to demonstrate how to model and simulate such system in Simulink. Note that solving and simulating differential equations via Simulink modeling is explained more in Chapter 8. Here, we put more emphasis on model building, adjusting block parameters, and interacting Simulink with MATLAB. Moreover, we address the issues of modeling continuous and discrete systems and of creating subsystems.

Given $m = 0.52$; $b = 0.00525$; $k = 165.5$; $f(t) = A\cos(\omega t)$; $\omega = 131$;

$A = 2.3$ and all initial conditions are "zero." The sampling time is $ts = 0.01$. The parameters of the system are m for mass, b for the damping coefficient, k for stiffness, f(t) for the input force, A for magnitude, and ω for frequency.

1. Collect all the necessary blocks from the Simulink Library by dragging and dropping them in a Blank Model window. Figure 5-29 shows the required blocks, taken from Simulink/Math Operations, Continuous, Discrete, Sources, Signal Routing, and Sinks. Also, the Bus collector block is taken from Simulink/Signal Routing.

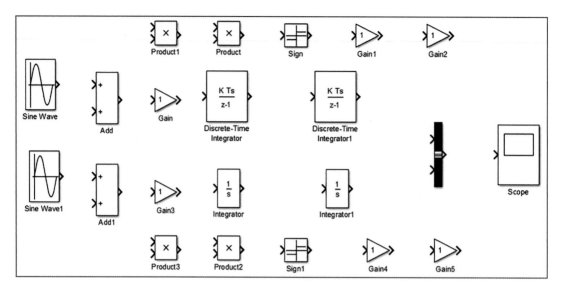

Figure 5-29. *Necessary the blocks for continuous and discrete systems*

2. Adjust the blocks in a more readable order by moving them around and rotating some of them by 180 degrees. The Product, Sign, and Gain blocks need to be rotated. You can do that by selecting a block and pressing the Ctrl+R keys on the keyboard or using the right mouse button's Rotate & Flip options. See Figure 5-30.

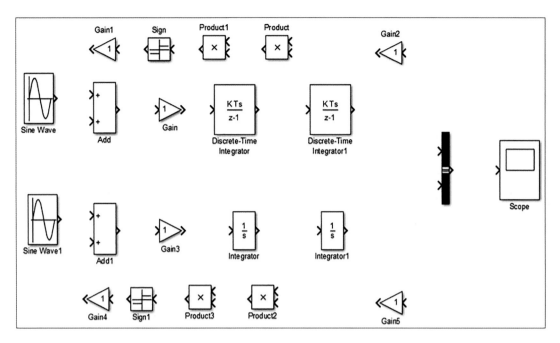

Figure 5-30. *Adjusted blocks*

3. Adjust the parameters of the blocks. First adjust the amplitude,
    frequency, and sampling time for the Sine Wave and Sine Wave1
    blocks, which are input signals (see Figure 5-31). Note that for Sine
    Wave1, the sampling time is set to 0 because by default it is used
    for continuous system modeling.

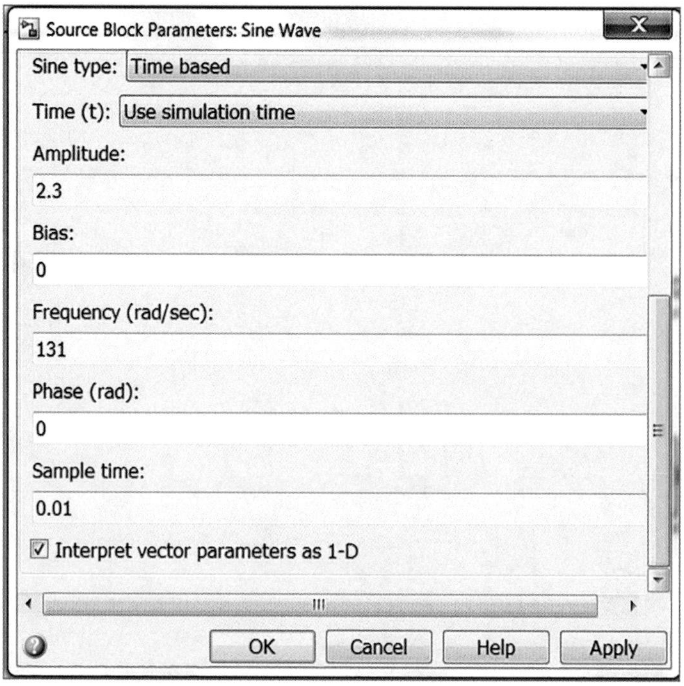

Figure 5-31. Adjusting Sine Wave block's parameters

4. Note how the sine wave sign in the Sine Wave block has changed
    from a smooth curve to a stairs curve when Sample time is set to
    be 0.01. Change the Gain values for Gain and Gain3 to 1/m, for
    Gain1 and Gain4 to b, and for Gain2 and Gain5 to k. Also, change
    the signs in the Add and Add1 blocks to -+-.

5. All blocks are connected, and complete models are attained. Two
    signals coming from the Discrete Time Integrator1 and Integrator1
    are connected to the Bus Creator block that subsequently is
    connected to the Scope block. In addition, two notations are
    added: Discrete System and Continuous System; see Figure 5-32.
    The complete model is saved (Ex4_Discrete_Continuous_
    Sys.slx).

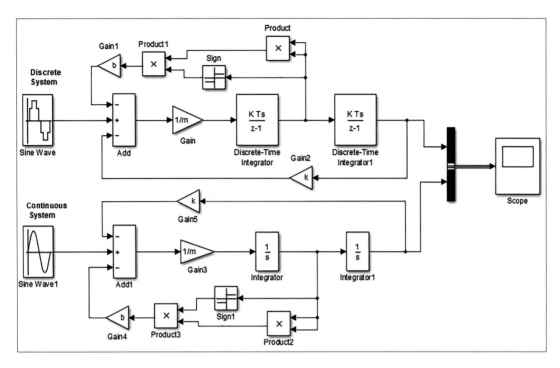

Figure 5-32. *Complete models of discrete and continuous systems*

6. You can execute these models by pressing the Ctrl+T keys on the keyboard or pressing the Run button.

Now the completed models (Ex4_Discrete_Continuous_Sys.slx) seem to be ready for simulation. However, if we execute them, error message windows will be launched, and no results will be attained. The reason for that is the values of three parameters—*m*, *k*, *b*—are not defined yet. You can define them in several different ways, one of which is to specify the values of the parameters in each block or in the MATLAB workspace. You can also specify this information in the Model properties/Callback/InitFcn, which can be accessed via File ➤ Model Properties. Enter the following in the Command window:

>> m=0.52; k=155; b=.00144.

Subsequently, click the 🔘 Run button in the Simulink model window. You can change the Scope block's Style settings (via ▶ drop-down options ➤ Style 🖼) and obtain the results of the simulation displayed in the Scope block; see Figure 5-33.

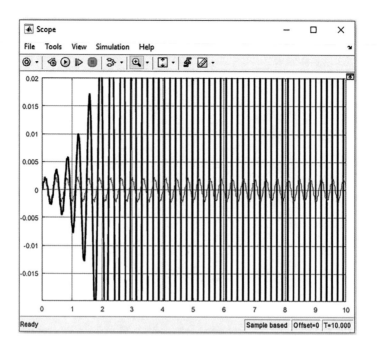

Figure 5-33. *Simulation results*

From the simulation results, you can see that one of the systems (the discrete one) is not stable and should be fixed. The problem with this discrete system modeling resides in the sampling time, which has to be adjusted. That can easily be fixed from the input signal block, i.e., the Sine Wave. In this block's sample, the time value is changed from 0.01 to 0.0001, which is 100 times smaller than initially set.

Note   If the sample time is set to -1 in a Discrete Integrator block, that makes the sample time be inherited automatically from an input signal source.

In addition to making the plot in the Scope more readable with legends displayed for the discrete and continuous system models, let's check the Legends option of the Scope block's properties ⚙. Second, click the signal going from Discrete-Time Integrator1 to Bus Creator and use the right mouse button's option to access Properties (Signal Properties). There, you specify the signal name to be Discrete Sys and then click OK. Similarly, change the signal name for the signal going from Integrator1 to Bus Creator and name the signal Continuous Sys. Then click OK, as shown in Figure 5-34.

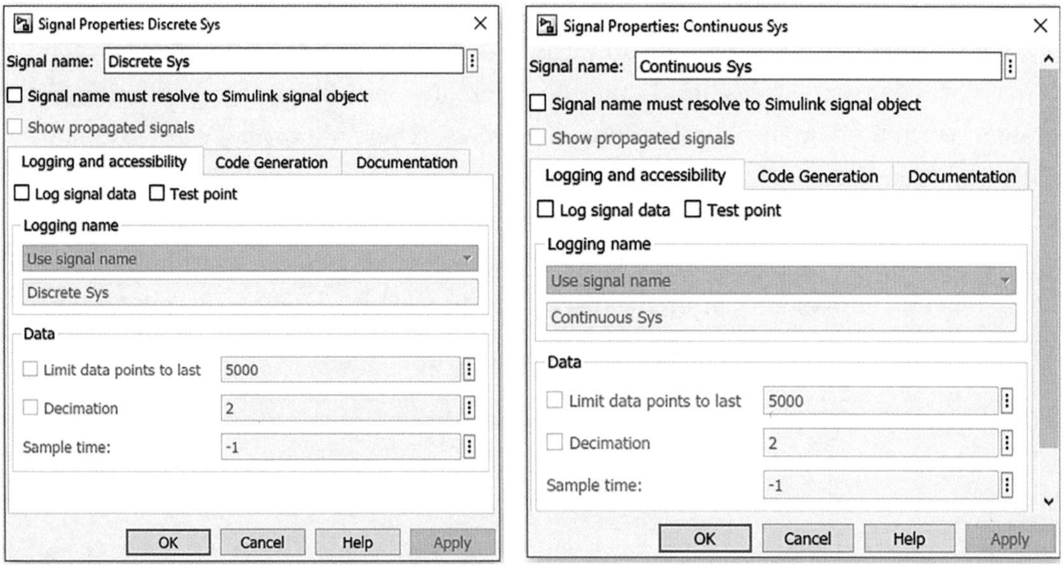

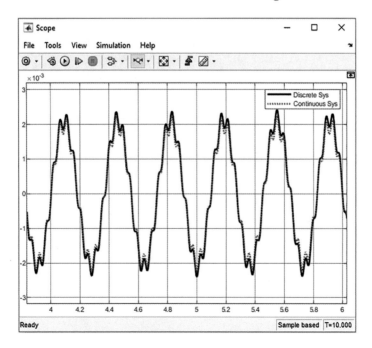

Figure 5-34. *Signal name change*

After these three changes, the whole model runs, and the next result is obtained, as shown in Figure 5-35. Note that it is a zoomed-in view along the horizontal axis.

Figure 5-35. *Simulation results displayed in Scope with its adjusted properties*

You have completed and verified the system models that can be simplified by using a subsystem option, as shown in the previous example (see Figure 5-25; Ex3_Function_ Compute_Simple.slx). Any system containing more than two blocks can be simplified or rather substituted by employing a subsystem block. There are several ways to create subsystems from system models. The easiest way is to select model blocks meant to be under one subsystem and then use the right mouse button options of Create Subsystem from Selection or press Ctrl+G on the keyboard. You create the two subsystems (Subsystem and Subsystem1) from the discrete and continuous system models (see Figure 5-36).

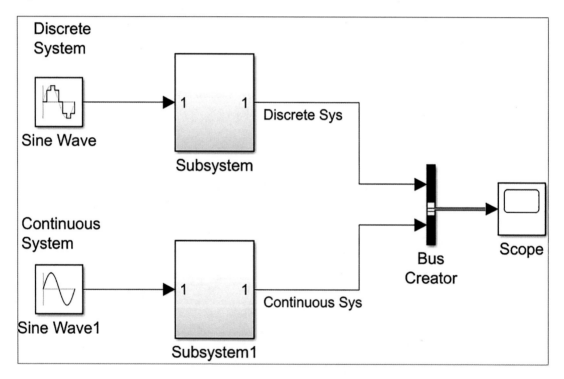

Figure 5-36. *The model containing two subsystem models*

A model under any subsystem can be accessed by double-clicking it. The subsystem representing the discrete system contains the model shown in Figure 5-37.

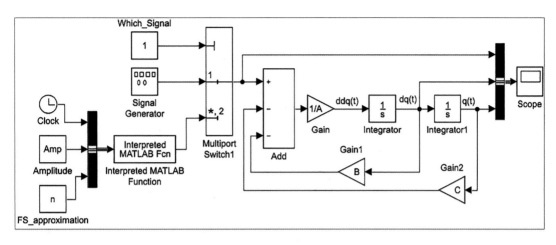

Figure 5-37. *Subsystem composed of this model*

Also, it is possible to reverse the process of subsystems by clicking a subsystem block and using the right mouse button option of Subsystem & Model Reference ➤ Expand Subsystem. You can also do this by pressing the Ctrl+Shift+G buttons after clicking a subsystem block.

In this exercise, you learned how to build discrete and continuous systems and how Simulink handles such composition of systems, but you have not made any adjustments to solver parameters. The accuracy and efficiency of simulation processes can be improved by adjusting solver type (variable step solver selected by default or fixed step solver) and parameter settings (solver, error tolerances, solver algorithm, step size if fixed step solver chosen, zero-crossings, and so forth).

Working with a Second-Order Differential Equation

Now let's build a Simulink model of the given system expressed by the second-order differential equation $A\ddot{q}(t) + B\dot{q}(t) + Cq(t) = F(t)$, where $F(t)$ is applied force (input signal to the system) associated with the MATLAB's function file. $F(t)$ is a rectangular pulse approximated by the Fourier series $F(t) = \sum_{n=1}^{N} \left(\frac{Amp}{\pi} \right) (1 - cos(n\pi)) sin(n\pi t)$, which is implemented in MATLAB via the function_pulse.m function file:

```
function F=function_pulse(t, Amp, n)
% HELP. Two input arguments, viz. t, Amp, and n are needed for
```

% simulation, where t is time vector, Amp is amplitude of a pulse and n % number of approximation terms in Fourier series.

```
F(1,:)=(Amp/pi)*(1-cos(pi))*sin(pi*t); for ii=2:n
    F=F+(Amp/(ii*pi))*(1-cos(ii*pi))*sin(ii*pi*t);
end
```

Let's build a Simulink model associated with the function file function_pulse.m. Moreover, you'll employ Simulink blocks—Repeating Sequence and Signal Generator— to generate a rectangular pulse signal. You'll explore the options with MATLAB-associated function files and Simulink blocks for input signal generation. In addition, you'll explore a few key Simulink modeling tools and aides, such as the model explorer, model advisor, code generation in C/C++, report generation, and so forth. For numerical simulations, the following values are used: $A = 2$; $B = 4$; $C = 200$; $Amp = 10$; $n = 25$.

Figure 5-38 shows the complete model of this exercise. It associates the given function file called function_pulse.m via the Interpreted MATLAB Fcn block. The input signal F(t) rectangular pulses are generated via two ways—one Interpreted MATLAB Fcn with three input variables (Amp, n, t) and a signal generator with two inputs (amp and frequency). You can compare the simulation results from the two input signal sources, i.e., the MATLAB associated input signal generation (pulse_function.m embedded via Interpreted MATLAB Fcn) versus Simulink Library block (the signal generator). Moreover, you'll see how to use Simulink's Model properties to enter the model parameters, such as Amp, n, A, B, C, and how to use Multiport Switch block.

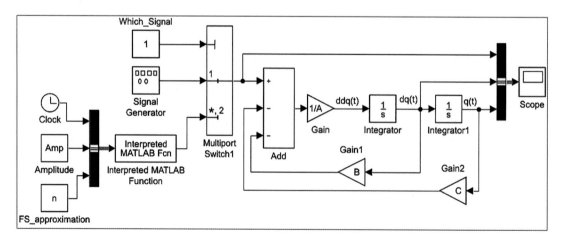

Figure 5-38.   *Complete model of the second-order ODE:* $A\ddot{q}(t) + B\dot{q}(t) + Cq(t) = F(t)$

The necessary blocks to model the given exercise are Add, Gain, Integrator, Bus Creator, and Scope, which you can drag and drop to a Blank Model window. The blocks are connected, as shown in Figure 5-38. Note that the three signals connected with Bus Creator and Scope blocks are called Input, dq(t), and q(t), and they represent input signal, pulse, velocity, and displacement. This is displayed via legends in the Scope block plot.

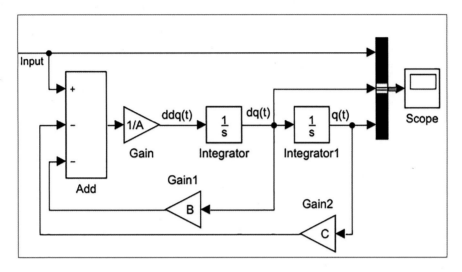

Figure 5-39. *Simulink model of the given second-order differential equation without Input force signal generators*

The function file (called function_pulse.m) is associated via the Interpreted MATLAB Fcn block with the Simulink model (see Figure 5-38). Interpreted MATLAB Fcn is modified to call the MATLAB function: function_pulse(u(1), u(2), u(3)), where u(1) calls the time signal, u(2) calls the amplitude Amp, and u(3) calls n number of the Fourier series approximation. Thus, the Interpreted MATLAB Fcn block requires three input signals simultaneously. That can be done via the Bus Creator block, as shown in Figure 5-40. Note that you can change block names by clicking the name of each block and typing the new name. Note that block tags are not considered during the model simulation, and thus, they have only an informative character for the user/programmer.

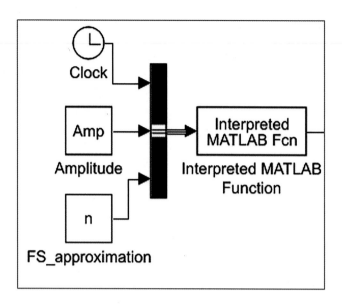

Figure 5-40. *MATLAB Fcn block with three inputs*

Subsequently, one Simulink block, called Signal Generator, is added to the model and its parameters to generate pulses (Amplitude = -Amp/2 and Frequency = 3.15 [rad/sec]). It's then adjusted according to the given pulse parameters, as shown in Figure 5-41.

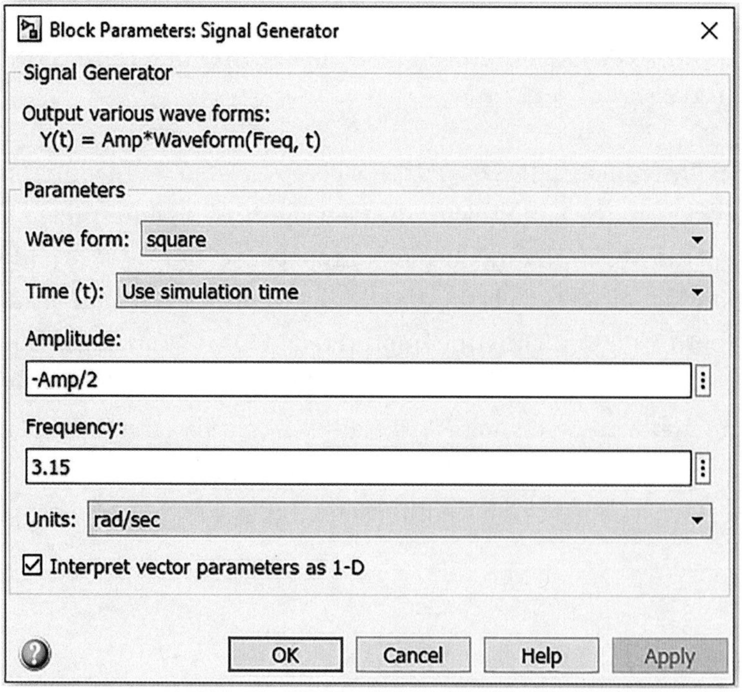

Figure 5-41. *Adjusted parameters of the Signal Generator block*

To connect two input signal sources (Signal Generator and MATLAB Interpreted Fcn), add the Multiport Switch block and connect it to another Constant block to specify a source signal block for selection. Finally, name the three signals going to the Scope block via the Bus selector block. Moreover, adjust the Scope block's Style options to make the output signals readable. Figure 5-42 shows a completed model.

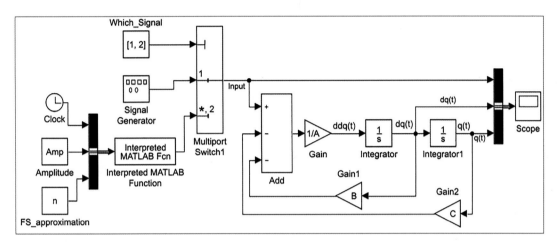

Figure 5-42.   *Complete Simulink model called Ex5_Function_PULSE.slx*

Before you start the simulation, you have to specify the values for A, B, C, Amp, and n. You can do that via the Callbacks option, from File ➤ Model Properties ➤ Model Properties ➤ Callbacks (Model callbacks) ➤ InitFcn. In the Model initialization function window (or alternatively, in the Command window), type A=2; B=4; C=200; n=25; Amp=10 and click OK. When you execute the model, both input signals are taken in the order of first and second with respect to the Constant block (called Which_Signal) values 1, 2. They correspond to Input 1 – Signal Generator and MATLAB Interpreted Fcn. The simulation results in Figure 5-43 show that the two Input signals and two pairs of Output signals (dq(t), q(t), which represent velocity $\dot{q}$ and displacement q, respectively, are well converged. Note that Input: 1 is Signal Generator and Input: 2 is MATLAB Interpreted Fcn.

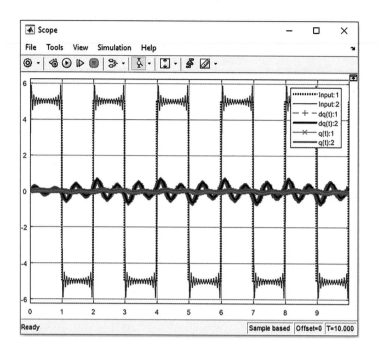

Figure 5-43. *Simulation results: a) from Signal Generator, b) from Repeating Sequence, and c) from Interpreted MATLAB Fcn*

It is clear from the simulation results displayed in Figure 5-43 that all three blocks generating pulse input signals have resulted in approximately the same excitation in the system. Simulink model blocks can be associated with MATLAB files if they are correctly modeled and adjusted. It should be noted that the second input signal generation approach via Interpreted MATLAB Fcn is less accurate. It approximates the Fourier Series from the M-file `function_pulse.m`. Moreover, it is slower since it calls an external M-file to generate the signal.

Subsystem in Simulink Modeling

Simulink has a handy function to create a subsystem from the existing Simulink model components or create model components inside the Subsystem block. The Subsystem block helps you make your created model in a more well-structured way. The Subsystem-based model and a model without it do not have any differences in terms of simulation speed.

Let's look at the following example to show how to employ the Subsystem block:

$$\ddot{u} - (t+1)\dot{w}^2 + 2uw - u^3 = \cos(t)\, 2\ddot{w} + \sin(t)\, \dot{u}\dot{w} - 6u = 2t + 3$$

With the initial conditions of $u(0)=1; w(0)=3; \dot{u}(0)=2; \dot{w}(0)=4$ and parameter values $a=2$, $b=10$. The u and w are the functions of time. To build a Simulink model of this exercise, we will follow the steps and procedures given in the previous two sections on solving second-order differential equations. Thus, all model building steps are skipped here.

To build a Simulink model of this given system of coupled differential equations, Sum, Integration, Gain, and Scope blocks are needed. Here again we follow the steps of building a Simulink model of a second-order ODE as explained previously.

In addition, the subsystem block will be used once the model is complete. The initial version of the complete Simulink model is Ex6_Coupled_ODE_ver1.slx, as shown in Figure 5-44.

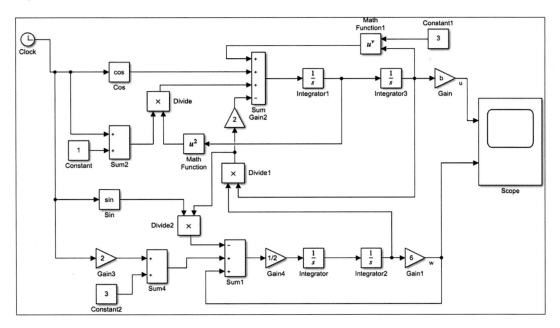

Figure 5-44. *Simulink model, Ex6_Coupled_ODE_ver1.slx*

The initial model called Ex6_Coupled_ODE_ver1.slx shown in Figure 5-44 is a bit complicated and not easy to read. Therefore, to make it more readable, we use the Subsystem block to re-create a new and simplified model out of this model.

You can create a subsystem out of the existing model in two different ways. The first way is to select blocks using the left mouse button as highlighted in Figure 5-45 and then press Ctrl+G keys on the keyboard or right-click and select Create Subsystem from

Selection. Once this step is completed, the subsystem is created from the selection; see Figure 5-46.

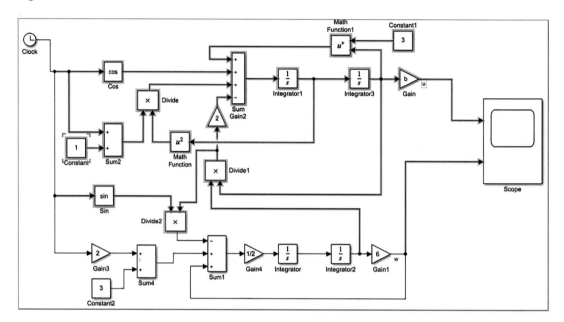

Figure 5-45. *Simulink model, Ex6_Coupled_ODE_ver1.slx, step 1: how to create a Subsystem*

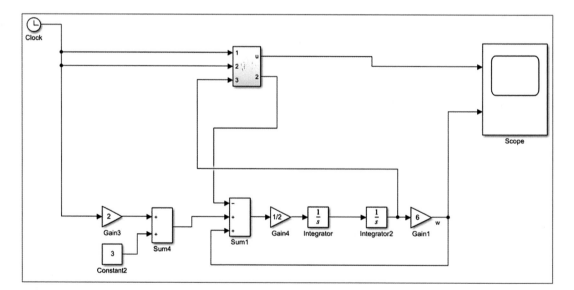

Figure 5-46. *Simulink model, Ex6_Coupled_ODE_ver1.slx, step 2: Subsystem created*

Similarly, you can select the blocks Gain3, Constant2, Sum4, Sum1, Gain4, Integrator, Integrator2, and Gain1 for w signal using the left mouse button and press Ctrl+G (Create Subsystem from Selection) on the keyboard. Subsequently, you obtain the simplified model with two subsystems, as shown in Figure 5-47, that you can save with a new file name: Ex6_Coupled_ODE_ver2.slx.

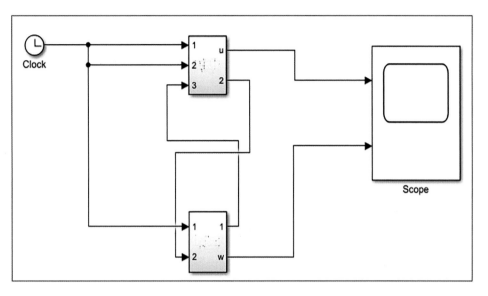

Figure 5-47. *Simplified Simulink model, Ex6_Coupled_ODE_ver2.slx with two subsystems*

To see or edit the model block parameters or connections, you can double-click the Subsystem block or select the subsystem block with the left mouse button and then use the right-click option of Open.

Let's simulate both models, Ex6_Coupled_ODE_ver1.slx and Ex6_Coupled_ODE_ ver2.slx, for t = [0, 5], from MATLAB using the sim() function and compare their simulation results. In addition, you should adjust the settings of the Scope block ☑ Log data to workspace and adjust the output data variable name ➤ OUT and format type ➤

Variable name:	OUT

dataset Save format: Dataset ▾ in both models: Ex6_Coupled_ODE_ ver1.slx and Ex6_Coupled_ODE_ver2.slx. Now, from MATLAB, you can recall these models and simulate them for five seconds, for instance, using the following short script (Sim_Models.m) for ten times to find out if there is any difference between their simulation time:

```
clc;
%%
clearvars
%%
for ii = 1:10
tic;
OUT2 = sim('Ex6_Coupled_ODE_ver2.slx', 5);
Sim_Time2(ii) = toc;
tic;
OUT1 = sim('Ex6_Coupled_ODE_ver1.slx', 5);
Sim_Time1(ii) = toc;
end
%% Compare simulation time
fprintf('Simulation time of Model 1: %f \n ', mean(Sim_Time1))
fprintf('Simulation time of Model 2: %f \n ', mean(Sim_Time2))
%% Compare simulation results
figure(1)
plot(OUT1.OUT{1}.Values.Time, OUT1.OUT{1}.Values.Data, 'r*')
hold on
plot(OUT2.OUT{1}.Values.Time, OUT2.OUT{1}.Values.Data, 'b-',
'linewidth', 2)
xlabel('Time, [s]')
ylabel('u(t)')
legend('Model 1', 'Model 2', 'location', 'NE')
figure(2)
plot(OUT1.OUT{2}.Values.Time, OUT1.OUT{2}.Values.Data, 'r*')
hold on
plot(OUT2.OUT{2}.Values.Time, OUT2.OUT{2}.Values.Data, 'b-',
'linewidth', 2)
xlabel('Time, [s]')
ylabel('w(t)')
legend('Model 1', 'Model 2', 'location', 'NE')
```

Note that these models are stored in your current MATLAB directory or you should have added their location directory to the MATLAB's path directory list using addpath().

Once the simulation is finished, the following output will be displayed in the Command window:

Simulation time of Model 1: 0.341179

Simulation time of Model 2: 0.384217

Also, these two plot figures shown in Figure 5-48 and Figure 5-49 will be displayed.

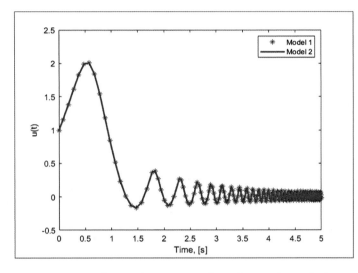

Figure 5-48. *Comparison of simulation results of Ex6_Coupled_ODE_ver1.slx and Ex6_Coupled_ODE_ver2.slx for u(t)*

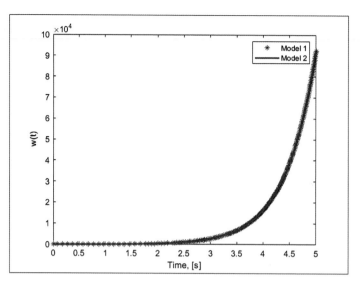

Figure 5-49. *Comparison of simulation results of Ex6_Coupled_ODE_ver1.slx and Ex6_Coupled_ODE_ver2.slx for w(t)*

The simulation results from the script `Sim_Models.m` for ten times show that there is not a significant difference in simulation time of these two models.

Simulink Model Analysis and Diagnostics

Simulink Model Analysis and Diagnostics tools provide good assistance to programmers for improving their models in terms of simulation speed, efficiency, and elimination of inaccurate and inefficient simulations. Therefore, it is recommended to perform analysis and diagnostics for efficiency and adequacy of employed blocks and combinations, chosen solver type, and many other options. All of these options can be explored via the

Model Explorer and the Model Advisor tools. Via the Model Explorer tools, you can generate C/C++code of a Simulink model, obtain a profile report of a model, start a model advisor, reset the configuration parameters of solver, view input/output, optimize the models, generate code, and much more. Let's look at some of the tools within the Model Explorer, considering the previous example.

Code Generation

Code generation (see Figure 5-50) can be accessed via the Model Explorer ▦ or Model Configuration Parameters ⚙ buttons.

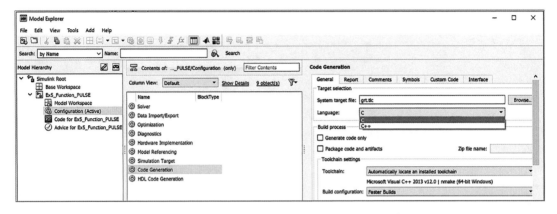

Figure 5-50. *Model Explorer tools*

After clicking Generate Code Only and Package Code and Artifacts, click the Generate Code button. Subsequently, the C code (C is the chosen language) will be generated. Note that there are some constraints in code generation; for instance, a chosen solver has to be a fixed step, and not all blocks used are compatible with code generation in C/C++. If these or other such requirements are not satisfied, the C/C++ code cannot be generated. Also, the code generation process depends on the installed compiler type and version.

Model Advisor

Model Advisor ⊘ tools (see Figure 5-51) can be helpful in identifying where problems have occurred within a model and where optimization is required. It identifies problems with code generation and model performance by product, by task types, or both.

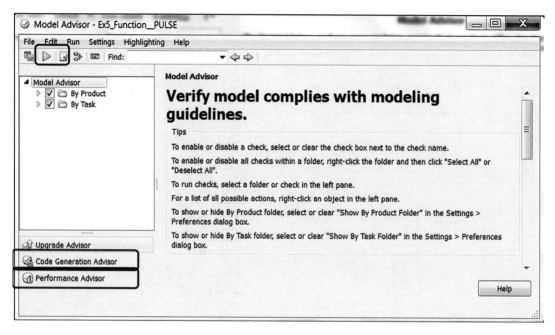

Figure 5-51. *Model Advisor tools*

You first choose which process to get help/advice from in the Model Advisor and then click the Run ▷ button. In this example, we chose the Model Advisor with By Product and By Task. Once the Model Advisor is launched, all diagnostic checks of the model (`Ex5_Function_PULSE.slx`) are run, and the report of all passed, failed, warning, and not run points is prepared. You can view the report by clicking the Generate Report button in the Model Advisor window. The Generate Report button opens the Generate Model Advisor Report window, from which you can select the directory (where to save the generated report), file name, file format (HTML by default, PDF, or Word), and check mark option to view the report after it's generated; see Figure 5-52. We chose HTML, which is the default report format of the Model Advisor.

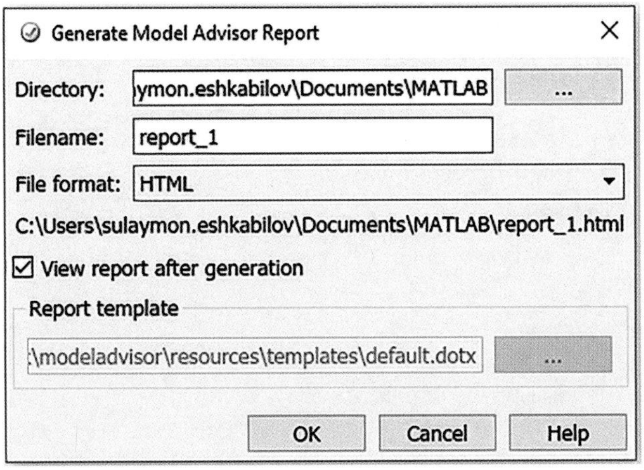

Figure 5-52. *The Generate Model Advisor Report window*

Moreover, there are warnings concerning double precision operations used by the blocks of the model. The blocks (Interpreted MATLAB Function) are not supported by code generation. In addition, the Clock, Integrator, Integrator1, Signal Generator, and Interpreted MATLAB Function blocks are not recommended for C/C++ production deployment.

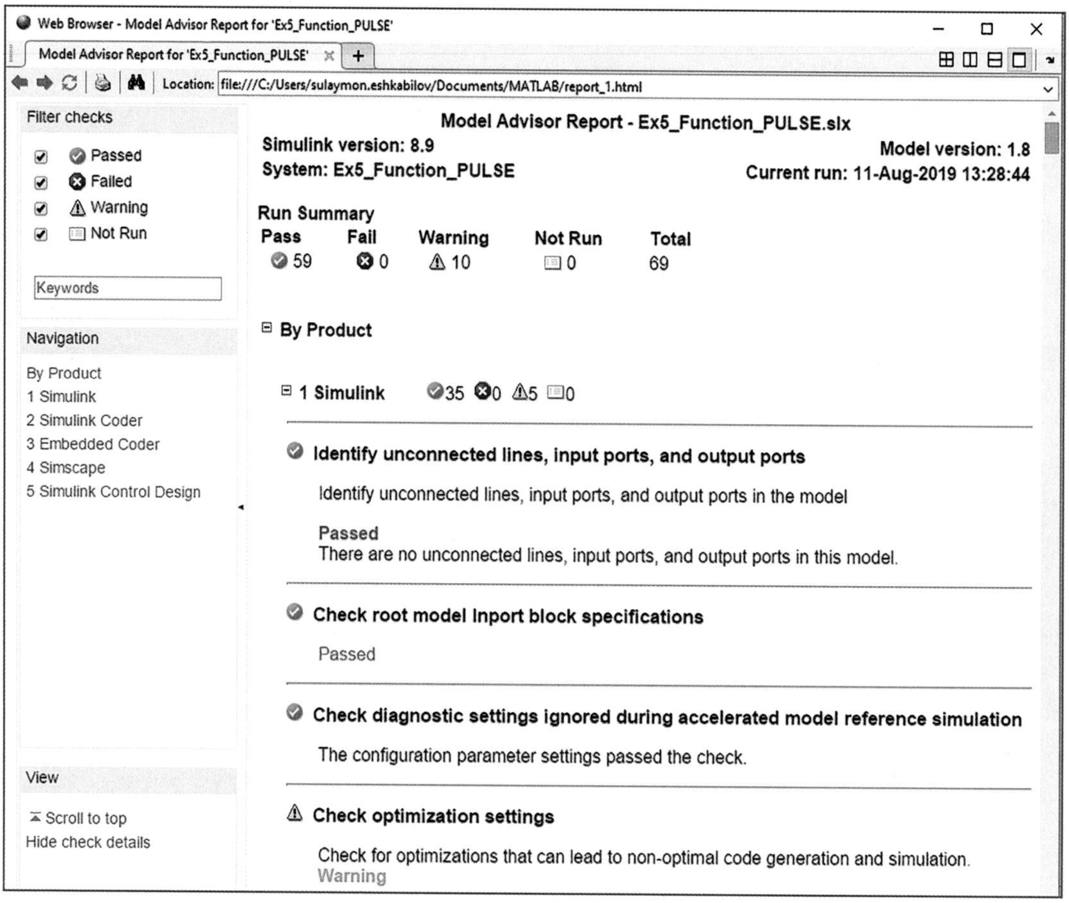

Figure 5-53. *Model Advisor Report*

Figure 5-53 shows part of the Model Advisor Report for Ex5_Function_PULSE. slx. The report is created in HTML format and shows 59 Pass, 0 Fail, 10 Warning, and 0 Not Run, for a total of 69 Run. By scrolling down the report, you can see where the model passed and where it had some warning issues, such as optimization settings. It's recommended to set the parameter of "Remove Code from Floating-Point to Integer Conversions That Warms Out-Of-Range Values (EfficientFloat2IntCast)" to on. Another recommendation is to set the parameter "Inline invariant signals (InlineInvariantSignals)" to on. Furthermore, another warning is Check Data Store Memory blocks for multitasking, strong typing, and shadowing issues. Duplicate data store names checking is not set to error. Duplicate usage of data store names can lead to unintended shadowing of data stores of higher model scope. For this reason, consider changing the duplicate data store names setting to error.

Another interesting warning is linked to bus signals. The warning says: "Check bus signals treated as vectors." The Bus signal was treated as a vector by the Simulink software. Identify bus signals in the model that are treated as vectors by the Simulink software.

Bus signal feeding input port 1 of the block: Ex5_Function_PULSE/Interpreted MATLAB Function. Bus signal is feeding input port 1 of the block in Ex5_Function_ PULSE/ Scope.

Recommended Action: The model contains bus signals that the Simulink software implicitly converts to vectors. However, the model is not configured to explicitly convert these signals to vectors. To fix this issue, insert Bus To Vector blocks at the imports of the blocks listed earlier.

You can do this automatically, by either pressing the modify button below or running the Simulink.BlockDiagram.addBusToVector function. You can do this manually using the Simulink ➤ Signal Attributes library.

By studying the Model Advisor's reports, you can improve your model by removing bugs, simulation bottlenecks, and unwanted warnings, and substituting some of the inefficient blocks in the model.

In addition to the Model Advisor, you can also employ the Optimization tools under Model Explorer or the Configuration parameters to optimize parameters and blocks in our model. In addition, to locate bugs or bottlenecks, you can use debugging tools. They can be accessed via the menu bar: Simulation ➤ Debug ➤ Debug Model. Another way to learn about the model's performance is from the menu bar: Analysis ➤ Performance Tools ➤ Show Profile Report (select) and then ➤ Performance Tools ➤ Performance Advisor. The options are displayed in Figure 5-54.

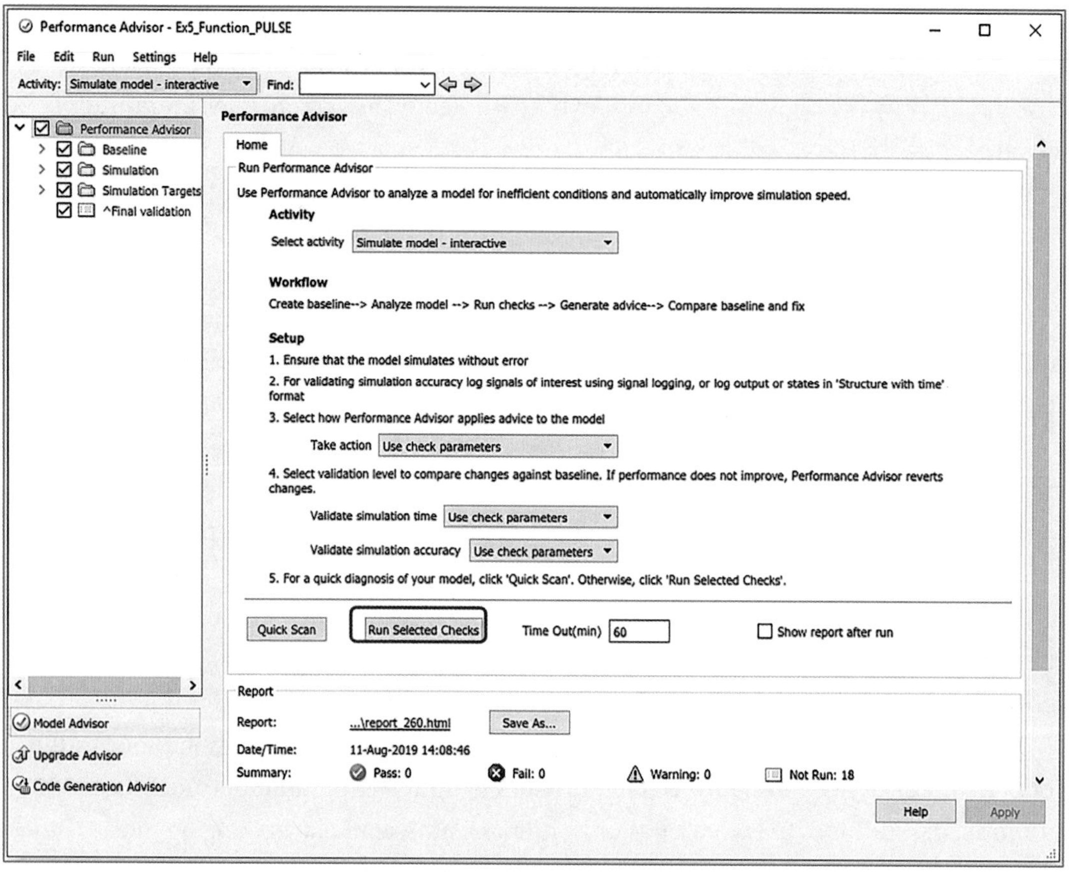

Figure 5-54. *Performance Advisor options*

After clicking Run Selected Checks (see Figure 5-55), the Simulink Profile Report is displayed. It's composed of the Summary and Simulink Performance Advisor Report, and it displays a complete picture of the model and its simulation processes.

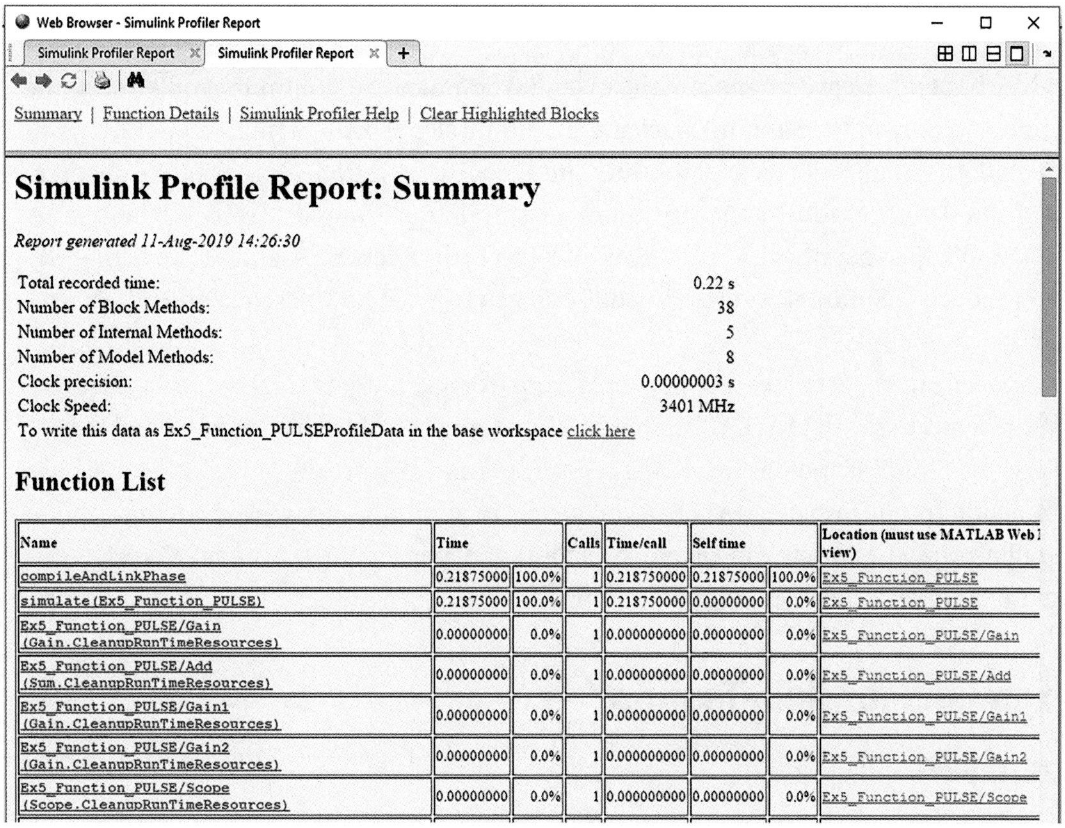
Web Browser - Simulink Profiler Report

Simulink Profiler Report | Simulink Profiler Report

Summary | Function Details | Simulink Profiler Help | Clear Highlighted Blocks

Simulink Profile Report: Summary

Report generated 11-Aug-2019 14:26:30

Total recorded time:	0.22 s
Number of Block Methods:	38
Number of Internal Methods:	5
Number of Model Methods:	8
Clock precision:	0.00000003 s
Clock Speed:	3401 MHz

To write this data as Ex5_Function_PULSEProfileData in the base workspace click here

Function List

Name	Time		Calls	Time/call	Self time		Location (must use MATLAB Web view)
compileAndLinkPhase	0.21875000	100.0%	1	0.218750000	0.21875000	100.0%	Ex5_Function_PULSE
simulate(Ex5_Function_PULSE)	0.21875000	100.0%	1	0.218750000	0.00000000	0.0%	Ex5_Function_PULSE
Ex5_Function_PULSE/Gain (Gain.CleanupRunTimeResources)	0.00000000	0.0%	1	0.000000000	0.00000000	0.0%	Ex5_Function_PULSE/Gain
Ex5_Function_PULSE/Add (Sum.CleanupRunTimeResources)	0.00000000	0.0%	1	0.000000000	0.00000000	0.0%	Ex5_Function_PULSE/Add
Ex5_Function_PULSE/Gain1 (Gain.CleanupRunTimeResources)	0.00000000	0.0%	1	0.000000000	0.00000000	0.0%	Ex5_Function_PULSE/Gain1
Ex5_Function_PULSE/Gain2 (Gain.CleanupRunTimeResources)	0.00000000	0.0%	1	0.000000000	0.00000000	0.0%	Ex5_Function_PULSE/Gain2
Ex5_Function_PULSE/Scope (Scope.CleanupRunTimeResources)	0.00000000	0.0%	1	0.000000000	0.00000000	0.0%	Ex5_Function_PULSE/Scope

Figure 5-55. *Simulink Model Profile Report Summary*

Thus, it is recommended that you run the Model Advisor and Performance Advisor options to obtain the many help hints to improve your model's performance. Also, the profile report generator can be recalled and executed using commands in order to locate inefficient operations and blocks of the model.

```
>> profile on; sim('Ex5_Function PULSE.mdl'); profile viewer
```

The profile report generator works well with all M-files and Simulink models and provides comprehensive reports including bottlenecks within a code/script/model in terms of computation and execution time spent on each command and operation.

Summary

In this chapter, we covered most of the essential graphical programming tools and some common blocks in the Simulink package, including signal sources, matrix operations, integration, visualization, signal routing, and C code generation. Moreover, the chapter highlighted and demonstrated, via numerical simulation examples, a few salient points on how to adjust parameters, use solver tools, set error tolerances, and improve the performance of Simulink models. In addition, you learned how to select and adjust solvers in Simulink.

You learned how to create subsystems from existing models and how to associate Simulink models with MATLAB scripts and function files. Moreover, you learned how to execute Simulink models from MATLAB scripts and acquire the Simulink model simulation results into the MATLAB workspace. In addition, you worked with the Simulink Model Analysis and Diagnostics tools and learned how to obtain Model Advisor and Performance Reports.

Exercises for Self-Testing

Exercise 1

Build a Simulink model to compute values of the cosine function $g(t) = \cos(\omega t)$ for $t = 0...3$ with 3,000 incremental steps and $\omega = [\pi, 2\pi, 3\pi, 5\pi, 7\pi]$ specified in MATLAB. Simulate your Simulink model using MATLAB with the `sim()` command using a `[for ... end]` or `[while .. end]` loop for all values of ω.

Exercise 2

The equation for charge in a resistor-inductance-capacitor (RLC) circuit (shown in the below figure) in a series is determined by Kirchhoff's law:
$$L\ddot{q} + R\dot{q} + q/C = \varepsilon_{max} \cos(\omega t)_.$$

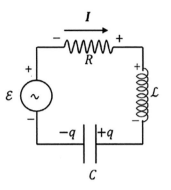

Create a Simulink model to simulate the given RLC system expressed by the second-order differential equation for $q(t)$ with the input arguments of R, L, C, ω, and t. Create a Simulink model to simulate the given RLC system.

$$\text{Take } R = 100 \ \Omega, \ L = 200 \ H, \ C = 0.02 \ \mu F, \ \omega = 60 \ rad/s.$$

Exercise 3

The acceleration of a skydiver is determined by the following:

$$a = g\left(1 - \frac{v^2}{3600}\right)$$

where $g = 9.81 \ m/\sec^2$.

Create a Simulink model to simulate the acceleration of a skydiver.

Exercise 4

A truck of mass m accelerates from rest at $t = 0$ with constant power P along a level road.

The speed of the truck as a function of time is given by $v(t) = \left(\dfrac{2P}{m}\right)^{\frac{1}{2}} \sqrt{t}$. If $x = 0$ at

time $t = 0$, the position function $x(t)$ is given by $x(t) = \left(\dfrac{8P}{9m}\right)^{\frac{1}{2}} \sqrt{t^3}$, where $P = 550 \ kW$ and $m = 15000$ kg.

1. Write an inline function to compute the position of the truck from the function $x(t)$ as a function of time t.

2. Create a Simulink model to obtain numerical values of $v(t)$, x(t) as a function of t in the MATLAB workspace and compare the results with the ones obtained from the function handle and inline function.

3. Build plots of $x(t)$ versus t and $v(t)$ versus t in two separate plot figures.

Exercise 5

Create a Simulink model with three subsystems to compute numerical solutions ($x_A(t)$, $x_B(t)$, $x_C(t)$) of the following second-order coupled differential equations.

$$kx_k + kx_A(t) + m\ddot{x}_A(t) = F_D(t) + kx_B(t)$$

$$2kx_B(t) + m\ddot{x}_B(t) = k\left(x_C(t) + x_B(t)\right)$$

$$m\ddot{x}_C(t) = k\left(x_k + x_B(t) - x_C(t)\right)$$

$$x_A(0) = 0, x_B(0) = x_k, x_C(0) = 2x_k, x_k = 1.5,$$

$$\dot{x}_A(0) = \dot{x}_B(0) = \dot{x}_C(0) = 0$$

$$F_D(t) = F_0 \cos(t\omega),\ F_0 = 5.75,\ \omega = 3.20;$$

$$t = [0,25], m = 2; k = 32.$$

Hint: For more information about ordinary differential equations, see Chapter 8.

CHAPTER 6

Plots and Data Visualization

The MATLAB package has numerous built-in functions for visualizing numerical data via plots, graphs, charts, and animations. There are two-dimensional (2D) and three-dimensional (3D) plots, charts, maps, etc. In addition, some of the MATLAB toolboxes have specific plot and visualization functions. Building plot figures is very straightforward and can be done in two different ways, one of which is using commands and writing scripts and the other is using GUI tools. In this chapter, we discuss and demonstrate some of the most essential tools and techniques used to build line, bar, pie, surface, mesh plots, graphs, and animated plots via examples.

Basics of Plot Building

On the PLOTS tab of the main menu (see Figure 6-1), you will see all the available plots in the current MATLAB package, including its installed toolboxes.

Figure 6-1. *PLOTS tab of the MATLAB Desktop menu*

If any data (variable) residing in the workspace is chosen, the applicable plot commands will become active. For instance, if the selected variable size is 20 by 1, then the plots shown in Figure 6-2 will become active, indicating that these plots can be used.

S. Eshkabilov, *Beginning MATLAB and Simulink*, https://doi.org/10.1007/978-1-4842-8748-4_6

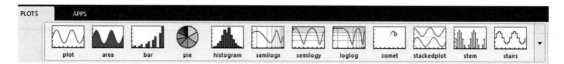

Figure 6-2. *Possible plots in the PLOTS tab for the selected data*

If the chosen data (variable) is 10 by 100, then the plots shown in Figure 6-3 will become active instead.

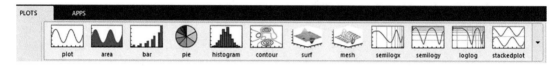

Figure 6-3. *Possible plots in the PLOTS tab for the selected data*

PLOT( )

Let's look at a few examples that create 2D line plots using the plot() command and GUI tools.

Example 1: Plotting Two Rows of Data

Say we're given these two sets of data: A=[1, 3 4.5 5 6.7 8.1 9.5 10.3] and B=[-2.32, 0.23, 2.14, 2.22, 3.92, 6.67, 7.41, 6.43].

Let's plot them using the plot() command.

```
>> A=[1, 3 4.5 5 6.7 8.1 9.5 10.3]; B=[-2.32, 0.23 , 2.14, 2.22, 3.92,
6.67, 7.41, 6.43]; plot(A, B)
```

Figure 6-4 shows the plot figure.

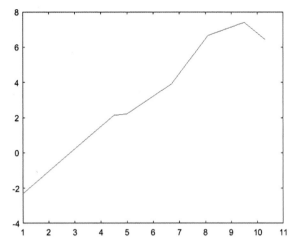

Figure 6-4. *Possible plots on the PLOTS tab for the selected data*

The same plot can be generated by using GUI tools and selecting the variables A and B in the workspace. Hold down Ctrl on the keyboard and use right mouse button (RMB) clicks to select multiple variables. Then with the left mouse button (LMB) options, you

can choose plot(A, B) or click the PLOTS tab and click the Plot icon:

Moreover, the A and B data can be plotted as a scatter, pie, or histogram plot.

Let's look at another example that plots computed data points.

Example 2: Plotting Function Values

Given: $y(t) = 1.2\ sin(2t + 10)$, $t = [-2\pi, 2\pi]$.

By using the following commands in the command window, you get a plot of $y(t)$ versus t, as shown here:

```
>> t=-2*pi:2*pi;  % value ranges for t are assigned
>> y=sin(2*t+10); % y values are computed
>> plot(t, y)     % t vs. y is plotted
```

These commands produce the plot shown in Figure 6-5.

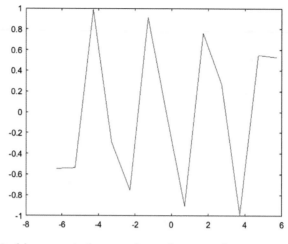

Figure 6-5. *Plot of y(t) = 1.2 sin(2t + 10), t = [−2π, 2π]*

The plot, shown in Figure 6-5, is not smooth even if it looks like a sine wave. So, where is the flaw?

The flaw is in the very first command: >> t=-2*pi:2*pi. It creates an array of *t* with a step size of $\Delta t = 1$, which is a default setting in the linear space definition. In this case, function values are computed at t = -6.28, -5.28, -4, 28 ... 5.28, and 6.28. To fix this problem, the step size must be specified and taken smaller than 1.0; for instance, $\Delta t = \pi/20$.

```
>> t=-2*pi:pi/20:2*pi; y=sin(2*t+10); plot(t, y)
```

After taking $\Delta t = \pi/20$, the plot of $y(t)$ becomes smooth and periodic, as shown in Figure 6-6.

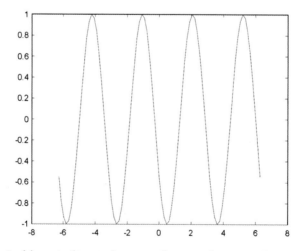

Figure 6-6. *Plot of y(t) = sin(2t+10) vs. t = [−2π, 2π], Δt = π/20*

Example 3: Building a Histogram

Say you are given 1,000 data points generated by a normally distributed pseudorandom number generator function called `randn()`. Let's build a histogram of the data points. See Figure 6-7.

```
>> F=randn(1,1000);
>> hist(F)
```

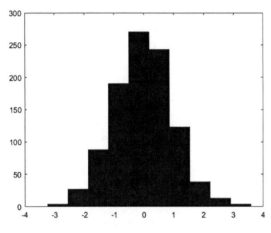

Figure 6-7. *Histogram of 1,000 normally distributed random numbers with the hist() command*

Example 4: Building a Bar Chart

Say you are given 5-by-3 data points generated by a normally distributed pseudorandom integer number 0 to 125 generator function called `randi()`. Let's build a bar chart of the data points. See Figure 6-8.

```
>> W= randi(125,5,3);
>> bar(W)
```

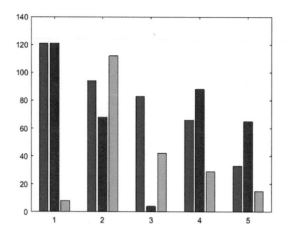

Figure 6-8. *Bar chart of 5-by-3 data points normally distributed integer pseudorandom numbers with the bar() command*

Example 5: Building a 3D Pie Chart

Say you are given A = [100, 121, 95 125 105, 111, 75, 55, 100]. Let's build a 3D pie chart of A. See Figure 6-9.

```
>> A = [100, 121, 95 125 105, 111, 75, 55, 100];
>> pie3(A)
```

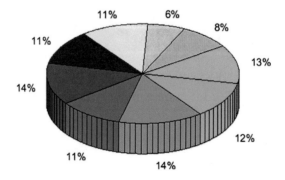

Figure 6-9. *3D pie chart of A*

TITLE, XLABEL, YLABEL, AXIS, GRID, and LEGEND

Plot titles, axis labels and scales, grid settings, and legend options can be implemented in plot figures by employing the appropriate commands, such as title(), xlabel(), ylabel(), axis(), grid, and legend(). Let's demonstrate how to use them via examples.

TITLE()

The title() command is used to set the plot title name tag. For example, title(Unit circle) gives the current plot figure a title of Unit circle.

The alternative syntax of this command is title 'Unit circle. That is compatible with later versions of MATLAB.

XLABEL, YLABEL, and ZLABEL

The xlabel, ylabel, and zlabel commands are used to assign the axis labels or titles for the x-, y-, or z-axis, respectively. The syntax is xlabel(t, [s]'), ylabel(f(t)). An alternative syntax for these commands is xlabel 't, [s]', ylabel' f(t), which assigns the labels for the x- and y-axes to t, [s], and f(t), respectively.

AXIS

By default, MATLAB scales the axes of a plot to fit the data. The axis command is used to set the scaling factor for the plot figure. The general syntax of the axis command is shown here.

For 2D plots:

```
xlim([xmin xmax])          % Min and Max limits set for x axis
ylim([ymin ymax])          % Min and Max limits set for y axis
axis([xmin xmax ymin ymax]) % Min and Max limits set for x & y
```

For 3D plots:

```
axis([xmin,xmax,ymin,ymax,zmin,zmax]) % Min and Max limits
```

In addition, axes can be set up in a few different ways with these commands:

```
axis equal  % Both or all 3 axes have equal tick marks
```

```
axis square % Current axis box square in size
axis normal % Restores current axis box to full size
axis off    % Removes all axis labels, tick marks, background
axis on     % Restores all axis labels, tick marks, background
axis tight  % Sets axis limits to the range of data
axis ij     % Puts MATLAB into its "MATRIX" axis mode
```

GRID

The grid on command is used to set grids on the plot figure to make it more legible, e.g., use grid on to set major grids and grid minor to set minor grids.

Let's use these additional plot commands.

Example 6: Plotting a Unit Circle with Plot Tools

Let's create a unit circle defined by $x = \sin(t)$, $y = \cos(t)$, $t = [0, 2\pi]$ (see Figure 6-10).

```
>>t=0:pi/100:2*pi; x=sin(t); y=cos(t); scatter(x, y);
>>title(Unit circle defined by x=sin(t) vs. y=cos(t)); axis tight;
>>grid on; axis tight; xlabel(sin(t)), ylabel(cos(t))
```

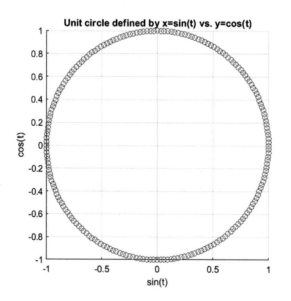

Figure 6-10. *Plot of a unit circle with plot tools*

There is an alternative solution in the polar coordinate system, using the polar() command, as shown in Figure 6-11.

```
>> t=0:pi/100:2*pi; x=sin(t);y=cos(t);r=sqrt(x.^2+y.^2);polar(t,r)
>> title(Unit circle defined by x=sin(t) vs. y=cos(t))
```

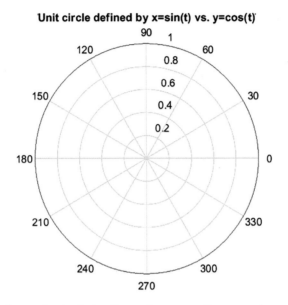

Figure 6-11. *Plot of a unit circle in the polar coordinate system*

Note   While plotting x versus y, the data size (i.e., the number of elements in x and y) has to be equal; otherwise, MATLAB will return an error. This is one of the most common mistakes.

LINE and MARKER Specifiers

Line and marker specifiers are useful for distinguishing a few data sets plotted in a single plot area. The specifiers can be line colors, types, and markers. Line specification (line specifier), marker specification (marker specifier), and color specification (color specifier) are explained in Tables 6-1, 6-2, and 6-3, respectively.

Table 6-1. *Line Style Specifiers*

Specifier	Line Style
-	Solid line (by default)
--	Dashed line
:	Dotted line (colon)
-.	Dash-dot line

Table 6-2. *Marker Specifiers*

Specifier	Marker Type
+	Plus sign
0	Circle
*	Asterisk
.	Point
x	Cross
'square' or s	Square
'diamond' or d	Diamond
^	Upward-pointing triangle
v	Downward-pointing triangle
>	Right-pointing triangle
<	Left-pointing triangle
'pentagram' or p	Five-pointed star (pentagram)
'hexagram' or h	Six-pointed star (hexagram)

Table 6-3. *Color Specifiers*

Specifier	Color
r	Red
G	Green
B	Blue
C	Cyan
M	Magenta
Y	Yellow
K	Black
W	White

Note   The color specifiers given in Table 6-3 can be lowercase or uppercase. The common practice is to use lowercase letters.

All plot functions (except for the family of easy plots, i.e., ezplot, ezsurf, and ezmesh, and implicit function plots, fimplicit, fimplicit3, and fsurf) accept a line specification argument that defines three components to specify lines with the following:

- Color
- Marker symbol
- Line style

Here's an example: >> plot(x, f, ':pb').

The command plots *x* versus *f* using a colon (:) place pentagram markers (p) at the data points, and it colors a line and a marker in blue (b). You can specify the components (in any order) as a quoted string after the data arguments. Note that line specifications are single strings, not property-value pairs.

This command: >> plot(x, h, 'd') plots *x* versus *h* with a marker diamond without any line. Note if the color of a plot line or marker is not specified, then the color by default is blue.

Moreover, while using the `plot` and `plot3` functions, you can specify other characteristics of lines using graphics properties.

> `LineWidth`—Specifies the width (in points) of the line, e.g., `'LineWidth, 2`

> `MarkerSize`—Specifies the size of the marker in points, e.g., `'MarkerSize, 5`

> `MarkerEdgeColor`—Specifies the color of the marker or the edge color for filled markers (e.g., circle o, square s, diamond d, pentagram p, hexagram h, and the four triangles <, >, ^, v)

> `MarkerFaceColor`—Specifies the color of the face of filled markers

In addition, you can specify the `LineStyle`, `Color`, and `Marker` properties instead of using the symbol string. This is useful when you want to specify a color that is not in the list. You can use the red-green-blue (RGB) values, e.g., to plot the line specifications.

`plot(t,y,'LineWidth, 1/25,'LineStyle,'--;Color,'magenta,'Marker,'v)`

This plots *t* versus *y* data with a line width of 1/25 inch, a line style of dashed lines, magenta dashed lines, and data points marked with triangle markers looking downward.

Let's look at another example.

Example 7: Plotting Sine Function Values with Plot Tools

Plot this function: $y(\alpha) = sin(2\alpha + 10)$, $\alpha = [-2\pi, 2\pi]$. The solution script (`Plot_E7.m`) contains the commands to compute $y(\alpha)$ with respect to $a=-2\pi : \pi/20 : 2\pi$. It has the plot commands specifying line color (b is blue), line type (: is colon), line width (2), marker type (d for diamond), marker size (7), and marker face color (m for magenta). Moreover, it displays the plot title text (`Plot of function y = sin(2a+10)`) with the font size of 13 and similarly, x- and y-axis label text (a, 'y(a)) with the font size of 13.

```
%% Plot_EX7.m
a=-2*pi:pi/20:2*pi;
y=sin(2*a+10); plot(a,y,'bd:;LineWidth,2;MarkerSize,7;MarkerFaceColor;m);
grid on; axis([-2*pi 0.5*pi -1 1]);
title(Plot of function y = sin(2a+10),'FontSize, 13) xlabel(a);
ylabel(y(a), FontSize, 13)
```

After executing the code (Plot_EX7.m), the plot figure shown in Figure 6-12 is created.

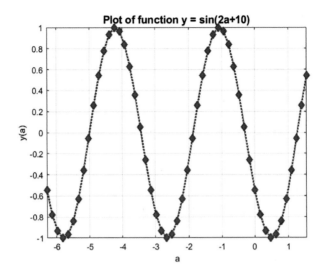

Figure 6-12. *Plot of y(α) = sin (2α + 10), α = [−2π, 2π]*

Special Characters

MATLAB can recognize various special characters by using TeX markup and LaTEX code syntax. It is handy to employ LaTEX in plots as well. The following examples display special characters and Greek letters in the plot figure in the title, axis label, plot text, or notation. You write special characters, such as the Greek letters, by using the backslash operator (\). Here are some examples:

- Title(\alpha) displays α on the plot title.

- xlabel(\beta) displays β on the x–axis label.

- text(0, 1, '\rightarrow) displays ➤ in the plot area at the coordinate points of (0, 1).

    - title(e^{at}) displays e^{at} in the plot title.

    - ylabel(\fontsize{13} y(x)) displays $y(x)$ on the y-axis with the font size of 13.

    - title(\it y(\alpha) vs. \alpha displays $y(\alpha)$ *vs.* α in italic in the plot title.

- xlabel(\bf f, [s^{-1}]) displays **f**, [**s⁻¹**] in bold font on the x-axis label.

- title(\copyright by SE) displays © by SE in the plot title.

- text(2,1,\Gamma \approx \pi/2) displays $\Gamma \approx \pi/2$ in the plot area at the coordinate points of (2,1).

For more extensive information and help on TeX and LaTEX, type the following in the command window:

```
>> help tex; help latex
```

Example 8: Plotting Sine Function Values with Plot Tools

Plot this function: $y(\alpha) = sin(2\alpha + 10)$, $\alpha = [-2\pi, 2\pi]$.

The answer script (Plot_EX8.m) has plot commands with line and marker specifiers, such as line type (- is a solid line), line width (2), line color (b for blue), marker type (o), marker size (7), and marker face color (y for yellow). Moreover, it includes commands for the title text, containing the Greek letter (α) in italic and a font size of 13.

Axis labels include the Greek letter (α) with a font size of 13. It includes the text note with a left arrow symbol $\leftarrow \alpha = 0^0$ displayed on the plot area, as shown in Figure 6-13.

```
%% Plot_EX8.m
a=-2*pi:pi/20:2*pi; y=sin(2*a+10);

plot(a,y,'b-o','LineWidth',2,'MarkerSize',7,'MarkerFaceColor','y'); grid on;
axis([-2*pi 0.5*pi -1 1]);
title('\it \fontsize{13} Plot of function y = sin(2\alpha+10))
ylabel('\fontsize{13} y(\alpha));
xlabel('\fontsize{13} \alpha) text(0,0, '\leftarrow \fontsize{13} \alpha
= 0^0);
```

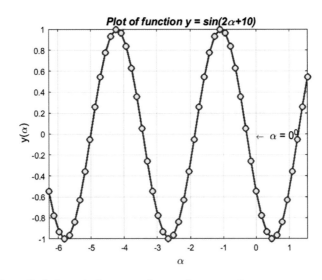

Figure 6-13. *Plot of y(α) = sin(2α + 10), α = [−2π, 2π]*

Plot Two Data Sets in Two Y–Y Axes

Sometimes you may need to plot two separate data sets or function values in one plot figure in two vertical (y-y) axes. This type of plot is relatively easy to implement, and there are two approaches to it. One approach is applicable to older versions of MATLAB, and the other works with more recent versions.

Let's look at the following example.

Example 9: Plotting Two Function Values on Y-Y Axes

Given:

- $y(\gamma) = 2.725e^{-0.1\gamma} \cos(2.725\,\gamma + 25); \gamma = 0...15\,\pi;$

- $z(\gamma) = 0.725\,\sin(0.725\gamma + 25)\,\ln(0.5\gamma); \gamma = 0...15\,\pi;$

The following code shows two solutions (Plot_EX9.m). Version 1 is for older versions of MATLAB, and version 2 is for recent versions. The main difference between the two versions are the commands plotyy and yyaxis (left and right). MATLAB is forward compatible in most cases, so the scripts and command syntax written for older versions are valid with newer versions.

```
%% Plot_EX9.m
% Version 1. For older versions of MATLAB clf

% Clean up an open figure gamma=0:pi/40:15*pi;
y=2.725*cos(2.725*gamma+25).*exp(-.1*gamma);
  z=0.725*sin(0.725*gamma+25).*log(0.5*gamma);
% gamma vs. y and gamma vs. z plotted for each value of gamma. [AX,Y1,Y2] =
plotyy(gamma, y, gamma, z,'plot);
% Title of the plot is in LaTEX
title([\fontsize{9} y=2.725*cos(2.725*\gamma+25)*e^{-0.1\gamma},...
'& z=0.725*sin(0.725*\gamma+25)*log(0.5*\gamma)]) xlabel(\fontsize{15} \
gamma);
set(get(AX(1),Ylabel),String,...'y=2.725*cos(2.725*\gamma+25).*e^{-^0.1\
gamma})
set(get(AX(2),Ylabel),String,...
'z=0.725*sin(0.725*\gamma+25)*log(0.5*\gamma)) grid on; axis tight
%% Version 2. For recent versions of MATLAB
clf % Clean up an open figure
gamma=0:pi/40:15*pi; y=2.725*cos(2.725*gamma+25).*exp(-.1*gamma);
  z=0.725*sin(0.725*gamma+25).*log(0.5*gamma);
yyaxis left % Select the left y-axis to plot the data plot(gamma,
y,'r-ò), ylabel(y=2.725*cos(2.725*\gamma+25).*e^{-^0.1\gamma}) yyaxis
right % Select the right y-axis to plot the data plot(gamma, z,'b--x);
ylabel(z=0.725*sin(0.725*\gamma+25)*log(0.5*\gamma)), grid minor;
title([\fontsize{9} y=2.725*cos(2.725*\gamma+25)*e^{-0.1\gamma},...
'& z=0.725*sin(0.725*\gamma+25)*log(0.5*\gamma)]); axis tight
```

The obtained plots are different as well. In version 1, there are some limitations, such as plot line, marker, and color specifications, that cannot be adjusted. Only the default colors are used (see Figure 6-14).

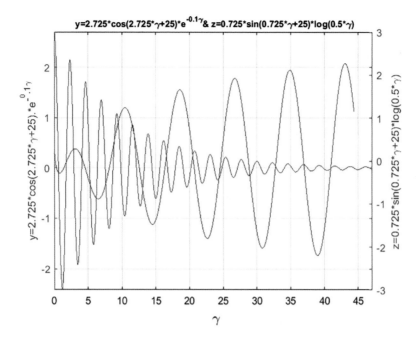

Figure 6-14. *Plot of two function values on y-y axes (version 1)*

The version 2 solution shown in Figure 6-15, for more recent versions of MATLAB, is more flexible. It includes adjustable plot specifiers, such as marker type, line type, color, size, etc.

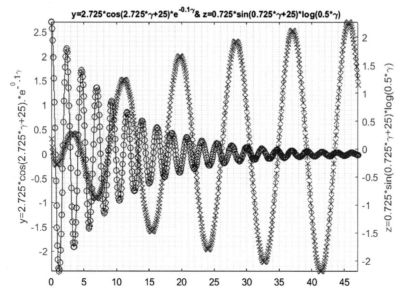

Figure 6-15. *Plot of two function values on y-y axes (Version 2)*

Subplots

Several plots can be plotted in a plot figure as subplots by employing the `subplot()` command.

Example 10: Building Subplots of Functions

Given these four functions:

$$z(\gamma) = 0.725 sin(2.725\gamma + 25)ln(0.5\gamma)$$

$$y(\gamma) = 2.725e^{-0.1\gamma}cos(2.725\gamma + 25)$$

$$x(t) = e^{2.725cos(2.725t+25)}$$

$$w(t) = 0.725 sin(0.725ln(0.5t) + 25)$$

$$\gamma = 0...15\pi; t = -5\pi...5\pi;$$

Let's plot them as subplots in a single plot figure. The solution script (`Plot_EX10.m`) computes and plots all four functions on one plot as four separate subplots (see Figure 6-16).

```
%% Plot_EX10.m
gamma=0:pi/40:15*pi; y=2.725*cos(2.725*gamma+25).*exp(-.1*gamma);
z=0.725*sin(0.725*gamma+25).*log(0.5*gamma); t=-5*pi:pi/40:5*pi;
x=exp(2.725*cos(2.725*t+25)); w=0.725*sin(0.725*log(0.5*t)+25);
figure % Creates a blank figure
subplot(2,2,1); plot(gamma, y,'r-'); grid on title(y=2.725*cos(2.725*\
gamma+25)*e^{-0.1\gamma}) subplot(2,2,2); plot(gamma, z,'b-'); grid on
title(z=0.725*sin(0.725*\gamma+25).*log(0.5*\gamma)) subplot (2,2,3)
semilogx(t, x, 'r-') % x axis in log. Scale for demo purposes grid on
axis([0.05 25 0 15.5]); % Assign min and max values for x & y axes
title(x=e^{2.725cos(2.725t+25)}) subplot(2,2,4)
semilogy(t, w,'b-'); grid on; axis([-17 17 0.025 5.5]); title(w=0.725*sin
(0.725*log(0.5*t)+25))
```

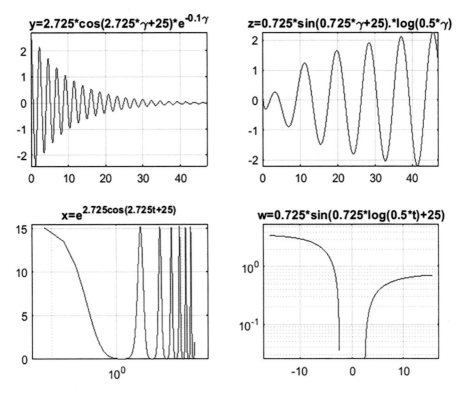

Figure 6-16. *Plot of four function values as subplots in one plot figure*

In addition, when all the plots are built, the following warning messages are displayed in the Command window:

```
Warning: Negative data ignored
Warning: Imaginary parts of complex X and/or Y arguments ignored Warning:
Negative data ignored
```

Imaginary parts of computed values are omitted when plotting them as complex numbers.

LEGEND

The legend command is used to specify legends of plotted data sets to recognize the data, e.g., legend(y(t), 'z(t), 'h(t). Any special notations of TeX and LaTEX can be implemented with the legend command. For example, legend(x(\alpha), 'y(\xi)) displays $x(\alpha)$ and $y(\xi)$ in the plot area. It is also possible to add a title to the list of legends. For example, a title ("Variables") for this list of legends $x(\alpha)$ and $y(\xi)$ can be attained with the following commands:

```
L =legend(x(\alpha),'y(\xi));
title(L, 'Variables')
```

HOLD

To hold and plot several plots in one plot figure, the hold on command is used. Once all the data sets are plotted, the held plot needs to be activated with the hold off command. Alternatively, when plotting sets of data in loops, you can use the hold all command to assign a unique color to each set of data.

Example 11: Plotting a Few Function Values in One Plot

Given: $y_1 = \dfrac{sec(x)}{x}, y_2 = \dfrac{10sin(x)}{x}, y_3 = \dfrac{10cos(x)}{x}, y_4 = \dfrac{10tan(x)}{x}, y_5 = \dfrac{exp(x)}{x}.$

Let's compute and plot numerical values of these five functions together in a single plot figure.

```
%% Plot_EX11.m
close all
x = linspace(-3*pi, 3*pi, 3333);
y=[sec(x)./x; 100*sin(x)./x;10*cos(x)./x;10*tan(x)./x; exp(x)./(x);];
plot( x, y(1,:), 'g:o'); hold on
plot(x, y(2,:),'k-', x, y(3,:), 'r:', x, y(4,:),'b-.', 'LineWidth',1.5)
plot(x, y(5,:),'m-','LineWidth',1.5)
title('[y_1, y_2, y_3, y_4, y_5] vs. x')
xlabel('\it x'); ylabel('\it y function values')
LEG =legend('sec(x)/x','100*sin(x)/x','10*cos(x)/x','10*tan(x)/x',
'e^x/x');
title(LEG, 'Plotted Equations:')
axis([-3*pi, 3*pi, -100, 100]); hold off; grid on
shg % Show a plot figure
```

After simulating the script, Figure 6-17 shows the result. Figure 6-17 shows how the title of a legend list is displayed.

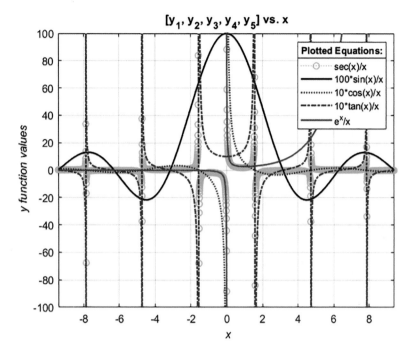

Figure 6-17. *Plotting several function values in one plot figure*

If there are different sets of data to be plotted in a single plot figure, loop control operators can be employed to handle line color, type, marker specifications, and legends. A line type, color, and marker type can be prespecified as a string and used later. Let's consider the following example.

Example 12: Plotting Function Values with Different Line Markers and Colors

Given: $F(\mu,t) = sin(2t) * \sqrt{sin\left(\mu t \sqrt{t + t^2 + t^3}\right)}$ for $t = 0...\pi$ ($\Delta t = \pi/1000$), $\mu = 1...5$.

If you take the stepwise approach that you used in the previous examples, your script becomes considerably long and untidy. Thus, you take a different approach to create a set of line and marker specifiers first and then use the loop control statements to choose from our predefined specifiers. You do this for every iteration while plotting the data sets with respect to every value of μ separately. The solution script (Plot_EX12.m) is rather compact.

399

```matlab
%% Plot_EX12.m
% Version 1
close all; clearvars
t=0:.001*pi:1*pi;
% Line color, type and marker specifiers are defined:
Labelit  = {};
Colorit  = 'krgbmk';
Lineit   = '--:-:--:-:-';
Markit   = '-+xdoshv<p>';
for ii=1:5
% Line and marker specifier are taken for each iteration:
    Stylo = [Colorit(ii) Lineit(ii) Markit(ii)];
    F     = sqrt(sin(ii*t.*sqrt(t+t.^2+t.^3))).*sin(2*t);
    subplot(211)
    plot(t, real(F), Stylo, 'markersize', 3)  % Real Part of F is plotted
    hold on; grid on
    Labelit{ii} = ['\mu = ' num2str(ii)];
    legend(Labelit{:})
    subplot(212)
    plot(t, imag(F), Stylo, 'markersize', 3) % Imaginary Part of F
    is plotted
    hold on
    Labelit{ii} = ['\mu = ' num2str(ii)];
    legend(Labelit{:})
end
axis tight; grid on
%% Version 2. No loop for calculations
close all; clearvars
t=0:.001*pi:1*pi;
[ts, mu]=meshgrid(t, 1:5);
F = sqrt(sin(mu.*ts.*sqrt(ts+ts.^2+t.^3))).*sin(2*ts);
%%
% Line color, type and marker specifiers are defined:
Labelit  = {};
Colorit  = 'krgbmk';
```

```matlab
Lineit   = '--:-:--:-:-';
Markit   = '-+xdoshv<p>';
for ii=1:5
% Line and marker specifier are taken for each iteration:
    Stylo = [Colorit(ii) Lineit(ii) Markit(ii)];
    subplot(211)
    plot(t, real(F(ii,:)), Stylo, 'markersize', 3)  % Real Part
    hold on; grid on
    Labelit{ii} = ['\mu = ' num2str(ii)];
    legend(Labelit{:})
    subplot(212)
    plot(t, imag(F(ii,:)), Stylo, 'MarkerSize',3) % Imaginary Part
    hold on
    Labelit{ii} = ['\mu = ' num2str(ii)];
    legend(Labelit{:})
end
axis tight;
grid on
shg             % Show the plotted graph
```

Versions 1 and 2 of the script (Plot_EX12.m) produce the same results and plot figure as shown in Figure 6-18. Note that similarly for a much larger number of μ values, a unified plot figure can be generated by specifying individual line type and color and marker type. Moreover, it is possible to add different marker size and line width values.

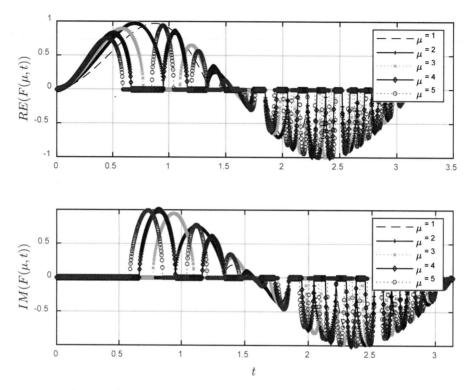

Figure 6-18. *Plotting function values with different line markers and colors*

Example 13: Bar Chart of Data with Standard Deviation

Given the data set (U = D±σ) with standard deviation values:

U= [22.3± 1.25; 25.7± 2.12; 21.2± 2.24 27.9± 1.21 25.5± 2.37 20.8± 2.31 19.3±1.13];

The given data set D can be plotted using the bar() chart and errorbar() plot functions including hold on. Here is the complete script (Bar_Plot_E13.m); see Figure 6-19:

```
%% Plot_EX13.m
D = [22.3   25.7    21.2  27.9  25.5  20.8  19.3];      % Data
S = [1.25    2.12   2.24  1.21  2.37   2.31  1.13];      % STD
n = 1:numel(D);
bar(n, D, 'y')                % Bar face color yellow
hold on
EB = errorbar(n,D,S,S, 'LineWidth', 2);
EB.Color = [1 0 0];        % STD line color
```

```
EB.LineStyle = 'none';       % No connection line for STD
hold off
legend('Data', 'std')
shg                          % Show the plotted graph
```

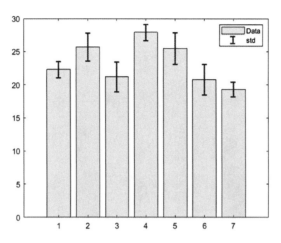

Figure 6-19. *Bar chart with standard deviations*

Example 14: Bar Chart of Data with Values Shown

Given the data (3-by- 5):

```
V = [12 13 16 17 15;
     25 21 23 26 24;
     31 33 30 35 37];
```

Let's plot bar chart and add the values of the plotted data points.

Here is the complete code (Plot_EX14.m):

```
%% Plot_EX14.m
V = [12 13 16 17 15;
     25 21 23 26 24;
     31 33 30 35 37];
n = 1:length(V);
b = bar(n,V);
for ii=1:3
xtips1 = b(ii).XEndPoints;
```

```
ytips1 = b(ii).YEndPoints;
labels1 = string(b(ii).YData);
text(xtips1,ytips1,labels1,'HorizontalAlignment','center',...
    'VerticalAlignment','bottom')
end
shg      % Sow the plotted graph
```

After executing the code (Plot_EX14.m), you will obtain the bar chart shown in Figure 6-20.

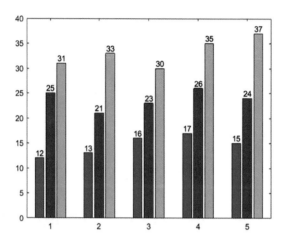

Figure 6-20. *Bar chart with data values shown*

Example 15: Bar Chart of Data with NaN Values Shown and Axis Tick Labels Off

Given: v = [3.7 4.3 nan 3 2 nan 5.5 2.9 NaN 4];

Let's plot [v], and display all NaN's and remove axis tick marks and labels using the axis off command.

Here is the final code (Plot_EX15.m):

```
%% Plot_EX15.m
clc; close all
D = [3.7  4.3 nan 3 2 nan 5.5 2.9 NaN 4];
H = bar(v);
```

```matlab
H.FaceColor = "g";  % Bar Face Color
H.EdgeColor ='r';   % Bar Edge Color
H.BarWidth = 0.75;  % Bar Width
grid on;
xlim([0, length(v)+1]);
yl = ylim;
for k = 1 : length(v)
    if isnan(v(k))
        y = 0;
        str = 'NaN';
    else
        y = v(k);
        str = sprintf('%.1f', y);
    end
    text(k, y, str, 'Color', 'b', 'FontSize', 13, 'FontWeight', 'bold', ...
        'VerticalAlignment', 'bottom', 'HorizontalAlignment', 'center');
end
axis off
title('Data with NaN')
% g = gcf;
% g.WindowState = 'maximized';  % To maximize Figure Window
```

After executing the script code (Plot_EX15.m), the plot in Figure 6-21 will be shown. Note that using the axis off command removes the tick mark labels of both axes.

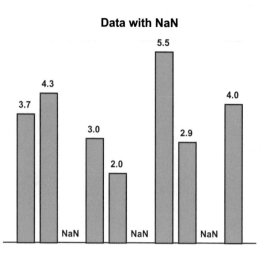

Figure 6-21. *Bar chart with data values shown including NaN*

EZPLOT, FPLOT, and FIMPLICIT with Function Handles (@)

EZPLOT and FPLOT are used to plot mathematical expressions with the function handle (@) and implicitly defined argument values. Here's an example:

```
fplot(@(t)([sin(t), cos(t), exp(cos(t))])); ezplot(@(x)(sin(x)));
fimplicit(@(x,y) (25-(x.^2+y.^2)));
```

Or with a range of given argument values. Here's an example:

```
fplot(@(t)([sin(t),cos(t),exp(cos(t))]),[-pi,pi]);
ezplot(@(x) (sin(x)),[0,2*pi]);
fimplicit(@(x,y)(25-(x.^2+y.^2)),[-5,5]);
```

The function handle (@) can be defined within fplot, ezplot, and fimplicit or outside of these plot functions:

```
G=@(t)([t.^2+2*t-13,sqrt(t.^3+3*t),(t.^5+t^(3/4))^(3/2)]); fplot(G,
[-2.5, 2]);
```

Example 16: Plotting a Mathematical Expression with ezplot( )

Given $f(x) = 0.5\cos\left(\dfrac{N\pi x}{180}\right) + 0.75\sin\left(\dfrac{N\pi x}{180}\right); N = 0:20:200$, with an argument of x, the function $f(x)$ is the summation of the cosine and sine functions within the frequency range of [0, 200] rad/sec with the increment of $\Delta N = 20$ rad/sec. Here is the solution script (Plot_EX16.m) with a loop control statement (see Figure 6-22):

```
%% Plot_EX16.m to demonstrate EZPLOT and FPLOT
for N=0:20:200
    FH=@(x)(0.5*cos(N*pi*x/180)+0.75*sin(N*pi*x/180));
    ezplot(FH, [-6.25, 6.25]); hold all % or fplot(FH, [-6.25, 6.25])
end
hold off
```

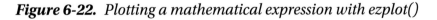

Figure 6-22. *Plotting a mathematical expression with ezplot()*

fplot, fimplicit, and ezplot are easy to employ with mathematical expressions that have one or two arguments to plot, with or without specifying their ranges. An important feature of fplot is that it can compute the values of the expression without plotting it, as follows:

```
[X, Y] = fplot(@(x)(0.5*cos(N*pi*x/180)+0.75*sin(N*pi*x/180)));
```

Note   In future releases of ATLAB, `ezplot()` will removed, and thus, it is recommended to use `fplot()` or `fimplicit()`.

GTEXT, TEXT, and GINPUT

The GTEXT and TEXT commands are used to add annotations and text strings to the plot figure. GTEXT is used to place text with a mouse. TEXT is used to add text descriptions to data points. The general syntax of the GTEXT command is as follows:

```
gtext('Text', 'PropertyName', PropertyValue,...)
gtext({'First line','Second line'},'FontName','Times','Fontsize',12)
```

The general syntax of the TEXT command is as follows:

```
text(x, y, 'Text Message')
```

Here are two examples of employing `text` and `gtext` to display text or notes on plot figures. The `gtext` and `text` graphical annotation commands are used:

```
>> text(0,0,'\alpha = \pi/2)
>> gtext([\alpha ='  num2str(13)' \pi])
```

The `text` command displays $\alpha = \pi/2$ in the current plot figure surface at $[0, 0]$, and `gtext` displays $\alpha = 13\,\pi$ where a user clicks with the cursor.

GINPUT is a graphical input from a mouse. It is a handy tool to locate the points of interest in a plot figure using a mouse pointer. The general syntax of this command is as follows:

```
[X,Y] = ginput(N); % N is number of points to select
```

This command returns N points of X and Y coordinates positioned/chosen using a mouse. The following command selects an unlimited number of points until the Enter key is pressed on the keyboard:

```
[X,Y] = ginput();
```

To locate points of interest on the plot or the local or global minima of the plotted data, the `ginput` is a good graphical tool.

```
>> [x, y] = ginput(3);
```

This command brings the cursor automatically to the surface of the current figure.

It selects three points, of which the x and y coordinates will be measured and saved in the workspace under the variable names x and y. If there is no open figure, ginput opens a blank plot figure. Let's look at an example of employing these commands in combination.

Example 17: Locate and Display Minimum Values of a Function Plot in a Plot Figure

Given a function: $f(x) = \sin(1.313^{-1.7*x}) - \cos(1.313^{1.7*x})$.

Let's compute and plot $f(x)$ and then find its minimum values within a region of $x = -6.5...6.5$. In addition, you can display the found value on the plot figure by employing text and gtext. You'll use two different ways to locate a minimum of a given function with ginput.

The solution script (Plot_EX17.m) uses a mouse cursor with ginput.

```
%% Plot_EX17.m
fun_fun=@(x)sin(1.313^(-1.7*x))-cos(1.313^(1.7*x));
fplot(fun_fun, [-6.5, 6.5]); grid on; hold on
[xm, ym]=ginput(3);     % Use mouse cursor carefully to locate minima
TXT1=['1st Local Min (square) @x= ', num2str(xm(1))];
gtext(TXT1, 'fontsize', 11)
text(-2,0,['2nd Local Min (circle) @x= ', num2str(xm(2))]);
text(-2, -0.5, ['3rd Local Min (star)   @x= ', num2str(xm(3))]);
plot(xm(1),ym(1),'rs', xm(2),ym(2), 'bo', xm(3),ym(3), 'kp', ...
'markersize',13,'markerfacecolor', 'm'); hold off
title('sin(1.313^{-1.7*x})-cos(1.313^{1.7x}')
xlabel('\it x'), ylabel('\it f(x)')
```

Within this script, the text and gtext commands are used to insert the three local minimum values of f(x) with the cursor clicks activated by ginput. After executing the script, you select the local minima of the plotted data from left to right and print the local minima on the plot figure. The plot figure shown in Figure 6-23 is the result.

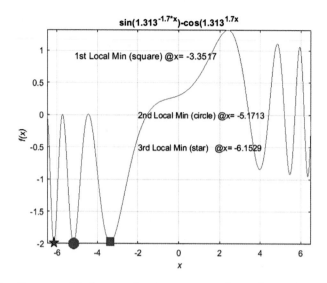

Figure 6-23. *Plot figure with located minimum values displayed*

Note that in Figure 6-23 the first local min (the square) @*x* = -6.1855 is displayed with the gtext command, and the other two local minima (the second and third minima) are displayed with the text command.

The GINPUT function is particularly useful when you need to locate several points in a plot with their corresponding coordinate points, but accuracy depends on how well or how precisely you can place a mouse cursor over the points of interest.

Axis Ticks and Tick Labels

In some cases, it might be necessary to display specific values of tick marks on the plot axis. In those cases, the xticks, xticklabels, yticks, yticklabels, zticks, and zticklabels commands will be employed, respectively, for the x-, y-, and z-axes.

Example 18: Display X-Axis Tick Labels

Given $G = [sin(2t), cos(2t), e^{sin(2t) + cos(2t)}]$, $t = [0, 2\pi]$, put the tick marks on the x-axis at $\left[0, \dfrac{\pi}{2}, \pi, \dfrac{3\pi}{2}, 2\pi\right]$.

Here is the solution script (Plot_EX18.m):

```
%% Plot_EX18.m to demonstrate Xticks and Xticklabels
H=@(t)([sin(2*t), cos(2*t), exp(sin(2*t)+cos(2*t))]); fplot(H, [0, 2*pi],'-*')
xticks([0:pi/2:3*pi]);
xticklabels({0,'\pi/2', '\pi', '3\pi/2', '2\pi'});
xlabel('\it t'), ylabel('\it H'),
title('\it \fontsize{9} Xticks & Xticklabels'), grid on
```

This solution script (Plot_EX18.m) results in a plot with the x-axis tick labels displayed, as shown in Figure 6-24.

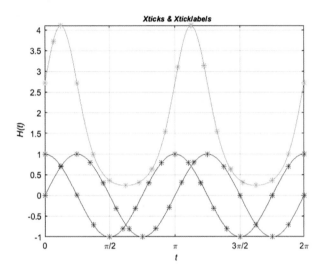

Figure 6-24. *Displaying x-axis tick labels*

In this example, as shown in Figure 6-24, we have demonstrated how to add specific tick marks and labels along the x-axis like can be done for the y-axis. In contrast, all axis labels and tick marks can be removed with a single command: axis off. Here's an example:

```
H=@(t)([sin(2*t), cos(2*t), exp(sin(2*t)+cos(2*t))]);
fplot(H, [0, 2*pi], '-*'); axis off
```

The previously shown commands produces the plot shown in Figure 6-24 with no axis labels (the resulted figure is not shown here).

Figure Handles

The plot figure properties and plot line and marker specifiers (such as line style, color, width, marker type, size, and face color) can be set and managed via a figure handle, e.g., FH = plot(x, y). The handle FH will contain all the figure properties of the plot. Let's look at the following example to see how to manage plot figure properties.

Example 19: Working with Figure Handles

Given: $G = [\sin(2t), \cos(2t), exp(\sin(2t) + \cos(2t))]$, $t = [0, 2\pi]$.

Here is the solution script (Plot_EX19.m):

```
%% Plot_EX19.m - Plot Handle.
% Part 1. Plot figure and get a handle: HG
clf                              % Clear current figure
t = 0:pi/30:2*pi;
GH = [sin(2*t); cos(2*t); exp(sin(2*t)+cos(2*t));];
HG = plot(t, GH); shg
%% Part 2. Change plot properties
% Set Line Color
HG(1).Color='r';                 % HG(1).color = [1 0 0]; % Alternative
HG(2).Color='g';                 % HG(2).color = [0 1 0]  % Alternative
HG(3).Color='b';                 % HG(3).color = [0 0 1]  % Alternative
% Set Line Type
HG(1).LineStyle = '-';
HG(2).LineStyle = '--';
HG(3).LineStyle = '-.';
% Set Line Width
HG(1).LineWidth = HG(1).LineWidth+0.5;
HG(2).LineWidth = HG(2).LineWidth+0.5;
HG(3).LineWidth = HG(3).LineWidth+0.5;
% Set Marker type
set(HG(1), 'marker', 'p')        % Marker is a Pentagon
set(HG(2), 'marker', 'd')        % Marker is a Diamond
set(HG(3), 'marker', 'o')        % Marker is a Circle
% Set Marker size
```

```
set(HG(1), 'MarkerSize', 5)
set(HG(2), 'MarkerSize', 7)
set(HG(3), 'MarkerSize', 9)
% Set Marker Face color
set(HG(1), 'markerfacecolor', [1 0 1])
set(HG(2), 'markerfacecolor', [0 1 1])
set(HG(3), 'markerfacecolor', [1 1 0])
title('\fontsize{11} G(\alpha) vs. \alpha. [\alpha = 0:\pi/30:2\pi]')
xlabel('\alpha'), ylabel 'F(\alpha)', grid on, axis tight
legend('sin(2\alpha)', 'cos(2\alpha)','e^{sin(2\alpha)+cos(2\alpha)}')
```

Part 1 of the script produces the plot figure shown in Figure 6-25, and Part 2 of the script changes the properties (the line and marker specifiers), as shown in Figure 6-24.

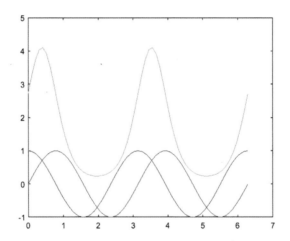

Figure 6-25. *Plot of three function values with default parameters of the plot() command*

The plot figure in Figure 6-26 shows all the changed properties (line and maker specifiers) and the added title, axis labels, and legends.

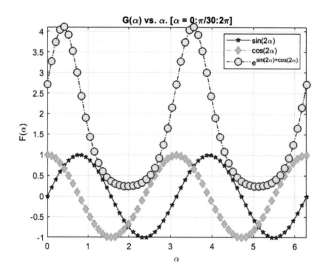

Figure 6-26. *Plot figure of three function values*

All of the 2D plot properties can be also altered using GUI tools (via the Property Inspector), which are accessed by activating the Plot Edit option with a single click on and then double-clicking the plot area.

3D Surface Plots

To create and edit 3D bars, charts, and plots, you use the bar3h, pie3, ezsurf, surf, surfc, mesh, contour3, ribbon, waterfall, and plot3 functions. Almost all of commands, tools, and functions that you use to build 2D plots are also applicable in building 3D plots. This includes title, axis label, axis scale, legend, and so forth.

Example 20: Creating a 3D Pie Plot with pie()

Given A =10; B = 15; C = 20; D = 17; E = 28; F = 10; H = [A, B, C, D, E, F], let's build a pie chart of the data set (see Figure 6-27).

```
pie3(1:6, H, {'A', 'B', 'C', 'D', 'E','F'}); title('Share Holders')
```

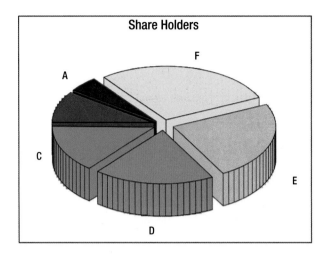

Figure 6-27. *3D pie plot*

Example 21: Creating a 3D Surface Plot with ezsurf( )

Given: $h(z, \beta) = \ln(0.01 * z) * cot(2.5 * \beta)$.

For this example, we define a function handle using the ezsurf( ) command (see Figure 6-28):

```
ezsurf(@(z, beta)log(0.01*z)*(1./tan(2.5*beta)))
```

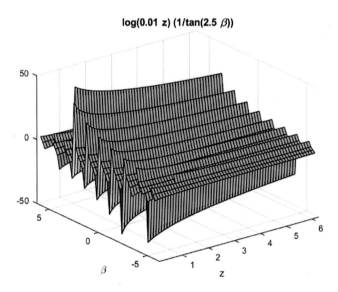

Figure 6-28. *3D surface plot*

An alternative and recommended solution here is to use the fsurf() plot function that is available in the recent versions of MATLAB.

```
Fsurf(@(z, beta)log(0.01*z)*(1./tan(2.5*beta)), [ 0 6, -2*pi, 2*pi])
zlim([-50, 50])
```

Example 22: Creating a 3D Mesh Plot with ezmesh()

Given: $h(\alpha, x) = cos(2.1\alpha)e^{0.1x}$

In this exercise, we create a symbolic math function of $h(\alpha, x)$ by using the function handle with a surface plot command: ezsurf() (see Figure 6-29).

```
%% Plot_EX22.m
% Part 1
ezmesh(@(alpha,x)cos(2.1*alpha)*exp(0.1*x))
```

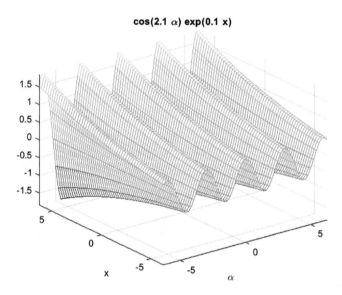

Figure 6-29. *3D mesh plot*

An alternative and recommended solution of this example is to use the fmesh() function with a function handle @.

```
%% Part 2. Alternative solution. Recommended one.
fmesh(@(alpha, x)cos(2.1*alpha)*exp(0.1*x),[-5.75 5.75 -5.75 5.75])
title('cos(2.1*\alpha)*exp(0.1*x)')
```

```
xlabel('\alpha')
ylabel('x')
```

The alternative and recommended solution of this example is to use the `fmesh()` function that produces the same plot as shown in Figure 6-29.

Note   `fmesh()` is recommended to use instead of `ezmesh()`, which is going to be obsolete in the future releases of MATLAB.

Example 23: Creating a 3D Surface-Contour Plot with ezsurfc(), fsurf(), and surfc()

Given $f(x) = 0.5 \cos\left(\dfrac{\omega \pi x}{180}\right) + 0.75 \sin\left(\dfrac{\omega \pi y}{180}\right); \omega = 50:5:150 ; x = [-\pi, \pi], y = [-\pi, \pi]$, with two arguments x and y, let's build a 3D plot $f(x, y)$ for certain frequency ranges with an animated simulation.

```
%% Plot_EX23.m
% Part 1
close all
for omega=50:5:150 handle=@(x,y)(0.5*cos(omega*pi*x/180)+0.75*sin(omega*y*
pi/180));
    ezsurfc(handle, [-pi, pi], [-pi, pi]);
end
```

This script builds a 3D plot (see Figure 6-30) of the given function $f(x, y)$ with an animated simulation with respect to the frequency ranges $\omega = 50 \dots 150$.

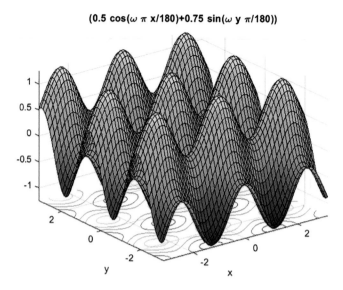

Figure 6-30. *3D surface-contour plot with* `ezsurfc()`

An alternative solution is to use the `fsurf()` function, which is a recommended function to use instead of `ezsurfc()`.

```
%%  Part 2. Alternative Solution
for omega=50:5:150
 handle=@(x,y)(0.5*cos(omega*pi*x/180)+0.75*sin(omega*y*pi/180));
    fsurf(handle, [-pi, pi]); shg
end
xlabel('x')
ylabel('y')
title('0.5*cos(\omega*\pi*x/180)+0.75*sin(\omega*\pi*y/180)')
```

Another alternative solution is to use the `surfc()` function:

```
%%  Part 3. Alternative Solution
[x, y]=meshgrid(linspace(-pi,pi, 50), linspace(-pi,pi, 50));
for omega=50:5:150
    F=(0.5*cos(omega*pi*x/180)+0.75*sin(omega*y*pi/180));
    surfc(x, y, F);
end
```

```
xlabel('x')
ylabel('y')
title('0.5*cos(\omega*\pi*x/180)+0.75*sin(\omega*\pi*y/180)')
```

This solution results in the same output as with `ezsurfc()` and `fsurf()`. Note that with the `surfc()` function all numerical values of the given expression f(x,y) are computed with respect to the equally spaced (generated) data points from the `meshgrid()` function for x and y.

Example 24: Creating a 3D Plot of an Electric Potential Field

The electric potential field V at a point, due to two charged particles, is given by

$$V = \left(\frac{1}{4\pi\epsilon_0} \right) \left(\frac{q_1}{r_1} + \frac{q_2}{r_2} \right)$$

where $q1$ and $q2$ are charges of the particles in Coulombs (C), $r1$ and $r2$ are the distances of the charges from the point (in meters), and $\epsilon0$ is the permittivity of free space, whose value is $\epsilon_0 = 8.854 * 10^{-12} \frac{C^2}{Nm^2}$.

Suppose the charges are $q1 = 2*10^{-10}C$ and $q2 = 4*10^{-10}C$. Their respective locations in the x-y plane are (0.3, 0) and (−0.3, 0). You will plot the electric potential field on a 3D surface plot, V plotted on the z-axis over the ranges of x and y, defined by $-0.25 \leq x \leq 0.25$ and $-0.25 \leq y \leq 0.25$, which correspond to r_1 and r_2. You can create a 3D plot in two ways.

- By using the `surf` function

- By using the `mesh` function

Here is the script (`Plot_EX24.m`) that computes the given electric potential field problem and plots its results:

```
%% Plot_EX24.m
% q1, q2 are charges of the particles in coulombs (C)
% r1, r2 are distances of the charges from the point in meters
% epsilon is permittivity of free space
% r1,r2, V(r1,r2) are coordinate systems for plotting close all
```

```
[r1, r2] = meshgrid(-0.25:0.01:0.25); epsilon = 8.854e-12;q1 = 2e-10;
q2=4e-10; coeff = (1./(4*pi*epsilon));
V = coeff.*(q1./r1 + q2./r2);
figure(1) % Surface plot
surface(r1,r2,V); xlabel('r_1'); ylabel('r_2');
zlabel('V(r_1,r_2)')
title(['Fig. 1. Electric potential field', ...
'of two charged particles with surface plot']) grid on, view(-15,15), axis
tight, colormap Jet figure(2) % Meshed plot with contour
mesh(r1,r2,V); xlabel('r_1'); ylabel('r_2');
zlabel('V(r_1, r_2)');
title(['Fig. 2. Electric potential field', ...
'of two charged particles with mesh plot']); axis vis3d; colormap hsv
```

After running this script, the 3D plots shown in Figure 6-31 and Figure 6-32 are obtained.

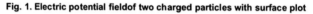

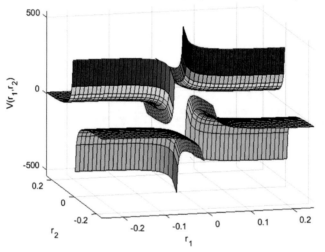

Figure 6-31. *3D plot surface plot of the potential field of the two charged particles*

Fig. 2. Electric potential field of two charged particles with mesh plot

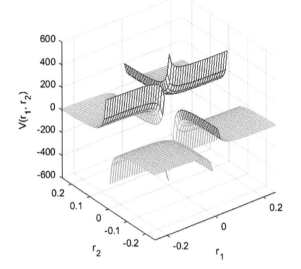

Figure 6-32. *Mesh plot of the electric potential field of two charged particles*

Example 25: Creating 3D Plots with waterfall(), ribbon(), meshc(), contour()

Given $G(t, \theta) = \ln(|\cos(0.5t + 5\theta) + \cosh(5t + 0.5\theta)|)$, $t = -0.2\pi...0.2\pi$; $\theta = -0.2\pi...0.2\pi$.

We define function variables as linearly spaced vectors (vector space) or arrays and then compute their function values according to the defined arrays. After that, you can build plots of the computed data. Here is the complete solution script (Plot_EX25.m):

```
% Plot_EX25.m
[t, theta]=meshgrid(linspace(-0.2*pi, 0.2*pi, 50));
G=log(abs(cos(0.5*t+theta*5)+sinh(5*t+0.5*theta)));
figure(1), waterfall(t,theta,G);
xlabel('t-axis'); ylabel('\theta-axis');
zlabel('G(t,\theta) function values');
title('G(t,\theta)=ln(abs((cos(0.5*t+\theta*5)+sinh(5*t+0.5*\theta)))');
figure(2)
ribbon(G), title('3D ribbon plot of the function G(t, \theta)');
xlabel('t-axis'); ylabel('\theta-axis');
zlabel('G(t,\theta) function'); axis vis3d
```

421

```
figure(3), meshc(t, theta, G)
title('3D meshed contour plot of the function G(t, \theta) ');
xlabel( 't-axis'); ylabel('\theta-axis');
zlabel('G(t,\theta) function'); axis vis3d
figure(4), contour(t, theta, G)
title('Contour plot of the function G(t, \theta)');
xlabel('t-axis'); ylabel('\theta-axis');
```

By executing the script (Plot_EX25.m), you get the plots of the $G(t, \theta)$ function shown in Figures 6-33, 6-34, 6-35, and 6-36.

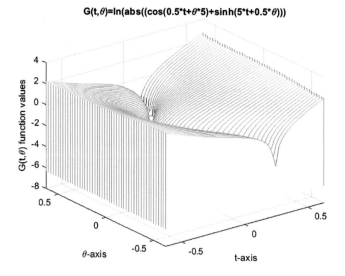

Figure 6-33. *3D surface-waterfall plot of*
$G(t, \theta) = \ln 3(|\cos(0.5t + 5\theta) + \cosh(5t + 0.5\theta)|)$

3D ribbon plot of the function G(t, θ)

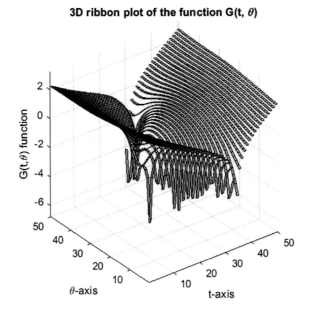

Figure 6-34. *3D ribbon plot of G(t, θ) = ln 3(| cos (0.5t + 5θ) + cosh (5t + 0.5θ)|)*

3D meshed contour plot of the function G(t, θ)

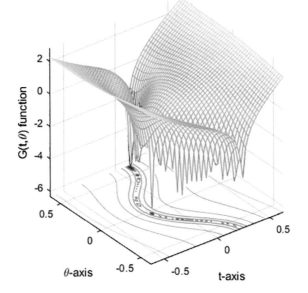

Figure 6-35. *3D meshed contour plot of*
G(t, θ) = ln 3(| cos (0.5t + 5θ) + cosh (5t + 0.5θ)|)

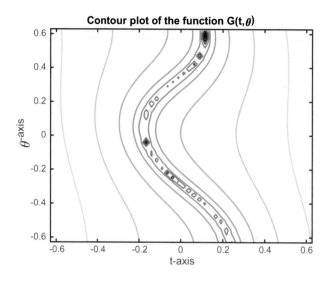

Figure 6-36. *Contour plot of G(t, θ) = ln 3(| cos (0.5t + 5θ) + cosh (5t + 0.5θ)|)*

With 3D plot tools of surfaces, we have demonstrated the mesh, surf (surface), waterfall, ezsurf, ribbon, meshc, and contour plot functions.

Save Plot Figure with saveas()

The created plot figures can be saved using the File ➤ Save As options and MATLAB's built-in function saveas(). The plot figure can be saved in more than a dozen file formats, such as *.png, jpg/.jpeg, .tif/tiff, .bmp, .eps,.pdf, .ps, and so forth. The general syntax of the saveas() command is as follows:

```
saveas(gcf, 'File Name.Extension')
saveas(gcf, 'File Name', 'File format')
```

Let's demonstrate a few examples of saving plot figures with saveas(). Here are the steps to write the complete code (SAVE_Plot.m) used in the examples:

(1)   Save the current plot figure in the MATLAB figure file format.

```
%% SAVE_Plot.m
% Part 1. Save in *.fig format
fplot(@(x)exp(sin(2*x)), [-pi, pi]), grid on;
  saveas(gcf, 'MY_fig.fig')  % Saves MY_fig.fig in current directory
```

(2)   Save the current plot figure in the *.png format.

```
% Part 2. Save in *.png format
[x, y]=meshgrid(linspace(-pi, pi, 75), linspace(-5, 5, 50));
F = exp(sin(x)+cos(y)); contour(x, y, F, '--')
  saveas(gcf, 'MY_Plot', 'png') % Saves MY_Plot.png in current directory
```

(3)   Save the current plot figure in the *.pdf format.

```
% Part 3. Save in *.pdf format
Theta = 0:0.01*pi:5*pi;
c = 2; a = 0.2; b=.5;
R = a + b*Theta.^(1/c);
polarplot(Theta, R, 'LineWidth', 2);
saveas(gcf, 'Spiral', 'pdf')
```

Here we have demonstrated how to use the saveas() function to save/write the created plot figure in different file formats.

Note   Note that the results from the saveas() function are not identical to the saved figure files from the File ➤ Save As options in the figure window.

Now you can recall already saved graphs in *.png and *.fig file formats and plot them as subplots using the imshow() command for the *.png and hgload(), copyobject(), allchild(), and get() commands for *.fig file formats. Here are the completed script (SAVE_Plot.m) commands:

```
% Part 4. Display saved graphs in subplots
H1=subplot(211); imshow('MY_Plot.png')
H2=subplot(212);
C=hgload('MY_Fig.fig'); copyobj(allchild(get(C, 'CurrentAxes')), H2);
title(H1, "MY plot.fig")
title(H2, "MY fig.fig")
```

3D Line Plots and Animations

There are several 3D space line plot tools (plot3, comet3, and scatter3), which are used to plot vector data points and space curves. The given vector data can be embedded in higher dimensions. 3D plotting methods are similar to plot tools employed in plot equations in 2D plots, which were described previously. The plot functions for 3D curves are plot3, comet3, ezplot3, and scatter3. They are implemented very much like the 2D plot functions, such as plot and comet. The comet and comet3 functions draw plots by evolving them in action/animation. Let's consider the next example.

Example 26: Building 3D Line Plots and Animated 3D Line Plots with plot3( ), comet3( ), and ezplot3( )

A popular amusement park ride known as the *corkscrew* has a helical shape. The parametric equations for a circular helix are defined by $x = a \cos t$; $y = a \sin t$; $x = bt$, where a is the radius of the helical path and b is a constant that determines the "tightness" of the path. In addition, if $b > 0$, the helix has the shape of a right-handed screw. If $b < 0$, then the helix is left-handed.

This example creates the three-dimensional plot of the helix in the following three cases and compares their appearance. Use $0 \le t \le 10\pi$ and $a = 1$.

a) $b = 0.1$; b) $b = 0.2$; c) $b = -0.1$;

```
%% Plot_EX26.m - Amusement ride - corkscrew plot in 3D
t=0:pi/15:10*pi; a=1; x=a*cos(t); y=a*sin(t);
% case (a) b1=0.1; z1=b1*t;
% case (b)
b2=0.2; z2=b2*t;
% case (c)
b3=-0.1; z3=b3*t;
subplot(311); plot3(x, y, z1, 'r*-'); legend('b_1=0.1') title('Corkscrew:
b_1=0.1; b_2=0.2, b_3= -0.1'); subplot(312); plot3(x, y, z2, 'bs-');
legend('b_2=0.2')
subplot(313); plot3(x, y, z3, 'ko--');legend('b_3=-0.1')
xlabel('X'); ylabel('Y'); zlabel('Z');
figure(2); X=[x,x,x]; Y=[y,y,y]; Z=[z1,z2,z3]; comet3(X,Y,Z); %
Animated plot
```

```
%% Alternative animated plot with ezplot figure(3);
ezplot3('cos(t)','sin(t)', '0.1*t', [0,10*pi], 'animate'); hold on
ezplot3('cos(t)', 'sin(t)', '0.2*t', [0, 10*pi], 'animate'),
ezplot3('cos(t)', 'sin(t)', '-0.1*t', [0, 10*pi], 'animate')
```

After executing the script called Plot_EX26.m, we obtain the plots displayed in Figure 6-37.

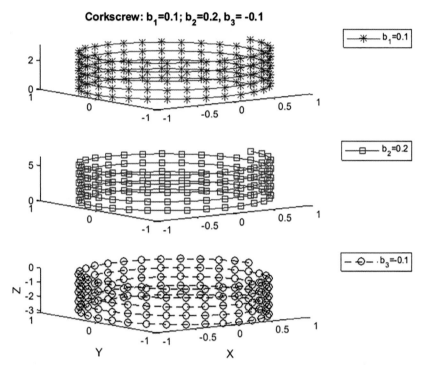

Figure 6-37. *Simulation plot of corkscrew amusement park rides*

With the 3D parametric curve plotter tools comet3 and ezplot3 shown in the previous script, we can obtain a simulation of the corkscrew amusement park ride in action. The comet3 and ezplot3 'animate' tools display simulated animation plots. There are also other 3D line and surface plot tools, including quiver, compass, feather, scatter3, stem3, and contour3.

Animated Plots

Plots can be animated with two functions, getframe and movie. There is also another tool called drawnow that can used to demonstrate this simulation process. It is a straightforward way to employ these tools and can be incorporated with other plot tools. Let's look at several exercises to see how to use these functions.

Example 27: Building an Animated Plot with getframe()

Given a function: $f(x, y) = (|x|)tan(0.5y/180)$, $x = -13...13$, $y = -13...13$.

We employ the getframe and movie tools to obtain an animated plot with the following script (Plot_EX27.m):

```
% Plot_EX27.m
    for m=1:20; [x, y]=meshgrid(linspace(-13, 13, 200)); z=log10(abs(x)).*t
    an(m*y*pi/180);
      mesh(x, y, z)
    M(m)=getframe(gcf);
end
title({['3D surface-contour plot of the function' ],... ['f(x,
y)=log10(abs(x))tan(0.5*y*\pi/180)']});
xlabel('X-axis'); ylabel('Y-axis');
zlabel('f(x, y) function values'); movie(M,2) % The plot movie is
played twice
```

Example 28: Building an Animated Plot with drawnow

Given a function: $f(t) = sin(t)\sqrt{|cos(t)|}$, $g(t) = cos(t)\sqrt{|sin(t)|}$, $x = 0...3\pi$.

For this example, we employ the drawnow function and write the following script (Plot_EX28.m) to obtain a 2D animated plot:

```
% Plot_EX28.m
for t=linspace(-2*pi, 2*pi, 200) f=sin(t)*sqrt(abs(cos(t)));
g=cos(t)*sqrt(abs(sin(t))); drawnow plot(t,f, 'o-',t, g, 'mp--');
hold all % used for colorful markers
end
```

```
title('Animated plot of the function')
xlabel('X-axis'); ylabel('Y-axis'); legend('f(t)', 'g(t)')
```

The drawnow command works very well with the plot (property) handle and the refreshdata commands.

Example 29: Building an Animated Plot with drawnow

Given a function: $y(t) = sinc\left(e^{sin(kt)}\right)$, $t = 0...5\pi$, $k = [1, 2]$.

```
% Plot_EX29.m
t= 0:pi/100:5*pi; y = sinc(exp(sin(t))); for k = 1:.01:2
    y = sinc(exp(sin(t.*k))); H = plot(t,y);H.LineStyle = ':'; H.Color =
[1 0 1];
    set(H, 'marker','o');
    set(H, 'markerfacecolor', [0 1 1]);
    set(H, 'markersize', 13); refreshdata(H,'caller') drawnow;
pause(.005), shg
end
```

Example 30: Building an Animated Plot of a Projectile with getframe()

The height and speed of a projectile (such as a thrown ball) launched at a speed of $v0$ at an angle θ to the horizontal are given by

$$h(t) = v_0 t sin(\theta) = 0.5gt^2$$

$$v(t) = \sqrt{v_0^2 - 2v_0 gt\, sin(\theta) + g^2 t^2}$$

Here, g is the acceleration due to gravity. The projectile will strike the ground when $h(t) = 0$, which gives the time to hit $t_{hit} = \dfrac{2v_0 sin(\theta)}{g}$. Suppose that $\theta = 30^0, v_0 = 40\dfrac{m}{s}$, and $g = 9.81\ m/s^2$. Let's animate the ball's trajectory with getframe (movie) and drawnow by choosing a sufficiently small step size for time.

Here is the solution script (plot_ex25.m):

```
% Plot_EX30.m
% Study the projectile trajectory and velocity clearvars; clc; close all
% Given data:
g=9.8; % acceleration due to Earth gravity in m/s^2 theta=30;   % angle to
the horizontal axis in degrees v0=40; % speed of the thrown ball in m/s
% t_hit is time needed for the thrown ball to hit the ground.
% h is height of the thrown ball relative to the ground in m.
% v is speed of the thrown ball over time.
% t is a time series for computation in [sec].
% dt is increment of total time. t_hit=2*v0*sin(theta*pi/180)/g; dt=t_
hit/80; t=(0:dt:t_hit); h=v0*t.*sin(theta*pi/180) - 0.5*g*t.^2;
v=sqrt(v0*v0-2*v0*g*t.*sin(theta*pi/180)+g*g*t.^2);
%% Animated plot of the projectile dt=t_hit/100; t=(0:dt:t_hit);
for ii=1:length(t)
    h(ii)=v0*t(ii).*sin(theta*pi/180) - 0.5*g*t(ii).^2; plot(t(ii), h(ii),
'o:', 'markerfacecolor', 'y') hold all
    M(ii)=getframe(gcf); end
% movie(M, 1)
%% Animated plot of the projectile velocity for ii=1:length(t)
v(ii)=sqrt(v0^2-2*v0*g*t(ii).*sin(theta*pi/180)+g^2*t(ii).^2); drawnow
plot(t(ii), v(ii), 'o:', 'markerfacecolor', 'c'), hold all end
```

Summary

This chapter discussed a few essential plot tools and functions. In particular, it demonstrated, with examples, how to employ the following plot commands and functions:

- 2D plot commands:

    - plot, plotyy, semilogx, semilogy, loglog, hist, bar,errorbar, title, axis, legend, grid, xlabel, ylabel, fplot, ezplot, xlim, and ylim

- 3D plot commands:
  - `ezsurf`, `surface`, `meshc`, `meshgrid`, `ribbon`, `pie3`, `waterfall`, `plot3`, `comet3`, `fsurf`, `surfc`,`contour`, `colormap`, `view`, and `zlabel`

- Additional plot-related commands:
  - `subplot`, `close all`, `axis`, `axis vis3d`, `set`, `gcf`, `grid`, `hold`, `axis`, `legend`, `clf`, `figure`, `title`, `text`, `gtext`, `leftarrow`, `ginput`,`fontsize`, `shg`, and `so forth`.

- Animated plot commands:
  - `getframe`, `movie`, `refreshdata`, and `drawnow`

Here are a few of the most common errors that occur while plotting:

- One of the most common errors made while plotting measured or computed data is when you try to plot two data sets with different sizes (e.g., as input versus output). A plot cannot be obtained because of the mismatched data points. Therefore, you need to be careful about the size of your data sets when plotting them in an x versus y plot.

- You can sometimes confuse or do not assign legends correctly, and as a result, you will get incorrect information from your plotted data sets.

- When you are dealing with large data sets composed of many rows and columns of data, you can sometimes confuse rows with columns, and vice versa.

- Sometimes you will try to plot complex numbers; however, MATLAB plots only the real part of complex numbers unless you plot them separately.

Exercises for Self-Testing

Exercise 1

Plot y versus $f(y)$ of the following polynomial for $y = -3 \dots 2$:
$$f(y) = y5 + 5y4 + 3y3 - 10y2 - (10^{0.5}) \text{ for } y = -3 \dots 2$$

1. Create an empty plot figure.

2. Plot y versus $f(y)$ with a 1.0 width solid line (style) in blue and with diamond markers in yellow using plot() by taking $\Delta y = 0.25$.

3. Hold on to the plot figure from step 2.

4. Plot y versus $f(y)$ with a 2.0 width dashed line (style) in magenta using fplot().

5. Add a plot title of "$y^5 + 5y^4 + 3y^3 - 10y^2 - (10^{0.5}) = f(y)$," axis labels, and grids.

Exercise 2

Plot the polynomial given in Exercise 1 using the fplot function.

1. Find the minimum value (f_{min}) of the polynomial by using ginput.

2. Plot the found values (minimums) with a diamond marker and circle in blue and red, respectively.

Exercise 3

Given:

$$f_1(y) = y5 + 5y4 + 3y3 - 10y2 - 10^{0.5}$$

$$f_2(y) = 0.001\, y5\, e^5 + 0.01 y4\, e^4 + 0.0125 y3\, e^3 - 0.0125 y2\, e^2 - 0.1 e^{0.5}$$

1.   Plot the two functions $f_1(y)$ and $f_2(y)$ in one plot area for $-3 \leq y \leq 2$ by using the fplot and plot functions. Compare the results of the two approaches. Insert all the necessary information (such as plot title, axis label, grid, line width, marker type, color, and so forth) to make the plot legible and informative for analysis.

2.   Plot these two functions: y versus $f_1(y)$, and y versus $f_2(y)$ in one plot area for $-1 \leq y \leq 1$ by using plotyy or yaxis right/left so that $f_1(y)$ and $f_2(y)$ are in two separate vertical axes.

3.   Locate the local minima of both plotted $f_1(y)$ and $f_2(y)$ functions in step 2 using ginput.

4.   Plot the local minimum values found in step 3 in the plot of step 2.

Exercise 4

A cable of length L_c supports a beam of length L_b so that it is horizontal when the weight W is attached to the beam end. The tension force T in the cable is given by

$T = L_c L_b \, W / D \sqrt{L_b^2 - D^2}$, where D is the distance of the cable attachment point to the

beam pivot.

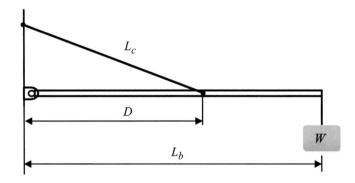

1.   Use the $W = 777\ N$, $L_b = 2.33\ m$, and $L_c = 3.77\ m$ element-wise operations and the function min() (hint: min(X, Y)) to compute the value of D that minimizes the tension value.

2. Check the sensitivity of the solution by plotting T versus D. How much can D vary from its optimal value before the tension T increases 13 percent above its minimum value?

Exercise 5

1. Use MATLAB to solve the following equations for x, y, and z as functions of the parameter c:

$$x - 5y - 2z = 11c$$

$$6x + 3y + z = 13c$$

$$7x + 3y - 5z = 10c$$

2. Plot the solutions for x, y, and z versus c on the same plot, for $-10 \leq c \leq 10$. Note that the incremental change of c is 1.

Exercise 6

Plot the following two polynomials over the interval of $-6 \leq x \leq 13$.

$$f(x) = 0.003x^4 - 1.5x^3 - 13x^2 - 0.08x + 13;$$

$$h(x) = -0.03x^3 - 1.5x^2 - 0.04x + 9.0;$$

1. Plot $f(x)$ versus x in a subplot 1 using fplot() and put a grid on the plot.

2. Plot $h(x)$ versus x in a subplot 2 using plot() with $\Delta x = 0.25$ and put a grid on the plot.

3. Using the ginput function, determine the coordinates of the peaks of f(x) and h(x) and add this information using text and gtext.

Exercise 7

Compute the following formulation: $h(\alpha) = sin\ sin\ (\alpha)/\alpha$, where $\alpha = 0 : \pi/40 : \pi$. Compute the values of π from the following formulations—(a) developed by Srinivasa Ramanujan in 1910 and (b) developed by Borwein-Plouffe in 1995:

$$\frac{1}{\pi} = \frac{2\sqrt{2}}{9801} \sum_{k=0}^{m} \frac{(4k)!1103 + 26390k}{(k!)^4 396^{4k}} \qquad (a)$$

$$\pi = \sum_{k=0}^{n} \frac{1}{16^k} \left[\frac{4}{8k+1} - \frac{2}{8k+4} - \frac{1}{8k+5} - \frac{1}{8k+6} \right] \qquad (b)$$

1. Compute π from (a) for m = 100; plot α versus $h(\alpha)$

2. Compute π from (b) for n = 10; plot α versus $h(\alpha)$

3. Build animated plots from steps 1 and 2 using `drawnow` and `getframe + movie`.

Exercise 8

The volume V and paper surface area A of a conical paper cup are given by

$$V = \frac{1}{3}\pi r^2 h \text{ and } A = \pi r \sqrt{r^2 + h^2}$$

Here, r is the radius of the base of the cone, and h is the height of the cone.

1. By eliminating h, obtain the expression for A as a function of r and V.

2. Create a user-defined function that accepts r as the only argument and computes A for a given value of V. Declare V to be global within the function.

Exercise 9

A torus is shaped like a doughnut. If its inner radius is a and its outer radius is b, its volume and surface area are given as follows:

$$V = \frac{1}{4}\pi^2(a+b)(b^2 - a^2)$$

$$A = \pi^2(b^2 - a^2)$$

1. Create a user-defined function that computes V and A from the arguments a and b.

2. Suppose that the outer radius is constrained to be 2 inches greater than the inner radius. Write a script file that uses your function to plot V and A versus a for $0.25 \le a \le 4$ [in].

Exercise 10

Create four anonymous functions to represent the function $6e^{3\cos x^2}$, which is composed of the functions $h(z) = 6e^z$, $g(y) = 3 \cos y$, and $f(x) = x^2$. Use `fplot` to plot $e^{3\cos x^2}$ over the range of $0 \le x \le 4$.

Exercise 11

Create an anonymous function for $20\, x^2 - 200x + 3$ and use it to plot the function in order to determine the approximate location of its minimum using `ginput`.

Exercise 12

Find the approximate roots of the equation $x^3 - 3x^2 + 5x$ with `fplot` and then use `ginput`.

Exercise 13

To compute the forces in structures, sometimes you must solve equations similar to the following. Use the `fplot` function to find all positive roots of this equation: $x \tan x = 7$.

Exercise 14

Cables are used to suspend bridge decks and other structures. If a heavy uniform cable hangs suspended from its two endpoints, it takes the shape of a catenary curve whose equation is $y = (x/a)$, where a is the height of the lowest point on the chain above some horizontal reference line, x is the horizontal coordinate measured to the right from the lowest point, and y is the vertical coordinate measured up from the reference line.

1. Let $a = 10 \; m$ and plot the catenary curve for $-20 \le x \le 30 \; [m]$. How high is each endpoint?

2. Let $a = 10 \; m$ with an increment of 1m and plot the catenary curve for $-20 \le x \le 30 \; [m]$ using the loop control statements and automatic labeling tools of plotted data points with strings.

3. Build an animated plot of the computation results from step 2 using `drawnow`.

Exercise 15

When a belt is wrapped around a cylinder, the relation between the belt forces on each side of the cylinder is as follows: $F_1 = F_2 e^{\mu \beta}$.

Here, β is the angle of the wrap of the belt and μ is the friction coefficient. Write a script file that first prompts a user to specify β, μ, and F2 and then computes the force, F1. Test your program with the values $\beta = 130^0$, $\mu = 0.3$, and F2 = 100 N. The output force (F1) has to be output in N in the Command window. (Hint: Be careful with β!)

1. Plot F1 versus β for $\beta = 30^0...360^0$ for 30 linearly spaced data points.

2. Add the plot title, axis labels, marker type (pentagram in cyan), and marker size (8.0), and then set a line width at 1.50.

3. Add text to the plot area "*F1* is dependent on wrap angle β!" with a font size of 13.

4. Build an animated plot of the output force $(F1)$ for $\beta = 30^0...360^0$ for 100 linearly spaced data points, $\mu = 0.1...0.3$ and $F2 = 100\ N$.

Exercise 16

Using estimates of rainfall, evaporation, and water consumption, the town engineer developed the following model of the water volume in the reservoir as a function of time:

$$V(t) = 10^9 + 10^8\left(1 - e^{-\frac{t}{100}}\right) - 10^7 t$$

Here, V is the water volume in liters and t is time in days. Plot $V(t)$ versus t. Use the plot to estimate how many days it will take before the water volume in the reservoir is 50 percent of its initial volume, 10^9 liters.

Exercise 17

Plot columns 2 and 3 of the following matrix A versus column 1. The data in column 1 is time (seconds). The data in columns 2 and 3 is force in N. Use the matrix axis mode to display the y-axis values in reverse order.

$$A = \begin{bmatrix} 0 & -8 & 6 \\ 5 & -4 & 3 \\ 10 & -1 & 1 \\ 15 & 1 & 0 \\ 20 & 2 & -1 \end{bmatrix}$$

Exercise 18

In certain kinds of structural vibrations, periodic force acting on the structure will cause the vibration amplitude to repeatedly increase or decrease with time. This phenomenon, called *beating*, also occurs in musical sounds. A particular structure's displacement is described by $y(t) = \dfrac{1}{f_1^2 - f_2^2}\left(\cos\left(f_2 t\right) - \cos\left(f_1 t\right)\right)$.

Here, y is the displacement in inches, and t is the time in seconds. Plot y versus t over the range $0 \leq t \leq 11\pi$ for $f_1 = 13$ rad/sec and $f_2 = 1.5$ rad/ sec . Make sure to select enough points to obtain an accurate plot of the process.

Exercise 19

A robot rotates around its base at two revolutions per minute while lowering its arm and extending its hand. It lowers its arm at the rate of 120^0 per minute and extends its hand at the rate of 5 m/min. The arm is 0.5 [m] long. The x-y-z coordinates of the hand are given as follows:

$$x = (0.5 + 5t) \sin\left(\frac{2\pi t}{3}\right) \cos(4\pi t)$$

$$y = (0.5 + 5t) \sin\left(\frac{2\pi t}{3}\right) \sin(4\pi t)$$

$$z = (0.5 + 5t) \cos\left(\frac{2\pi t}{3}\right)$$

Here, t is time in minutes.

1. Obtain a 3D plot of the path of the hand for $0 \leq t \leq 0.2$ [min].

2. Simulate the arm's trajectory using `comet3` and `drawnow` and analyze which tool results in better visualization.

Exercise 20

Obtain surface and contour plots for the function $z = 9x2 + 2\ xy + 3y2$. This surface has the shape of a saddle. At its saddle-point at $x = y = 0$, the surface has zero slope, but this point does not correspond to either a minimum or a maximum. What types of contour lines correspond to the saddle-point?

Exercise 21

The following function describes oscillations in some mechanical structures and electric circuits:

$$z(t,\tau) = e^{-\frac{t}{\tau}} \sin(\omega t + \varphi)$$

In this function, t is time, and ω is the oscillation frequency in radians per unit time. The oscillations have a period of $\frac{2\pi}{\omega}$, and their amplitudes decay in time at a rate determined by τ, which is called the *time constant*. The smaller τ is, the faster the oscillations die out.

Suppose that $\varphi = 0$, $\omega = 2.5$, and τ can have values in the range of $0.5 \le \tau \le 10$ *sec*. Then the proceeding equation becomes the following:

$$z(t,\tau) = e^{-\frac{t}{\tau}} \sin(\omega t)$$

Obtain a surface plot with `waterfall` and a contour plot of this function with `contour` to help visualize the effect of τ for $0.5 \le t \le 10$ sec. Label the axes by function name $z(t, \tau)$, with variable names of t and τ.

Exercise 22

The following equation describes the temperature distribution in a flat rectangular metal plate. The temperature on three sides is held constant at T_1 and T_2 on the fourth side (see the following figure). The temperature $T(x, y)$ as a function of the x-y coordinates shown is given as follows:

$$T(x,y) = (T_2 - T_1) w(x,y) + T_1$$

Here, $w(x,y) = \dfrac{2}{\pi} \sum_{n=1}^{\infty} \dfrac{\frac{2}{n} \sin\left(\frac{n\pi x}{L}\right) \sinh(n\pi y)}{\sinh\left(\frac{n\pi W}{L}\right)}$.

The given data for this problem is $T_1 = 22^\circ C$, $T_2 = 75^\circ C$, $W = L = 3$ [m].

Using a spacing of 0.05 for x and y, generate a surface mesh plot and a contour plot of the temperature distribution.

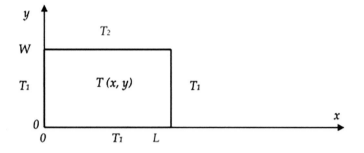

Exercise 23

Given a function $f(x, y) = (|x|)\, tan\,(0.5y/180)$, create a 3D plot of $f(x, y)$ by using meshc and ribbon.

Exercise 24

Given a function $H(\theta, \varphi) = log\,10(|\theta|)\, tan\,(0.5\varphi)$, $\theta = -\pi...\pi$; $\varphi = -\pi...\pi$, do the following:

1. Plot the given function $H(\theta, \varphi)$ by using the 3D surface and waterfall plot functions on two subplots.

2. Build a 3D animated plot of $H(\theta, \varphi) = log10(|\theta|)\, tan\,(a * \varphi)$, with $\theta = -\pi...\pi$; $\varphi = -\pi...\pi$; and $a = -1.5...1.5$.

Exercise 25

Given these functions:

$h_1 = cos\,(2t)$; $h_2 = sin\,(3t)$; $h_3 = cosh\,(t/10)$; $t = 0...8\pi$, $\Delta t = \pi/10$

Do the followings:

– Plot the given functions with plot3 and comet3.

– Build an animated plot of the given function by using getframe and movie.

– Build an animated plot of the given function by using drawnow.

CHAPTER 7

Linear Algebra

This chapter introduces linear algebra. It discusses some of the essential approaches to solving systems of linear equations, as well as various matrix operations (matrix inverse, determinant, sum, subtraction, division, multiplication, power, exponential, elementwise and array-wise operations, and so forth). It covers eigen-value problems and matrix factorizations/decompositions, such as Cholesky, Schur, LU, QR, and singular value decomposition. It also includes built-in functions and scripts in MATLAB and Simulink models. Moreover, the chapter explains the standard matrix generator functions of MATLAB, how to create vector spaces, how to solve polynomials, and the logical indexing of matrices, all via examples in MATLAB and Simulink.

Introduction to Linear Algebra

Linear algebra is one of the more important branches of mathematics. It deals with vectors, vector spaces, linear spaces, matrices, and systems of linear equations. There is a wide range of linear algebra applications in engineering and scientific computing, including many fields of natural and social studies. Linear algebra starts with a system of linear equations for underdetermined, overdetermined, and well-defined systems.

If a given system is composed of m-linear equations with n-unknowns and $m \geq n$, that is solvable for unknowns. Consider the following linear system, formulated by the system of equations (Equation 7-1):

$$\begin{cases} a_{11}x_1 + a_{12}x_2 + .. + a_{1n}x_n = b_1 \\ : \quad + : \quad + \quad \quad : \quad = : \\ a_{m1}x_1 + a_{m2}x_2 + .. + a_{mn}x_n = b_m \end{cases} \qquad \text{(Equation 7-1)}$$

The system of linear equations (Equation 7-1) is solvable directly for all cases when $m \geq n$. If $m < n$, there are more unknowns than the number of linearly independent equations, and such a system is called *underdetermined* and not solvable directly.

443

© Sulaymon Eshkabilov 2022
S. Eshkabilov, *Beginning MATLAB and Simulink*, https://doi.org/10.1007/978-1-4842-8748-4_7

If $m > n$, there are more linearly independent equations, and such a system is called *overdetermined* and is solvable directly.

For the sake of simplicity, let's take $m = n$ and rewrite Equation 7-1.

$$\begin{cases} a_{11}x_1 + a_{12}x_2 + .. + a_{1n}x_n = b_1 \\ \ : \quad + \ : \quad + \ : \ =: \\ a_{n1}x_1 + a_{n2}x_2 + .. + a_{nn}x_n = b_n \end{cases} \qquad \text{(Equation 7-2)}$$

The given system of linear equations in Equation 7-2 can also be written in matrix notation form.

$$[A]*\{X\} = [B] \qquad \text{(Equation 7-3)}$$

Here, A and B are matrices and X is a vector of unknowns.

$$\text{Where } [A] = \begin{bmatrix} a_{11} & \cdots & a_{1n} \\ : & : & : \\ a_{n1} & \cdots & a_{nn} \end{bmatrix}, \ \{X\} = \{x_1, x_2, \ldots, x_n\}, [B] = \begin{bmatrix} b_1 \\ : \\ b_n \end{bmatrix}$$

Equation 7-3 can also be rewritten in the form of column matrices.

$$x_1 \begin{bmatrix} a_{11} \\ : \\ a_{n1} \end{bmatrix} + x_2 \begin{bmatrix} a_{12} \\ : \\ a_{n2} \end{bmatrix} + \ldots + x_n \begin{bmatrix} a_{1n} \\ : \\ a_{nn} \end{bmatrix} = \begin{bmatrix} b_1 \\ : \\ b_n \end{bmatrix} \qquad \text{(Equation 7-4)}$$

The system in Equation 7-3 or 7-4 can be solved for X (unknowns) with the next formulation:

$$\{X\} = [A]^{-1}*[B] \qquad \text{(Equation 7-5)}$$

Here, $[A]^{-1}$ is the inverse of the matrix [A].

Matrix Properties and Operators

Matrices have several important properties and operators, such as determinant, diagonal, transpose, inverse, singularity, rank, and so forth.

The *determinant* of a matrix can be computed only if the given matrix is a square. Here's an example:

$$M = \begin{bmatrix} a & b & c \\ d & e & f \\ g & h & i \end{bmatrix}$$

The determinant of M will be computed with the following expression:

$$det(M) = aei + bfg + dhc - ceg - dbi - hfa$$

The MATLAB command for the determinant computation is det(). Here's an example:

```
>> A=[  8   1   6;   3   5   7;   4   9   2]
A =

     8     1     6
     3     5     7
     4     9     2
>> det(A)
ans =
-360
```

The *diagonal* of a matrix is composed of its element along its diagonals. For example, in the previous example, the diagonals are *aei* and *ceg*.

The MATLAB command for diagonal separation is diag(). Here's an example:

```
>> A = [  8   1   6;   3   5   7;   4   9   2];
>> diag(A)
ans =
     8
     5
     2
```

The *transpose* of a matrix can be determined by the counterclockwise rotation of a matrix by 90^0 (degrees). The transpose properties are as follows:

$$\left(M^T\right)^T = M$$

$$\left(M + B\right)^T = M^T + B^T$$

$$\left(kM\right)^T = kM^T$$

$$\left(MB\right)^T = B^T M^T$$

$$\left(M^{-1}\right)^T = \left(M^T\right)^{-1}$$

Here, M and B are matrices of the same size, k is a scalar, and ^{T} and $^{-1}$ are the transpose and inverse operators.

The MATLAB command for the transpose operation is transpose(), or ' .

Here's an example:

```
>> A =[   8   1   6;   3   5   7;   4   9   2];
>> transpose(A)
ans =

      8      3      4
      1      5      9
      6      7      2
>> A'
ans =

      8      3      4
      1      5      9
      6      7      2
```

Simulink Blocks for Matrix Determinant, Diagonal Extraction, and Transpose

Simulink has blocks that you can use to compute the matrix determinant, extract the matrix diagonal elements, and obtain the matrix transpose. The determinant block ([det(A) (3x3)]) is present in Simulink's Aerospace Blockset/Utilities/Math Operations, and it has a constraint and can only compute the determinant of 3-by-3 matrices.

Note   The block [det(A) (3x3)] from the aerospace blockset is limited; it can only compute the determinant of 3-by-3 matrices.

The block to extract the diagonal elements of a matrix is available in the DSP System Toolbox/MATH Functions/Matrices and Linear Algebra/Matrix Operations. The block to compute the matrix transpose is present in Simulink/Math Operations, and the block name is Math Function. It has a few math functions embedded in it, including exp (by default), log, 10^u, magnitude^2, square, pow, and transpose. Any of these math functions in the Math Function block can be chosen. You simply click the Apply and OK buttons of the block, and the chosen math function becomes available. Figure 7-1 shows these three blocks.

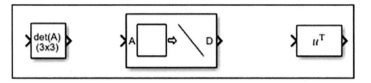

Figure 7-1. *Simulink blocks used for determinant calculation, diagonal extraction, and transpose operation, from left*

These blocks have one input and one output port. Therefore, you need to add two additional blocks, specifically, one Constant block for input entry and one Display block, to obtain/see the computation results. The Constant block can be taken from the Simulink Library Simulink/Sources or DSP System Toolbox/Sources. Similarly, the Display block can be taken from Simulink/Sinks or DSP System Toolbox/Sinks. Alternatively, with the latest versions of MATLAB starting from 2018a, you can obtain all the necessary blocks by double-clicking (with the left mouse button) and typing the block name in the search box. As discussed in the previous chapters, in any Simulink model one signal source can be used as many times as necessary. There is no need to generate that signal within one model to use it with other blocks as an input signal. Moreover, to optimize the Simulink model, it is strongly advised you build a Simulink model with fewer blocks to make your models more readable, comprehensive, and easy to edit. Therefore, this example uses one Constant block for input source [A]. Figure 7-2 shows the primary version of the Simulink model.

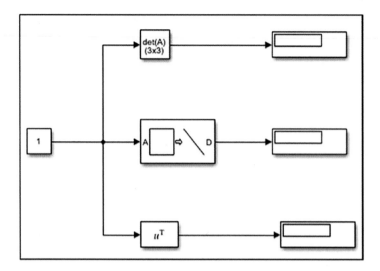

Figure 7-2. *Simulink model to compute the determinant of a matrix, extract diagonal elements of a matrix, and perform a transpose on a matrix*

Let's use example matrix [A] to demonstrate these three Simulink blocks. The elements of the matrices [A] can be entered in two different ways:

- By typing all elements in the Constant block's Constant Value box, as shown in Figure 7-3. Click the Apply and OK buttons.

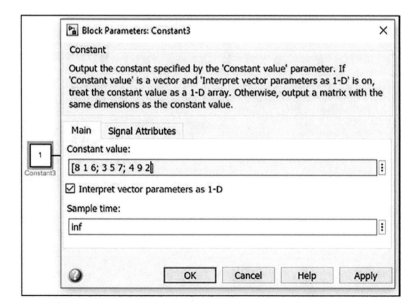

Figure 7-3. *Entering matrix elements in a Constant block*

- By defining [A] via MATLAB's Command window and workspace:

```
>> A =[  8   1   6;   3   5   7;   4   9   2];
```

Then provide the variable name A in the Constant block's Constant Value box for [A], as shown in Figure 7-4.

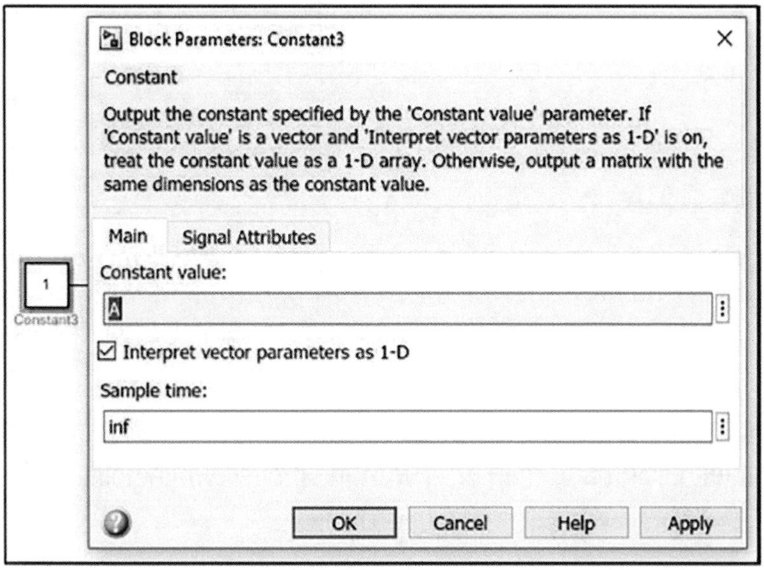

Figure 7-4. *Matrix [A], defined in the MATLAB workspace, called via the Constant block*

Then click the Apply and OK buttons. Note that we are not going to use the second method (see Figure 7-4) of defining matrix [A] elements in this example; it's just shown here for explanation purposes.

Finally, you'll get the complete model in which the matrix [A] elements are entered in the Constant block directly, as shown in Figure 7-5. After you complete the model, by pressing Ctrl+T on the keyboard or clicking the Run ⊙ button in the Simulink model window, the complete model with its computed results will be created.

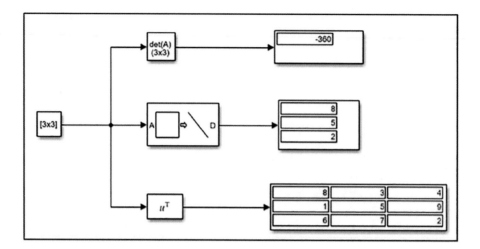

Figure 7-5. *Completed Simulink model that computes the determinant, extracts diagonal elements, and performs the transpose operation on the 3-by-3 matrix*

Note   To see the simulation results in the Display block, it has to be resized/ stretched. You left click it and then drag with the mouse while holding the button.

The simulation results of the Simulink models match the ones from the MATLAB commands, such as det(), diag(), and transpose(), or '.

Matrix Inverse or Inverse Matrix

The *inverse matrix* has the following important property:

$$[A]*[A]^{-1}=[I]$$

Here, [*I*] is the identity matrix.

For example, $A = \begin{bmatrix} 1 & 1 \\ 3 & 4 \end{bmatrix}$ has its inverse $A^{-1} = \begin{bmatrix} 4 & -1 \\ -3 & 1 \end{bmatrix}$ that is computed from the following:

$$A^{-1} = \frac{1}{\det(A)} adjugate(A) = 1/(4*3-(-1*-3))* \begin{bmatrix} 4 & -1 \\ -3 & 1 \end{bmatrix}$$

The MATLAB command to compute the inverse of a matrix is inv(). Here's an example:

```
>> A =[   8   1   6;   3   5   7;   4   9   2];
>> inv(A)
ans =
     0.1472   -0.1444    0.0639
    -0.0611    0.0222    0.1056
    -0.0194    0.1889   -0.1028
```

A given matrix is *singular* if it is square, if it does not have an inverse, and if it has a determinant of 0.

Simulink Blocks for Inverse Matrix

The *matrix inverse* can also be calculated via several Simulink blocks with respect to a given matrix size, i.e., square matrix or rectangular. The inverse matrix or matrix inverse computing blocks are present in the DSP System and Aerospace Blockset Toolboxes of Simulink and can be accessed via the Simulink Library: the DSP System Toolbox/Math Functions/Matrices and Linear Algebra/Matrix Inverses, and the Aerospace Blockset/ Utilities/Math Operations. Let's test the available blocks of this toolbox to compute the inverse of the matrix [A] shown in the previous example. Open a blank Simulink model and drag and drop the block from the libraries of the DSP System and Aerospace Blockset Toolboxes shown in Figure 7-6.

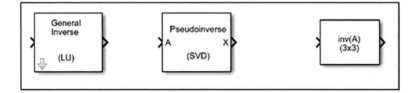

Figure 7-6. *Simulink blocks for computing the inverse matrix*

They are as indicated on the top of each block—General Inverse (LU), Pseudoinverse (SVD), and inv(A)—used to compute the matrix inverses based on LU factorization for square matrices, and pseudoinverse for rectangular matrices (i.e., m>n, or the number of rows is larger than the number of columns or vice versa). Theoretical aspects of the LU, SVD, and other matrix decomposition and transformation operations are highlighted in the "Matrix Decomposition" section.

The three blocks have one input port for the entry matrix and one output port for the computed inverse. Add two additional blocks—one Constant block and one Display—by following the procedures. Figure 7-7 shows the primary version of the Simulink model.

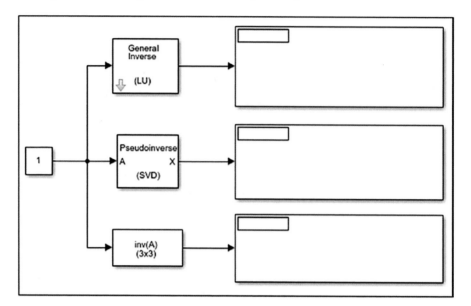

Figure 7-7. *Simulink model to compute the inverse matrix via three different blocks*

The elements of the matrices [A] can be entered in two ways: (1) by typing all the elements in the Constant block's Constant Value box and then clicking the Apply and OK buttons; or (2) by defining [A] via the MATLAB's Command window and workspace.

Finally, you'll get the following complete model in which the matrix [A] elements are entered in the Constant block directly. After you complete the model, by pressing Ctrl+T on the keyboard or clicking the Run ⊙ button in the Simulink model window, the finalized model with its computed results is created, as shown in Figure 7-8.

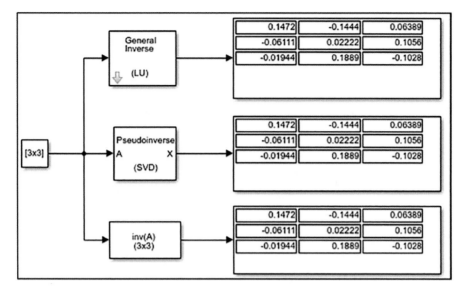

Figure 7-8. *The inverse matrix computed via three different blocks*

The computed inverse matrix (A^{-1}) values match the ones computed using MATLAB's inv() command, within four correct decimal places.

Another important operator of matrices is its *rank*. The *rank* of a matrix (e.g., [A]) is the maximum number of linearly independent row vectors of the matrix, which is the same as the maximum number of linearly independent column vectors. The [A] matrix is considered to have a full rank if its rank equals the largest possible for a matrix of the same dimensions. The [M] matrix is considered to be rank deficient if it does not have full rank. A matrix's rank determines how many linearly independent rows the system contains. The MATLAB command to compute the rank of a matrix is rank(). Here's an example:

```
>> A =[8,   1,   6;   3,   5,   7;   4,   9,   2];   % Full rank matrix
>> rank(A)
ans = 3
>> M  =[8   0   6;   -3,   0,   7;   0   0   2]   % Rank deficient matrix
M =

      8     0     6
     -3     0     7
      0     0     2
>> rank(M)

ans =

      2
```

Based on the rank, the systems (system matrices) can be full rank, overdetermined, and underdetermined.

Example 1: Solving a System of Linear Equations

The following example shows you how to solve a linear equation by using these formulations:

$$\begin{cases} 2x+3y+5z=1 \\ -3x-2y+5z=2 \\ 4x-7y+6z=3 \end{cases}$$

To solve this problem for unknowns, such as x, y, z, you apply Equations 7-3, 7-4, and 7-5 directly and then use the following operations:

$$\begin{bmatrix} 2 & 3 & 5 \\ -3 & -2 & 5 \\ 4 & -7 & 6 \end{bmatrix} * \begin{Bmatrix} x \\ y \\ z \end{Bmatrix} = \begin{bmatrix} 1 \\ 2 \\ 3 \end{bmatrix}$$

That can be written as follows:

$$\begin{Bmatrix} x \\ y \\ z \end{Bmatrix} = \begin{bmatrix} 2 & 3 & 5 \\ -3 & -2 & 5 \\ 4 & -7 & 6 \end{bmatrix}^{-1} * \begin{bmatrix} 1 \\ 2 \\ 3 \end{bmatrix}$$

$$\begin{Bmatrix} x \\ y \\ z \end{Bmatrix} \cong \begin{bmatrix} 0.0754 & -0.1738 & 0.0820 \\ 0.1246 & -0.0262 & -0.0820 \\ 0.0951 & 0.0852 & 0.0164 \end{bmatrix} * \begin{bmatrix} 1 \\ 2 \\ 3 \end{bmatrix} \cong \begin{bmatrix} -0.0262 \\ -0.1738 \\ 0.3148 \end{bmatrix}$$

$$Solution: \begin{Bmatrix} x \\ y \\ z \end{Bmatrix} \cong \begin{bmatrix} -0.0262 \\ -0.1738 \\ 0.3148 \end{bmatrix}$$

Let's solve this exercise using the reduced row echelon method in MATLAB.

```
% Step 1. Write an augmented matrix: AU = [A, b]
A = [2 3 5; -3 -2 5; 4 -7 6;]; b = [1;2;3];
AU=[A, b];
% Step 2. Row1 = Row1 - Row2
```

```
AU(1,:)=AU(1,:)-AU(2,:);
% Step 3. Row3 = Row3-4*Row1/5
AU(3,:)= AU(3,:)-4*AU(1,:)/5;
% Step 4. Row2 = Row2+3*Row1/5
AU(2,:)= AU(2,:)+3*AU(1,:)/5;
% Step 5. Row3 = Row3+11*Row2
AU(3,:)= AU(3,:)+11*AU(2,:);
% Step 6. Row2 = Row2+5*Row3/61
AU(2,:)= AU(2,:)-5*AU(3,:)/61;
% Step 7. Row1 = Row1/5-Row2
AU(1,:)= AU(1,:)/5-AU(2,:);
% Step 8. Row3 = Row3/61
AU(3,:)= AU(3,:)/61;
% Step 9. Solution:
x= AU(:, end)
x =
        -0.0262295081967213
        -0.173770491803279
         0.314754098360656
```

Alternative ways of solving this example include Gauss elimination and graphical methods. There are a number of operators and built-in functions in MATLAB that can be used to solve a linear system of equations. They are as follows:

- `inv()`, which computes the inverse of a given matrix or the pseudo-inverse of the given system (used for overdetermined systems).

- `\`, the backslash operator, which solves the system of linear equations directly. It's based on the Gaussian elimination method. This is one of the most powerful MATLAB operators (tools) for handling matrices.

- `mldivide()`, which is a built-in function similar to the \ backslash operator.

- `linsolve()`, which is a built-in function similar to the \ backslash operator.

- `lsqr()`, which is a built-in function based on the least squares method.

- `lu()`, which is a built-in function based on the Gauss elimination method.

- `rref()`, which is a built-in function based on the reduced row echelon method.

- `svd()`, which is a built-in function based on the singular value decomposition.

- `chol()`, which is a built-in function based on the Cholesky decomposition.

- `qr()`, which is a built-in function based on the orthogonal triangular decomposition.

- `decomposition()`, which is a built-in function that automatically choses the decomposition method.

- `bicg()`, `cgs()`, `gmres()`, `pcg()`, `symmlq()`, and `gmr()`, which are built-in functions that are based on gradient methods.

- `solve()`, which is a built-in function from the Symbolic MATH toolbox.

Note   Among these listed functions/commands and operators, some of them use the same computing algorithm and are alternatives to each other. For example, the \ backslash operator is an alternative to `mldivide()`.

First, denote the given system with the following notations:

$$A = \begin{bmatrix} 2 & 3 & 5 \\ -3 & -2 & 5 \\ 4 & -7 & 6 \end{bmatrix}, \ B = \begin{bmatrix} 1 \\ 2 \\ 3 \end{bmatrix}$$

The entries of $[A]$ matrix (coefficients of the unknowns x, y, z) are defined, and the elements of $[B]$ matrix are defined in the Command window.

```
>> A = [2 3, 5; -3, -2, 5; 4, -7, 6]
A =
      2     3     5
     -3    -2     5
      4    -7     6
```

```
>> B = [1;2;3]
B =
    1
    2
    3
```

Using inv() and (*), we can compute the solutions of the system.

```
>>Ai=inv(A)    %  [B] matrix is an  inverse  matrix  of [A]  matrix.
Ai =0.0754    -0.1738    0.0820
0.1246    -0.0262    -0.0820
0.0951    0.0852    0.0164
>> XYZ1=Ai*B    %   Solutions of the problem
Ai =-0.0262
-0.1738
0.3148
```

The next example uses the backslash \ operator based on the Gaussian elimination method. This approach is quite simple and efficient in terms of computation time.

```
>> XYZ2=A\B
Ai=-0.0262
-0.1738
0.3148
```

Using mldivide():

```
>>XYZ3=mldivide(A,B)
-0.0262
-0.1738
0.3148
```

Using linsolve():

```
>>XYZ4=linsolve(A,B)
-0.0262
-0.1738
0.3148
```

   Using lsqr():

```
>>XYZ5=lsqr(A,B)
lsqr converged at iteration 3 to  a  solution  with  relative
residual  6.6e-17.
-0.0262
-0.1738
0.3148
```

   Using lu():

```
>>[L, U, P] = lu(A); %L-lower; U-upper triangular; P-Permutation matrix
>> y = L\(P*B);
>> XYZ6 =  U\y
XYZ6 =
-0.0262
-0.1738
0.3148
```

   Using rref():

```
>>   MA   = [A,  B];   % Augmented matrix
>> xyz = rref(MA);
>> XYZ7= xyz(:,end)
-0.0262
-0.1738
0.3148
```

   Using svd() and inv():

```
>> [U, S, V]= svd(A);
>> XYZ8 = V*inv(S)*U'*B
-0.0262
-0.1738
0.3148
```

Using chol():

```
>>   [U,  L]  =  chol(A);    % A has to be Hermitian positive definite
>>    XYZ9  =  U\(U'\B)    % U'*U = A
-0.0262
-0.1738
0.3148
```

Using qr():

```
>> [Q, R] = qr(A);
>>   XYZ10    = R\Q.'*B
-0.0262
-0.1738
0.3148
```

Using decomposition():

```
>> XYZ11 = decomposition(A)\B
-0.0262
-0.1738
0.3148
```

Using bicg() gradient methods:

```
>> XYZ12 = bicg(A, B)
bicg converged at iteration 3 to a solution with relative  residual  3.1e-14.
-0.0262
-0.1738
0.3148
```

Using solve(), which is a Symbolic Math Toolbox function:

```
>> syms x y z
>> sol=solve(2*x+3*y+5*z-1, -3*x-2*y+5*z-2, 4*x-7*y+6*z-3);
>> XYZ13=[sol.x; sol.y; sol.z]
-8/305
-53/305 96/305
```

```
>> XYZ13=double([sol.x; sol.y; sol.z])
-0.0262
-0.1738
0.3148
```

All of the computed solutions are accurate within four decimal places of the employed operators and functions. In fact, the accuracy of the solutions and the computation time of each operator or function will differ. For instance, the inverse matrix calculation is not only costly in terms of computation time but is also less accurate. Moreover, among the studied methods, the last function of the Symbolic Math Toolbox, solve(), is the slowest and least efficient method.

Note   The decomposition() function is available in the recent versions of MATLAB starting from MATLAB 2018b.

Simulink Modeling

In addition to the MATLAB commands demonstrated, Simulink has several blocks by which the linear system of equations, such as [A]{x} = [B], can be solved. All of the solver blocks are present in the DSP System Toolbox and can be accessed via the Simulink Library: the DSP System Toolbox/Math Functions/Matrices and Linear Algebra/Linear System Solvers. Let's test some of the blocks here to solve the previous example, called Example 1. Open a blank Simulink model and drag and drop the block from the DSP System Toolbox library, as shown in Figure 7-9.

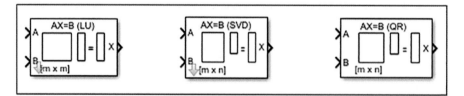

Figure 7-9. *Simulink blocks used to solve a system of linear equations*

They are as indicated on the top of each block—LU, SVD, QR factorization and decomposition operation-based solvers. All of them have two input ports for [A] and [B] and one output port for a solution, {x}. Therefore, you need to add three additional blocks—two Constant and one Display block—which you add as explained previously in building Simulink models to compute determinant, transpose, and inverse of matrices. Figure 7-10 shows the primary version of the Simulink model.

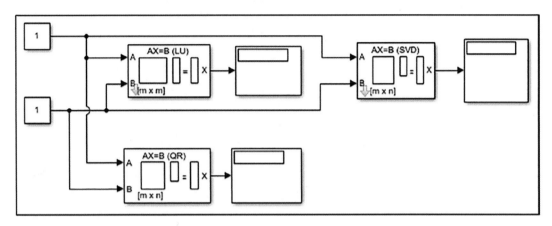

Figure 7-10. *Simulink model to solve a system of linear equations*

The elements of the matrices [A] and [B] can be inserted, as shown in Figure 7-3, directly in the Constant block's Constant Value window. Or you can define the elements of [A] and [B] via MATLAB's Command window and workspace.

```
>> A=[2, 3, 5; -3, -2, 5; 4, -7, 6]
>> B=[1; 2; 3];
```

The variable names A and B are then entered in the first and second Constant block's Constant Value box for [A] and [B], respectively, as shown in Figure 7-11. Click Apply and OK to complete the model.

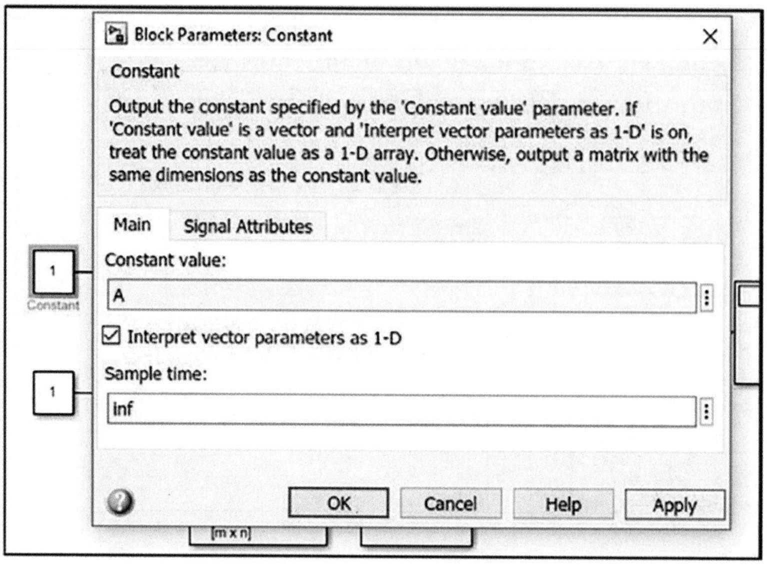

Block Parameters: Constant ✕

Constant

Output the constant specified by the 'Constant value' parameter. If 'Constant value' is a vector and 'Interpret vector parameters as 1-D' is on, treat the constant value as a 1-D array. Otherwise, output a matrix with the same dimensions as the constant value.

Main   Signal Attributes

Constant value:

A

☑ Interpret vector parameters as 1-D

Sample time:

inf

OK   Cancel   Help   Apply

Figure 7-11. *The variable names defined in the Constant block*

By pressing Ctrl+T on the keyboard or clicking the Run ⊙ button in the Simulink model window, you'll obtain the complete model with its simulation results (see Figure 7-12). The computed results/solutions match the MATLAB solutions to four decimal places.

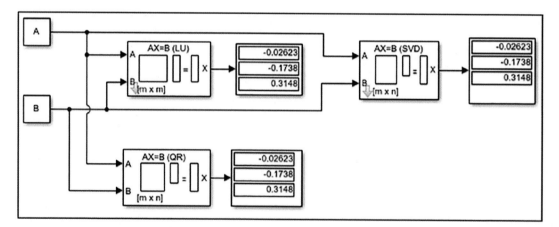

Figure 7-12. *Complete model with computed results*

Note that the variables (matrices) A and B are defined via MATLAB's Command window.

To obtain more decimal places of the computed results with the Display block, the block parameters (Format Type) need to be tuned by selecting long_e, as shown in Figure 7-13.

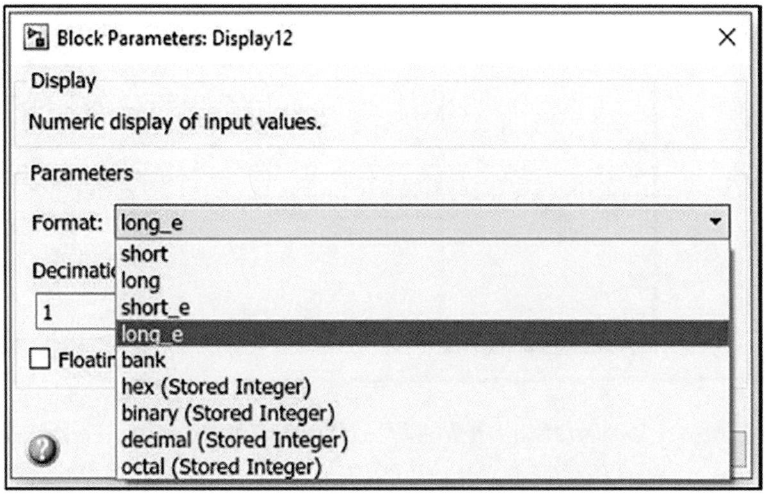

Figure 7-13. *Adjusting the Display block's Format parameter*

Example 2: Embedding a MATLAB Function Block to Compute the Determinant and Solve Linear Equations

All of the aforementioned MATLAB functions/commands used for computing matrix determinants, matrix inverses, or solutions of linear systems can be embedded in

Simulink via the MATLAB Function block  . Let's take two MATLAB functions/ commands used for computing a determinant of a matrix of any size with det() and solving with linsolve() and embed them into a Simulink model. Here's an example:

$$A1 = \begin{bmatrix} 16 & 2 & -3 & 13 \\ -5 & 11 & 10 & -8 \\ 9 & 7 & -6 & 12 \\ -4 & 14 & 15 & 1 \end{bmatrix}, B1 = \begin{bmatrix} 3 \\ 2 \\ 4 \\ 5 \end{bmatrix}$$

Here are completed Simulink models. Figure 7-14 is built with three Constant, two MATLAB Function, and two Display blocks.

463

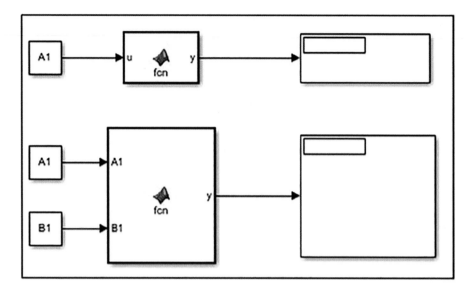

Figure 7-14. *Simulink models with MATLAB Function blocks to compute the determinant and solve a linear system of equations*

The input variables/entries for A1 and B1 are defined via the Command window and MATLAB workspace in this model. To edit and type in the necessary script, you have to open the MATLAB Function block. It can be opened by double-clicking it, which opens the MATLAB editor window. The following function file scripts for the MATLAB Function blocks are typed in the MATLAB editor for the upper MATLAB Function block (with one input) and the lower one (with two inputs A1 and B1) models, respectively. After editing the codes of the blocks, save them. They will be saved under the created Simulink model and not as a separate MATLAB function file.

```
function y  =  fcn(u)
y = det(u);
end

function y =  fcn(A1,  B1)
y = linsolve(A1, B1);
```

The model is then completed, and the finalized model is executed. Figure 7-15 shows the completed model with its computed results in the Display blocks. The upper Display block shows the determinant, and the lower one shows the solution of the given system.

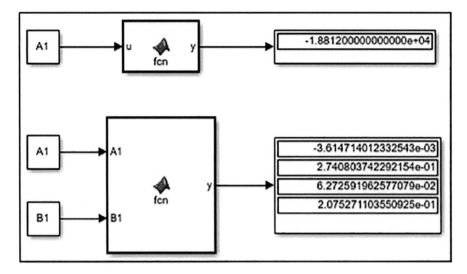

Figure 7-15. *Completed models with computed results*

The computed results of the Simulink model can be compared with MATLAB.

```
>> A1=[16 2 -3 13 ; -5 11 10 -8; 9 7 -6 12; -4 14 15 1 ];
>> B1 = [3; 2; 4; 5];
>> det(A1)
ans =
-18812

>> linsolve(A1, B1)
-3.614714012332536e-03 2.740803742292154e-01
6.272591962577077e-02
2.075271103550925e-01
```

The computed results from the determinant calculation and linear MATLAB solver match the Simulink model's results to 13 decimal places.

Example 3: Accuracy of Solver Functions of Linear Equations

Let's find out which one of the functions/tools (methods) highlighted in Example 1 is more accurate in computing the solutions. For this exercise, you'll take the following 13-by-13 [A] and 13-by-1 [B] matrices generated by the magic() and randi() (random

465

integer) matrix generator functions of MATLAB. Moreover, the norm() function is used to compute the norm of the given linear system with its computed solutions. LA_Ex3.m is the complete solution script.

```
%% Given 13-by-13 system of linear equations
A    =    magic(13);
B = randi([-169,169], 13,1); % Elements of B vary within [-169, 169]
%% 1-Way: inv()   or  pinv()   %% INVERSE matrix method
x1a = inv(A)*B; Err_INV  =   norm(A*x1a-B)/norm(B)     %#ok: ERROR checking
x1a = inv(A)*B; Err_PINV  =   norm(A*x1b-B)/norm(B)    %#ok: ERROR checking
%% 2-Way: \   %% backslash
x1a = inv(A)*B; Err_BACKSLASH = norm(A*x2-B)/norm(B)      %#ok: ERROR checking
%%  3-Way: mldivide()   %% Left divide function
x1a = inv(A)*B; Err_MLDIVIDE  =  norm(A*x3-B)/norm(B)    %#ok: ERROR checking
%% 4-Way: Using linsolve();
x1a = inv(A)*B; Err_LINSOLVE  =   norm(A*x4-B)/norm(B)       %#ok: ERROR
checking
%% 5-Way: Using lsqr()
x1a = inv(A)*B; Err_LSQR  =   norm(A*x5-B)/norm(B)           %#ok: ERROR
checking
%% 6-Way: Using lu()
x1a = inv(A)*B; y = L\(P*B); x6 = U\y;
Err_LU  =  norm(A*x6-B)/norm(B)   %#ok: ERROR checking
%% 7 - Way: Using rref()
x1a = inv(A)*B; xyz = rref(MA); x7= xyz(:,end);
Err_RREF  =   norm(A*x7-B)/norm(B)    %#ok: ERROR checking
%%   8 - Way: Using svd()
x1a = inv(A)*B; x8 = V*inv(S)*U'*B;
Err_SVD  =   norm(A*x8-B)/norm(B)    %#ok: ERROR checking
%% 9 - Way: Using chol()
x1a = inv(A)*B; x9 = U\(U'\B);
Err_CHOL  =   norm(A*x9-B)/norm(B)    %#ok: ERROR checking
%% 10 - Way: Using qr()
x1a = inv(A)*B; x10   =   R\Q.'*B;
Err_QR  =  norm(A*x10-B)/norm(B)   %#ok: ERROR checking
%% 11 - Way: Using decomposition()
```

```
x1a = inv(A)*B; Err_DECOMPOSITION = norm(A*x11-B)/norm(B) %#ok: ERROR checking
%% 12 - Way: Using bicg()
x1a = inv(A)*B; Err_BICG  =  norm(A*x12-B)/norm(B)   %#ok: ERROR checking
%% 13-Way: solve()   %% SOLVE() symbolic math method
x = sym('x', [1, 13]); x=x.'; Eqn = A*(x); Eqn = Eqn - B;
Solution = solve(Eqn); SOLs = struct2array(Solution); SOLs = double(SOLs);
 x13 = SOLs';
Err_SOLVE  =  norm(A*x13-B)/norm(B)   %#ok: ERROR checking
```

Here are the errors that were made while computing the solutions of the system with the employed methods:

```
Err_INV =
5.8087e-16
Err_PINV =
3.7982e-15 Err_BACKSLASH = 3.0569e-16 Err_MLDIVIDE = 3.0569e-16 Err_
LINSOLVE = 3.0569e-16
```

lsqr converged at iteration 7 to a solution with relative residual 3.5e-07. Err_LSQR =

```
3.4959e-07
Err_LU =
3.0569e-16
Err_RREF =
1.1576e-05
Err_SVD =
3.7982e-15
Err_CHOL  =
2.2400
Err_QR =
7.0615e-16
Err_DECOMPOSITION =
3.0569e-16
```

bicg stopped at iteration 13 without converging to the desired tolerance 1e-06 because the maximum number of iterations was reached.

The iterate returned (number 13) has relative residual 9.6e-06.

```
Err_BICG = 9.5856e-06
Err_SOLVE = 1.5109e-16
```

From the computed errors, it is clear that the RREF(), BICG(), and LSQR() functions make errors within the margin of $10^{-5}...10^{-7}$ and all other methods make errors within the margin of $10^{-15}...10^{-16}$ while computing the solutions of this given system.

Example 4: Efficiency of Solver Functions of Linear Equations

This example demonstrates which one of the shown ways is more efficient in terms of computation time. For this demonstration, you'll consider two large matrices of 1000-by-1000 and 1000-by-1, generated by the random integer number generator function randi() to generate the elements of matrices [A] and [B]. In addition, to record the elapsed time of each computation method, the [tic, toc] functions are used. Here is the complete solution script, called LA_Ex4.m:

```
clearvars
A=randi([-100,100],1000); B=randi([-100,  100],  1000,  1);
%% 1) inv() or pinv()
tic; Ai = inv(A); xyz1=Ai*B; T_inv=toc
%% 2) bacslash operator: \
clearvars
A=randi([-100,100],1000); B=randi([-100,  100],  1000,  1);
tic; xyz2 = A\B; T_backslash = toc
%% 3) mldivide()
clearvars
A=randi([-100,100],1000); B=randi([-100,  100],  1000,  1);
tic; xyz3= mldivide(A, B); T_mld = toc
%% 4) linsolve()
clearvars
A=randi([-100,100],1000); B=randi([-100,  100],  1000,  1);
tic; xyz4 = linsolve(A, B); T_linsolve =   toc
%% 5) lsqr()
clearvars
A=randi([-100,100],1000); B=randi([-100,  100],  1000,  1);
tic; xyz5 = lsqr(A, B); T_lsqr = toc
%% 6) lu()
clearvars
A=randi([-100,100],1000); B=randi([-100, 100], 1000, 1);
tic; [L, U, P]=lu(A); y=L\(P*B); xys6=U\y; T_lu=toc
```

```
%% 7) rref()
clearvars; A=randi([-100,100],1000); B=randi([-100, 100], 1000, 1);
tic; MA = [A, B];xyz7 = rref(MA); XYZ7=xyz7(:, end); T_rref=toc
%% 8) svd()
clearvars; A=randi([-100,100],1000); B=randi([-100, 100], 1000, 1);
tic; [U S V] = svd(A); xyz8 = V*inv(S)*U'*B; T_svd=toc
%% 9) chol()
clearvars; A=randi([-100,100],1000); B=randi([-100, 100], 1000, 1);
tic; [U L]= chol(A); xyz9 = U\(U'\B); T_chol=toc
%% 10) qr()
clearvars
A=randi([-100,100],1000); B=randi([-100, 100], 1000, 1);
tic; [Q R] = qr(A); xyz10 = R\Q.'*B ; T_qr=toc
%% 11) decomposition()
clearvars; A=randi([-100,100],1000); B=randi([-100, 100], 1000, 1);
tic; xyz11 = decomposition(A)\B; T_decom = toc
%%  12)  bicg()   Gradient methods
clearvars; A=randi([-100,100], 1000); B=randi([-100, 100], 1000, 1);
tic; xyz12 = bicg(A, B); T_bicg=toc
%% 13) solve()
A=randi([-100,100],100); B=randi([-100, 100], 100, 1);
tic;
x = sym('x', [1, 100]); x=x.';
Eqn = A*(x); Eqn = Eqn - B;

Solution = solve(Eqn); SOLs = struct2array(Solution); SOLs = double(SOLs);
x13 = SOLs';
T_solve=toc
```

Here are the elapsed computation time values from the simulations:

```
T_inv =
0.0390
T_backslash = 0.0173
T_mld =
0.0171
T_linsolve =
0.0171
```

lsqr stopped at iteration 20 without converging to the desired tolerance 1e-06 because the maximum number of iterations was reached.

The iterate returned (number 20) has relative residual 0.24.

```
T_lsqr = 0.0236
T_lu =
0.0235
T_rref =
10.1406
T_svd =
0.4263
T_chol =
0.0330
T_qr =
0.1045
T_decom =
0.0459
```

bicg stopped at iteration 20 without converging to the desired tolerance 1e-06 because the maximum number of iterations was reached.

The iterate returned (number 0) has a relative residual of 1.

```
T_bicg =
0.0195
T_solve =
14.6306
```

From these computations, it is clear that linsolve(), mldivide, and \ (the backslash operator) (Gaussian elimination method) are the fastest among all the tested methods. The slowest and computationally costliest one is the solve() operator of the Symbolic MATH even when the size of the system was 10 times smaller. It is worth noting that the reduced row echelon method called rref() is the next slowest, after the solve() operator.

Let's consider another example to solve these four different methods, which are \, linsolve(), inv(), and solve(), discussed previously.

Example 5: Solving Linear Equations ([A]{x} = [b]) by Changing Values of [b]

This exercise is composed of two parts:

[1]. Solve the given linear system for unknowns a, b, and c.

$$\begin{cases} -0.072a - c = -12 \\ 0.12b - c = -9 \\ a + b = 50 \end{cases}$$

[2]. Solve the given system for unknowns a, b, and c. The third equation's value changes in the range of 50...250.

$$\begin{cases} -0.072a - c = -12 \\ 0.12b - c = -9 \\ a + b = 50...250 \end{cases}$$

The system is rewritten in a matrix form as $[A]\{x\} = [B]$ and then solved directly for unknowns a, b, and c. Here is the solution script (LA_Ex4.m):

```
% PART 1.
% The given system is written from the Ax=B as [A]*[abc]=[B]
A=[.072, 0, -1; 0, .12, -1; 1 1 0];
B=[-12, -9, 50];

abc1=A\B'               %#ok   % BACKSLASH \
abc2  =   linsolve(A,B')   %#ok   % LINSOLVE()
abc3  =   inv(A)*B'        %#ok   % INV
% SOLVE() in symbolic MATH
syms a b c; abc4=solve(0.072*a-c+12, 0.12*b-c+9, a+b-50);
abc4=double([abc4.a;   abc4.b;   abc4.c])   %#ok
% SOLVE()
%% Part II. %%%%%%%%%%%%%%%%%%%%%%%%%%%%%%%%%%%%%%%%%%%%%%%%%%%%%%%%
% BACKSLASH \ ; LINSOLVE(); INV
tic; Bk=50:250;
a=zeros(numel(Bk),1); b=zeros(numel(Bk),1);
c=zeros(numel(Bk),1); A=[.072, 0, -1; 0, .12, -1; 1 1 0];
```

```
for ii=1:numel(Bk)
B=[-12; -9; Bk(ii)];
abc=A\B;
a(ii)=abc(1,:);
b(ii)=abc(2,:);
c(ii)=abc(3,:);
end Time1=toc;
fprintf('Computation time with  BACKSLASH: %3.3f  \n', Time1); clearvars
tic; Bk=50:250;
a=zeros(numel(Bk),1); b=zeros(numel(Bk),1);
c=zeros(numel(Bk),1); A=[.072, 0, -1; 0, .12, -1; 1 1 0];
for ii=1:numel(Bk)
B=[-12; -9; Bk(ii)];
abc=linsolve(A,B);
a(ii)=abc(1,:);
b(ii)=abc(2,:);
c(ii)=abc(3,:);
end
Time2=toc;
fprintf('Computation  time with LINSOLVE:   %3.3f   \n', Time2) clearvars
tic Bk=50:250;
a=zeros(numel(Bk),1); b=zeros(numel(Bk),1); c=zeros(numel(Bk),1); A=[.072,
0, -1; 0, .12, -1; 1 1 0];
for ii=1:numel(Bk)
   B=[-12; -9; Bk(ii)];
   abc=inv(A)*B;
   a(ii)=abc(1,:); b(ii)=abc(2,:); c(ii)=abc(3,:);
end
Time3=toc;
fprintf('Computation  time with  INV:   %3.3f  \n', Time3)
%% SOLVE() from symbolic math
clearvars; tic;
Bk=50:250;
a1=zeros(numel(Bk),1);b1=zeros(numel(Bk),1); c1=zeros(numel(Bk),1);
syms a b c
for ii=1:numel(Bk)
```

```
    abc=solve(0.072*a-c+12,0.12*b-c+9,a+b-Bk(ii));
    a1(ii)=double(abc.a);
    b1(ii)=double(abc.b);
    c1(ii)=double(abc.c);
end
Time4=toc;
fprintf('Computation  time with SOLVE:   %3.3f  \n', Time4)
```

Here are the results of the calculations from Part 1:

```
abc1 =
    15.6250
    34.3750
    13.1250
abc2 =
15.6250
34.3750
13.1250
abc3 =
15.6250
34.3750
13.1250
abc4 =
15.6250
34.3750
13.1250
```

Here are the results of the script from Part 2:

```
Computation time with BACKSLASH: 0.002
Computation time with LINSOLVE:  0.002
Computation time with INV:       0.002
Computation time  with  SOLVE:   22.066
```

From the computation time spent to compute solutions of the given linear system with three variables and 201 possible cases using four ways, it is clear that the least efficient way of solving linear equations is using the Symbolic Math toolbox's solve()

function. The backslash operator (\) and `linsolve()` and `inv()` methods all performed similarly. The solver `linsolve()`, \, and `inv()` methods are more than 11,033 times more efficient and faster than the `solve()` function.

Example 6: Linear Equations ([A]{x} = [b]) Applied for the Least Squares Method

This exercise demonstrates how to apply the principles of solving linear equations in the form of [A]{x} = [b] to solve the least squares problem to find best-fit model coefficients. In this exercise, we introduce the Vandermonde matrix expression to determine the polynomial fit models.

Here is the N-th order polynomial:

$$f(x) = a_n x^n + a_{n-1} x^{n-1} + \ldots + a_3 x^3 + a_2 x^2 + ax + a_0$$

To compute the fit model f(x_i), we set it equal to the measured data y_i: f(x_i) = [y_i].

$$a_n x_1^n + a_{n-1} x_1^{n-1} + \ldots + a_3 x_1^3 + a_2 x_1^2 + a_1 x_1 + a_0 = y_1$$

$$a_n x_2^n + a_{n-1} x_2^{n-1} + \ldots + a_3 x_2^3 + a_2 x_2^2 + a_1 x_2 + a_0 = y_2$$

$$a_n x_3^n + a_{n-1} x_3^{n-1} + \ldots + a_3 x_3^3 + a_2 x_3^2 + a_1 x_3 + a_0 = y_3$$

$$\vdots \quad \vdots \quad \ldots \quad \vdots \quad \vdots \quad \vdots \quad \vdots = \vdots$$

$$a_n x_m^n + a_{n-1} x_m^{n-1} + \ldots + a_3 x_m^3 + a_2 x_m^2 + a_1 x_m + a_0 = y_m$$

These expressions can be written as follows:

$$[V]\{a_i\} = [y_i]$$

Here, [V] is the Vandermonde matrix, {a_i} is the coefficients of the n-th order polynomial, and [y_i] is the measured data points.

$$V = \begin{bmatrix} 1 & x_1 & x_1 & \cdots & x_1 \\ 1 & x_2 & x_2^2 & \cdots & x_2^n \\ \vdots & \vdots & \vdots & \vdots & \vdots \\ 1 & x_m & x_m^2 & \cdots & x_m^n \end{bmatrix}; \ a_i = \begin{Bmatrix} a_0 \\ a_1 \\ a_2 \\ \vdots \\ a_n \end{Bmatrix}; \ y_i = \begin{bmatrix} y_1 \\ y_2 \\ y_3 \\ \vdots \\ y_m \end{bmatrix}$$

Or

$$
V = \begin{bmatrix} x_1^n \cdots & x_1^2 & x_1 & 1 \\ x_2^n \cdots & x_2^2 & x_2 & 1 \\ & \vdots & & \\ x_m^n \cdots & x_m^2 & x_m & 1 \end{bmatrix}_{;} \quad a_i = \begin{Bmatrix} a_n \\ \vdots \\ a_1 \\ a_0 \end{Bmatrix}; \quad y_i = \begin{bmatrix} y_1 \\ y_2 \\ y_3 \\ \vdots \\ y_m \end{bmatrix}
$$

Here, x_i and y_i are known, and a_i polynomial fit coefficient values are needed to be computed. Therefore, we can compute a_i from the next expression:

$$
\{a_i\} = [V]^{-1} * [y_i]
$$

Let's consider the following example.

Given test data:

Test #	Test1	Test2	Test3	Test4	Test5	Test6	Test7
Applied Load, [N]	10	20	30	40	50	60	70
Deflection, δ[m]	0.145	0.435	0.505	0.765	1.025	1.199	1.430

The task is to compute the fit model using Hooke's law formulation for linear elastic materials. The Hooke's law formulation is $F = k\delta$, where F is applied force in [N] and δ is a dependent variable, which is the deflection of an elastic material when F force is applied. And k is the stiffness coefficient of a material. Thus, the unknown variable here is k that will be computed using the least squares criterion.

First, we express the test data with respect to the system of linear equations $[A]\{x\} = [b]$. Here the applied force is the dependent variable [b], and the independent variable $\{x\}$ corresponds to the resulted deflection δ. Therefore, in this exercise, the unknown variable is k, which is stiffness of the material. In this exercise, a first tricky point is how to compute the values of [A]. To compute the elements of [A], we use the Vandermonde matrix approach. According to Hooke's law, it is a first-order polynomial, i.e., $F(\delta) = k\delta$, that can be also written as $k = F(\delta)/\delta$. Using the given data in this exercise, we can define the Vandermonde matrix and load matrix.

$$V = \begin{bmatrix} \delta_1 & 0 \\ \delta_1 & 0 \\ \vdots & \vdots \\ \delta_n & 0 \end{bmatrix} = \begin{bmatrix} 0.145 & 0 \\ 0.435 & 0 \\ \vdots & \vdots \\ 1.430 & 0 \end{bmatrix} F = \begin{bmatrix} 10 \\ 20 \\ \vdots \\ 70 \end{bmatrix}$$

Here, V is the Vandermonde matrix. Note the size of the Vandermonde matrix is 7-by-2 and the size of the applied load is 7-by-1. Therefore, the size of the stiffness matrix will be 1-by-2. The reason of having zeros in the second column of [V] is that according to Hooke's law, the linear relationship between the applied load and deflection of a linear elastic material is in the form of $f(x) = a_1 * x + a_0$ and $a_0 = 0$. Therefore, the unknown stiffness is found from the following:

$$k = [V]^{-1} * [F]$$

Note that to compute the values of [k] in a more efficient and exactly, we employ the backslash (\) operator. An alternative solution function to the backslash operator is linsolve() or mldivide().

The final solution script (LA_Ex6.m) is shown here:

```
% LA_Ex6.m
% Part 1. Vandermonde matrix
clc; clear variables
F = (10:10:70).';                                    % Applied Load
d = [0.145  0.435 0.505 0.765 1.025 1.199 1.430].';  % Deflection
scatter(F, d, 'filled')
ylim([0, max(d)+.2]),shg
A = [F zeros(size(F))];
FM =A\d;
FM_values = FM(1)*F;
hold on
plot(F, FM_values, 'k-', 'linewidth', 2)
gtext(['Fit model: F = '  num2str(FM(1)) '*\delta'])
gtext(['Stifness is: '  num2str(FM(1))])
grid on
xlabel('Applied Load, F [N]')
ylabel('Deflection, \delta [m]')
```

Figure 7-16 shows the resulted plot of the calculations from the script.

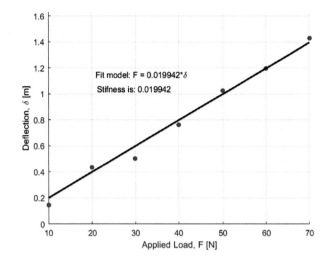

Figure 7-16. *Fit model is computed using the least squares method*

There are a few functions (polyfit, fitlm, fit) in Curve Fitting and Statistics and Machine Learning Toolboxes, which can be used easily to compute approximation polynomials. Let's look at the previous example of how to employ these functions:

```
% Part 2. Polynomial Approximation Fcn: Curve Fitting Toolbox
FM2 = polyfit(F,d, 1);
fprintf('CFTOOL Fit Model: F(d) = %f*d \n', FM2(1));
% Part 3. Polynomial Approximation Fcn: Stats and ML Toolbox
FM3 = fitlm(F,d, 'linear');
fprintf('Stats and ML Fit Model: F(d) = %f*d \n', FM3.Coefficients.
Estimate(2));
```

Parts 2 and 3 of the code (LA_Ex6.m) produce close approximation coefficients of the first-order polynomial. The following results will be displayed in the Command window:

```
CFTOOL Fit Model: F(d) = 0.021082*d
Stats and ML Fit Model: F(d) = 0.021082*d
```

Note that there is a small difference between the Vandermonde approach and polyfit() and fitlm() functions. The reason for the difference is the intercept value is set equal to "0" with the Vandermonde matrix, and with the other two functions, the intercept is considered.

Example 7: Linear Equations ($[A]\{x\} = [b]$) Applied for the Least Squares Method

The following data table gives the stopping distance y as a function of initial speed v, for certain car model. Find the quadratic polynomial coefficients that fit the data.

$v(km/h)$	20	30	40	50	60	70
$y(m)$	45	80	130	185	250	330

The Vandermonde matrix of this exercise for the quadratic fit model is computed from the following:

$$V = \begin{bmatrix} v_1^0 & v_1 & v_1^2 \\ v_2^0 & v_2 & v_2^2 \\ \vdots & \vdots & \vdots \\ v_n^0 & v_n & v_n^2 \end{bmatrix}$$

Note that $v_1^0, v_2^0, \ldots v_n^0 = 1$ corresponds to a_0. Therefore, V can be also expressed as follows:

$$V = \begin{bmatrix} 1 & v_1 & v_1^2 \\ 1 & v_2 & v_2^2 \\ \vdots & \vdots & \vdots \\ 1 & v_n & v_n^2 \end{bmatrix}$$

Note that V can be also expressed as follows:

$$V = \begin{bmatrix} v_1^2 & v_1 & 1 \\ v_2^2 & v_2 & 1 \\ \vdots & \vdots & \vdots \\ v_n^2 & v_n & 1 \end{bmatrix}$$

The Vandermonde matrix of the data from this exercise is equal to the following:

$$V = \begin{bmatrix} 1 & 20 & 20^2 \\ 1 & 30 & 30^2 \\ \vdots & \vdots & \vdots \\ 1 & 70 & 70^2 \end{bmatrix} \text{ or } V = \begin{bmatrix} 20^2 & 20 & 1 \\ 30^2 & 30 & 1 \\ \vdots & \vdots & \vdots \\ 70^2 & 70 & 1 \end{bmatrix}$$

The measured data points in this exercise are as follows:

$$y_i = \begin{bmatrix} 45 \\ 80 \\ \vdots \\ 330 \end{bmatrix}$$

The unknown coefficient of the quadratic polynomial is found from the following, depending on which way [V] is defined:

$$a = [a_0, a_1, a_2] \text{ or } a = [a_2, a_1, a_0]$$
$$a = V^{-1} * [y_i]$$

Note that in this exercise, the size of the Vandermonde matrix is 6-by-3.

The complete code of this exercise is LA_Ex7.m.

```
% LA_Ex7.m
clc; clear variables; close
% Part 1. Vandermonde matrix
v = (20:10:70).';                       % Velocity, [km/h]
y = [45  80 130 185 250 330].';         % Braking distance, [m]
scatter(v, y, 'filled')
ylim([0, max(y)+.2])
A = [v.^2, v,  ones(size(v))];
FM =A\y;
FM_values = FM(1)*v.^2+FM(2)*v+FM(3);
hold on
plot(v, FM_values, 'k-', 'linewidth', 2)
gtext(['Fit model: s(v) = '  num2str(FM(1))  'v^2 +' num2str(FM(2)) '*v +',
num2str(FM(3))])
grid on
```

```
xlabel('\it Velocity, v [km/h]')
ylabel('\it Braking Distance, s [m]')
% Part 2. Polynomial Approximation Fcns: Curve Fitting Toolbox
FM2 = polyfit(v,s, 2);
fprintf('CFTOOL Fit Model: s(v) = %f*v.^2 + %f*v + %f \n', FM2);
% Part 3. Polynomial Approximation Fcn: Stats and ML Toolbox
FM3 = fitlm(v, s, 'poly2');
fprintf('Stats and ML Fit Model: s(v) = %f*v.^2 + %f*v + %f \n', flip(FM3.
Coefficients.Estimate));
```

Figure 7-17 shows the simulation results of LA_Ex7.m.

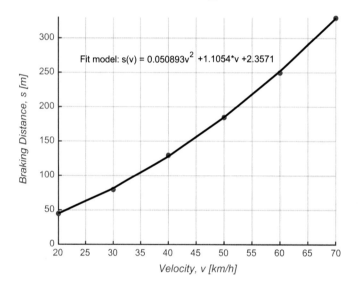

Fit model: s(v) = 0.050893v² +1.1054*v +2.3571

Figure 7-17. *Quadratic fit model is computed using the least squares method*

Also, in the Command window, the following outputs will be displayed after executing the script: LA_Ex7.m:

```
FMM =

    0.0509
    1.1054
    2.3571
CFTOOL Fit Model: s(v) = 0.050893*v.^2 + 1.105357*v + 2.357143
Stats and ML Fit Model: s(v) = 0.050893*v.^2 + 1.105357*v + 2.357143
```

The results from the three approaches are identical, which proves that the Vandermonde approach is well correlated with the functions of the two toolboxes.

Example 8: Linear Equations ([A]{x} = [b]) Applied for the Least Squares Method Using Simulink Modeling

The following data table gives the stopping distance y as a function of initial speed v, for a certain car model. Find the quadratic polynomial coefficients that fit the data.

$v(km/h)$	20	30	40	50	60	70
$y(m)$	45	80	130	185	250	330

Let's build a Simulink model to solve this exercise and apply the least squares polynomial solver block. A Simulink model of this exercise is relatively simple and composed of three blocks: Constant, Least Squares Polynomial Fit, and Display blocks, as shown in Figure 7-18.

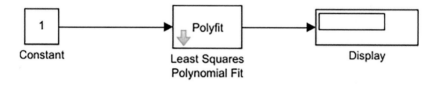

Figure 7-18. *Simulink model, the least squares method*

The Simulink model shown in Figure 7-18 is not complete yet. There are two more adjustments to be made in the Constant and Least Squares Polynomial Fit blocks. The Constant should be opened by double-clicking it, and the data for y, i.e., [45 80 130 185 250 330].' should be entered. Note the data has to be a column vector. Then the next block parameters should be adjusted, as shown in Figure 7-19. Note that Control Parameter (X) values are v values in a column vector form, and Polynomial order (N) is 2 because we are looking for a quadratic polynomial fit.

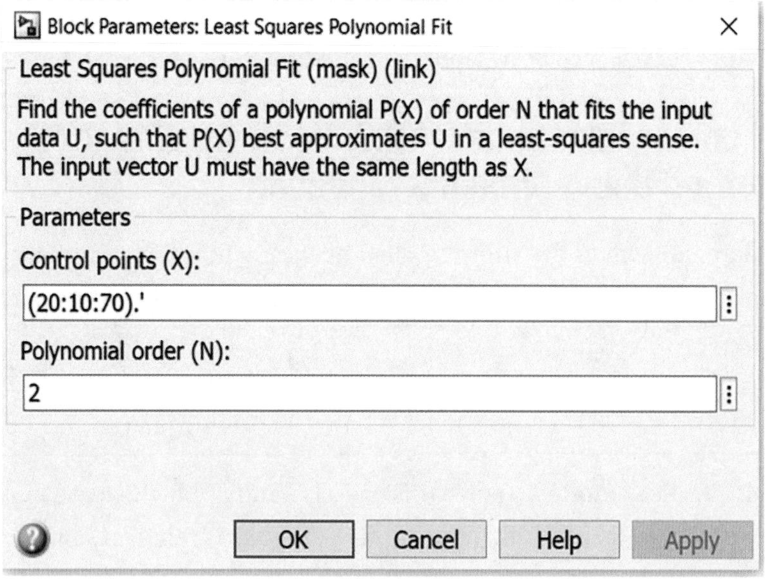

Block Parameters: Least Squares Polynomial Fit   ✕

Least Squares Polynomial Fit (mask) (link)

Find the coefficients of a polynomial P(X) of order N that fits the input
data U, such that P(X) best approximates U in a least-squares sense.
The input vector U must have the same length as X.

Parameters

Control points (X):

(20:10:70).'

Polynomial order (N):

2

OK   Cancel   Help   Apply

Figure 7-19. Least squares Polynomial Fit block parameters adjustment

Once all adjustments are made and values are entered, the model is ready to
simulate. The completed model (LA_Ex8.slx) with simulation results after resizing the
Display block to see all results is shown in Figure 7-20.

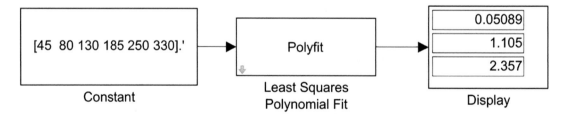

[45 80 130 185 250 330].'	Polyfit	0.05089
		1.105
Constant	Least Squares Polynomial Fit	2.357
		Display

Figure 7-20. Simulink model, LA_Ex8.slx

Note that the found results from the Simulink model LA_Ex8.slx match perfectly
well with the ones found using the Vandermonde matrix, polyfit() and fitlm().

Matrix Operations

This section covers general mathematical operations and computations of matrices,
vectors, and eigen-vectors. Many numerical examples are used to explain the matrix
operations. Table 7-1 lists the matrix operations their command syntax.

Table 7-1. *Matrix Operators in Two Equivalent Formulations*

Operation Name	MATLAB First Way	MATLAB Second Way
Matrix multiplication	A*B	mtimes(A,B)
Array-wise multiplication	A.*B	times(A,B)
Matrix right division	A/B	mrdivide(A,B)
Array-wise right division	A./B	rdivide(A,B)
Matrix left division	A\B	mldivide(A,B)
Array-wise left division	A.\B	ldivide(A,B)
Matrix power	A^B	mpower(A,B)
Array-wise power	A.^B	power(A,B)
Complex transpose	A'	ctranspose(A)
Matrix transpose	A.'	transpose(A)
Binary addition	A+B	plus(A,B)
Unary plus	+A	uplus(A)
Binary subtraction	A-B	minus(A,B)
Unary minus	-A	uminus(A)
Determinant	det(A)	det(A)
Rotate by 90^0	rot90(A)	rot90(A)
Replicate and tile an array n times	repmat(A, n)	repmat(A, n)
Flip matrix left/right	fliplr(A)	fliplr(A)
Flip matrix in up/down	flipud(A)	flipud(A)

Basic MATLAB unit data is in the array type format. Matrices and vectors can be employed in many cases to define input and output, local data, and function inputs and outputs. Moreover, they can be used to combine separate scalars into one signal and process multidimensional input and output signals. An array is defined by a single name and a collection of data arranged by rows and columns, as shown here.

Row # 1 ⇨	A_{11}	A_{12}	A_{13}
Row # 2 ⇨	A_{21}	A_{22}	A_{23}
Row # 3 ⇨	A_{31}	A_{22}	A_{33}
Row # 4 ⇨	A_{41}	A_{42}	A_{43}
	⇧	⇧	⇧
	Column # 1	Column # 2	Column # 3

Let's look at some numerical examples. They perform matrix operations with scalars, such as addition, subtraction, power, multiplication, and division, including array-wise (elementwise) operations in the Command window.

```
>>   A=[8,1,6;   3,5,7;   4,9,2]   % Matrix 3-by-3
A =

     8     1     6
     3     5     7
     4     9     2
>> a = 2; b = 2+3i; c = 5j;
>> B=A^a  % Note the difference between ^ and .^
B =

    91    67    67
    67    91    67
    67    67    91

>> C=A.^a  % Elementwise. Note the difference between ^ and .^
C =

    64     1    36
     9    25    49
    16    81     4
```

```
>> D = A*a+B/b
D =
  30.0000 -21.0000i  12.3077 -15.4615i  22.3077 -15.4615i
  16.3077 -15.4615i  24.0000 -21.0000i  24.3077 -15.4615i
  18.3077 -15.4615i  28.3077 -15.4615i  18.0000 -21.0000i
>> E =  C./c
E =
  0.0000 -12.8000i   0.0000 - 0.2000i   0.0000 - 7.2000i
  0.0000 - 1.8000i   0.0000 - 5.0000i   0.0000 - 9.8000i
  0.0000 - 3.2000i   0.0000 -16.2000i   0.0000 - 0.8000i
>> F = C/c
F =
  0.0000 -12.8000i   0.0000 - 0.2000i   0.0000 - 7.2000i
  0.0000 - 1.8000i   0.0000 - 5.0000i   0.0000 - 9.8000i
  0.0000 - 3.2000i   0.0000 -16.2000i   0.0000 - 0.8000i
```

Example: Performing Matrix Operations

Given six arrays: A (4 − by − 3), B(3 − by − 4), C(4 − by − 4), D(4 − by − 3), E(3 − by − 3), and F(3 − by − 3).

Let's perform several matrix operations—such as summation, subtraction, multiplication, power, scalar multiplication, square root, mean, round, standard deviations, and replicate/rotate/flip matrix—from the Command window.

```
>> A=[2 -3 1; 3 2 5; 1 3 4; -3 -2 3] ;
>> B=[3,4,-2 1;2,5,4,-6;4,-3, 1,2] ;
>> C=[16,2,3,13;5,11,10,8;9 4 7 14;6 15 12 1] ;
>> D=[1 2 3; 2 3 4; 4 3 1; -2 -3 1] ;
>> E=[8, 1, 6; 3, 5, 7; 4, 9, 2];
>> F=[3 7 3; 3 2 8; 9 2 1];
>> M_AB = A*B
M_AB =
     4   -10   -15    22
    33     7     7     1
    25     7    14    -9
    -1   -31     1    15
```

```
>> M_BA = B*A
M_BA =
    13     -9     18
    41     28     25
    -6    -19     -1
>> M_S = M_AB-C
M_S =
   -12    -12    -18      9
    28     -4     -3     -7
    16      3      7    -23
    -7    -46    -11     14
>> M_S= M_BA-C
```

Matrix dimensions must agree.

```
>> CM=C*M_S    % Not equivalent to M_S*C
CM =
  -179   -789   -416    243
   352   -442   -141   -150
    18   -747   -279     88
   533   -142    -80   -313
>> CM1=M_S*C    % Not equivalent to C*M_S
CM1 =
  -360    -93   -174   -495
   359   -105    -61    283
   196   -252   -149    307
  -357   -354   -390   -599
>> CM2=M_S.*C    % Elementwise  operation:  NOT  equivalent  to  M_S*C
CM2 =
  -192    -24    -54    117
   140    -44    -30    -56
   144     12     49   -322
   -42   -690   -132     14
```

```
>> MDE=M_S./C    % Elementwise operation:  NOT  equivalent  to  M_S/C
MDE =
    -0.7500    -6.0000    -6.0000     0.6923
     5.6000    -0.3636    -0.3000    -0.8750
     1.7778     0.7500     1.0000    -1.6429
    -1.1667    -3.0667    -0.9167    14.0000

>> MD=M_S/C    % Not equivalent to M_S./C
MD =
     1.9275     8.5704    -5.6271    -5.8414
     1.4496    -6.4076     1.5420     3.8277
    -1.0389   -10.8246     5.0116     6.9401
    -4.9118   -12.9118    12.6765     3.6765
>> M_AD =A.*D % Elementwise operation: matrix multiplication
M_AD =
     2    -6     3
     6     6    20
     4     9     4
     6     6     3
>> MM_AD= A*D    % Error due to size mismatch of [A] and [D]
```

Error using * Incorrect dimensions for matrix multiplication. Check that the number of columns in the first matrix matches the number of rows in the second matrix. To perform elementwise multiplication, use '.*'. Related documentation

```
>> M_EF=E.*F    % Elementwise multiplication of square matrices
M_EF =
    24     7    18
     9    10    56
    36    18     2
>> MM_EF=E*F    % Square matrices can be multiplied matrix-wise
MM_EF =
    81    70    38
    87    45    56
    57    50    86
```

```
>> Csqrt=sqrt(C) % Not equivalent to sqrtm(C)
Csqrt =
     4.0000      1.4142      1.7321      3.6056
     2.2361      3.3166      3.1623      2.8284
     3.0000      2.0000      2.6458      3.7417
     2.4495      3.8730      3.4641      1.0000
>> Csqrt=sqrtm(C) % Not equivalent to sqrt(C)
Csqrt =
   3.8335 - 0.0167i    0.0738 + 0.7839i    0.1262 + 0.3666i    1.7975 - 1.1337i
   0.3251 + 0.0011i    2.6850 - 0.0526i    1.6850 - 0.0246i    1.1359 + 0.0761i
   1.3123 - 0.0237i    0.7322 + 1.1107i    1.9687 + 0.5194i    1.8178 - 1.6064i
   0.5925 + 0.0373i    2.0922 - 1.7477i    1.7997 - 0.8172i    1.3466 + 2.5276i
>>  C_E1 = expm(C)   % Matrix exponential not equal to  exp(C)
C_E1 =
   1.0e+14 *
     1.5718      1.3711      1.3622      1.5295
     1.5718      1.3711      1.3622      1.5295
     1.5718      1.3711      1.3622      1.5295
     1.5718      1.3711      1.3622      1.5295
>> C_E2 = exp(C) % Exponential  of  a  matrix:  not  equal  to  expm(C)
C_E2 =
   1.0e+06 *
     8.8861      0.0000      0.0000      0.4424
     0.0001      0.0599      0.0220      0.0030
     0.0081      0.0001      0.0011      1.2026
     0.0004      3.2690      0.1628      0.0000
>> S=[A(1,1:3); B(2,1:3);C(3,2:4)]; % Created from the existed
>> Y=[A(1), 1.3];                   % Created from the existed
>> Arot90=rot90(A)                  % Matrix rotate
```

```
Arot90 =
     1     5     4     3
    -3     2     3    -2
     2     3     1    -3
>> Crep=repmat(C,  2,1)    % Matrix replication/copy
Crep =
    16     2     3    13
     5    11    10     8
     9     4     7    14
     6    15    12     1
    16     2     3    13
     5    11    10     8
     9     4     7    14
     6    15    12     1
>> Bflip=fliplr(B)    % Matrix flip
Bflip =
     1    -2     4     3
    -6     4     5     2
     2     1    -3     4
Cud=flipud(Crep)    % Matrix flip up or down
Cud =
     6    15    12     1
     9     4     7    14
     5    11    10     8
    16     2     3    13
     6    15    12     1
     9     4     7    14
     5    11    10     8
    16     2     3    13
```

Many of these matrix operations can also be performed in the Simulink environment. Let's use the previous examples to demonstrate how and what Simulink uses for matrix operations and manipulations.

The Simulink Library contains the blocks for sum, multiplication/division, power, exponent, and concatenation, as shown in Figure 7-21.

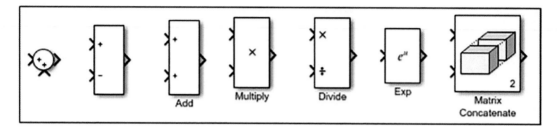

Figure 7-21. *Matrix operation blocks in the Simulink Library*

First define the [A] and [D] matrices in the Command window.

```
>> A=[2 -3 1; 3 2 5; 1 3 4; -3 -2 3] ;
>> D=[1 2 3; 2 3 4; 4 3 1; -2 -3 1] ;
```

Now compute the sum and subtraction of matrices [A] and [D], as shown in Figure 7-22.

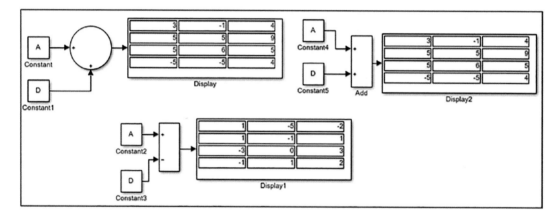

Figure 7-22. *Matrix sum and subtraction operations in Simulink*

Note that matrices [A] and [D] are defined via the Command window and workspace. The computed sums match the ones calculated using MATLAB's Command window.

Here are the results of multiplication (see Figure 7-23), exponent, and square (see Figure 7-24) of the matrices.

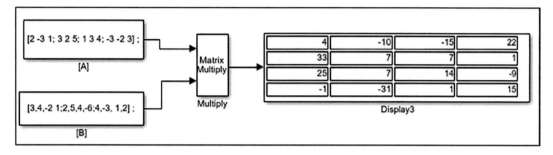

Figure 7-23. *Matrix multiplication in Simulink*

Note that for the matrix multiplication operation shown in Figure 7-23, the Multiply block changes from element-wise (.*) multiplication to matrix (*) multiplication, as shown in Figure 7-24.

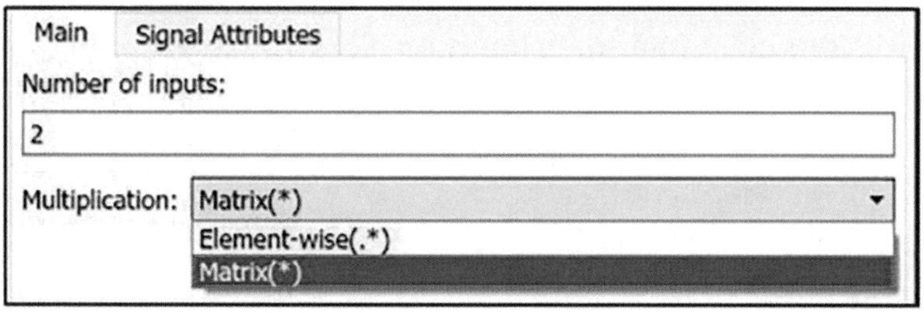

Main	Signal Attributes
Number of inputs:	
2	
Multiplication:	Matrix(*) ▼
	Element-wise(.*)
	Matrix(*)

Figure 7-24. *Setting up the Matrix Multiply block for matrix multiplication (*) or element-wise multiplication (.*)*

Otherwise, the multiplication operation will not be performed due to the mismatched sizes of [A] and [B]. Again, the computed results match the ones from MATLAB.

C
[C]
3
power
u^v
Exp
4096 8 27 2197
125 1331 1000 512
729 64 343 2744
216 3375 1728 1
Display5
C
[C]
u^2
Exp1
256 4 9 169
25 121 100 64
81 16 49 196
36 225 144 1
Display6

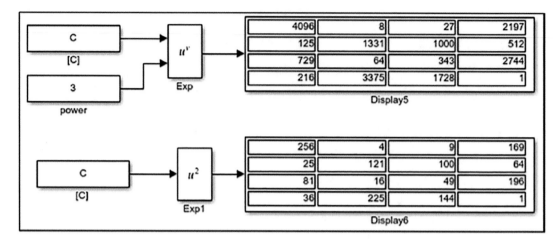

Figure 7-25. *Matrix exponential and square operation blocks*

Note that in the operations in Figure 7-25, the exponential and power operations are performed with one block (one Math Function block), by choosing its Function type [pow] in u^v and [square] in u^2 (see Figure 7-26).

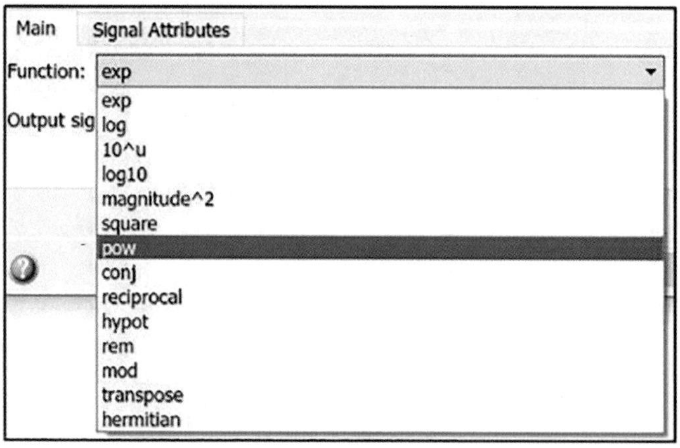

Main
Signal Attributes
Function:
exp
Output sig
exp
log
10^u
log10
magnitude^2
square
pow
conj
reciprocal
hypot
rem
mod
transpose
hermitian

Figure 7-26. *How to set the Math Function block for matrix operations*

Now by using the Matrix Concatenate block, we create a new matrix (4-by-10) from the computed the matrix sum (4-by-3), square (4-by-4), and matrix division (4-by-3). See Figure 7-27.

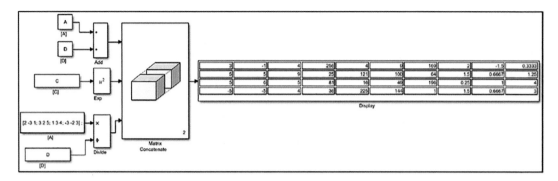

Figure 7-27. *The Matrix Concatenate block performs matrix concatenation.*

As demonstrated, Simulink blocks perform various matrix operations, much like MATLAB functions. However, there are computationally costly simulations with matrix and array operations in which Simulink models might be slower than MATLAB scripts. For example, when computing discrete Fourier transforms, Simulink models are much slower than MATLAB. For some matrix and array operations, the MATLAB Fcn block or the Interpreted MATLAB Fcn block can be used in Simulink modeling.

In addition to these matrix operations, there are a few other operations by which you can create new matrices. For instance, you can take out diagonals of existing matrices with diag(A) or take out selected elements of matrices and create a new matrix.

```
>> E=[8, 1, 6; 3, 5, 7; 4, 9, 2];
>> F=[3 7 3; 3 2 8; 9 2 1];
>> EF = [diag(E),  diag(F)]
EF =
8   3
5   2
2   1
```

Standard Matrix Generators

MATLAB has numerous standard array and matrix generators, which can be used to generate a wide range of matrices. For instance, eye(n), eye(k, m), ones(m), ones(m, k), zeros(1), zeros(1,k), magic(k), pascal(k), pascal(k, m), rand(m), rand(k, m), randi(n,m,k), repmat(A, r, c), blkdiag(A, B, C), sparse(m,n), and many more. Here's an example:

```
>> eye(3)
ans =
     1     0     0
     0     1     0
     0     0     1
>> magic(5)  %   Magic   matrix in a size of 5 by 5
ans =
    17    24     1     8    15
    23     5     7    14    16
     4     6    13    20    22
    10    12    19    21     3
    11    18    25     2     9
>>  A=pascal(4)   % Pascal   matrix   in   a   size   of   4   by   4
A =
     1     1     1     1
     1     2     3     4
     1     3     6    10
     1     4    10    20
>> A=pascal(4,2)   % Pascal matrix in a size of 4 by 4
A =
    -1    -1    -1    -1
     3     2     1     0
    -3    -1     0     0
     1     0     0     0
>> zeros(3) % Zero matrix 3-by-3
ans =
     0     0     0
     0     0     0
     0     0     0
>> zeros(2,3) % Zero matrix 2-by-3
ans =
     0     0     0
     0     0     0
```

```
>> ones(3)   % Ones matrix 3-by-3
ans =
    1    1    1
    1    1    1
    1    1    1
>> ones(2,3)  % Ones matrix 2-by-3
ans =
    1    1    1
    1    1    1
>> eye(3,4)   % Unit diagonal matrix of size  3  - by - 4
ans =
    1    0    0    0
    0    1    0    0
    0    0    1    0
>> eye(4,5)   % Unit diagonal matrix of size 4 - by - 5
ans =
    1    0    0    0    0
    0    1    0    0    0
    0    0    1    0    0
    0    0    0    1    0
>> rand(2)    % Uniform random matrix 2-by-2
ans =
    0.8147    0.1270
    0.9058    0.9134
>> rand(2, 4)    % Uniform random matrix 2-by-4
ans =
    0.6324    0.2785    0.9575    0.1576
    0.0975    0.5469    0.9649    0.9706
>> randn(3)    % Normally distributed random matrix 3-by-3
ans =
    0.7254   -0.2050    1.4090
   -0.0631   -0.1241    1.4172
    0.7147    1.4897    0.6715
```

```
>> A = round(randn(3))    % Round up to the nearest 0
A =
    -1     0     0
     1     1     0
     2     1    -1
>> A_rep=repmat(A, 2, 3) % replicating the matrix A by making its
% replication 2 times of rows and 3 times of columns
A_rep =
    -1     0     0    -1     0     0    -1     0     0
     1     1     0     1     1     0     1     1     0
     2     1    -1     2     1    -1     2     1    -1
    -1     0     0    -1     0     0    -1     0     0
     1     1     0     1     1     0     1     1     0
     2     1    -1     2     1    -1     2     1    -1

>> C=eye(2); B=magic(3); A=ones(4);
>> D=blkdiag(A,B,C)    % combine matrices in diagonal directions to
% create a block diagonal matrix.
D =
     1     1     1     1     0     0     0     0     0
     1     1     1     1     0     0     0     0     0
     1     1     1     1     0     0     0     0     0
     1     1     1     1     0     0     0     0     0
     0     0     0     0     8     1     6     0     0
     0     0     0     0     3     5     7     0     0
     0     0     0     0     4     9     2     0     0
     0     0     0     0     0     0     0     1     0
     0     0     0     0     0     0     0     0     1
>> randi([-13, 13], 5) % Random  integers within [-13, 13]
ans =
     0     7    12     9    -4
    12    -7     1    -7    -8
    -4     0   -10     8    -7
     2     5    -9    -7     3
    -7    11    -7    12    -1
```

```
>> K=reshape(randperm(9),  3,3)  % Change  the  size  (reshape)
of  array  %  to  make 3 by 3 matrix by random permutation
K =
      6      4      7
      1      3      2
      9      8      5
```

In addition, there are a few dozen matrix generation functions. They are the gallery of test matrices, such as binomial, cauchy, clement, invol, house, krylov, leslie, lesp, neumann, poisson, ris, rando, smoke, wilk, and many more. In general, the command syntax of these matrices is as follows:

```
[A, B, C,...] = gallery(matname,P1,P2,...);
[A, B, C,...] = gallery(matname,P1,P2,..., classname);
A=gallery(3);
B=gallery(5);
```

To get more information about the gallery of matrices, type this in the Command window:

```
>> help gallery
>> doc gallery
```

Here are several examples of how to employ gallery matrices:

```
>> S=[3 2 7]; X=[2 2];
% This is the 3-by-3 Leslie population matrix taken from the model with
average birth numbers S(1:n) and survival rates X(1:n-1)
>> L=gallery('leslie',  S,  X)
L =
      3      2      7
      2      0      0
      0      2      0
% Chebyshev spectral differentiation matrix of order 3
>> C = gallery('chebspec', 3,1)
C =
   -0.3333    -1.0000     0.3333
    1.0000     0.3333    -1.0000
   -1.3333     4.0000    -3.1667
```

```
% Cauchy matrix 3-by-3, C(I, j) = 1/(S(i)+Y(j)). The arguments S and Y are
vectors of length 3.
% If you pass in scalars for S and Y, they are interpreted as vectors 1:S
and 1:Y.
>> S = [3 2 6]; Y = [1 3 2];
>> C = gallery('cauchy', S, Y)
C =
    0.2500    0.1667    0.2000
    0.3333    0.2000    0.2500
    0.1429    0.1111    0.1250
>> % Krylov matrix of size 5-by-5.
>>  B  =  gallery('krylov',  randn(5))
B =
    1.0000    2.4392    3.9250   26.5823   24.9976
    1.0000    1.2031    7.8039    6.5275   61.9487
    1.0000   -1.3094   -7.4622   11.6113  -14.5418
    1.0000    0.3038   -3.8311  -16.0811  -10.1830
    1.0000   -3.7454    0.3824    5.7186  -65.2352
>> % House-holder matrix of size 3-by-1.
>> A = [3;2;5];      % Must be a column matrix
>> H =   gallery('house', A)
H =
    9.1644
    2.0000
    5.0000
>> % Hankel matrix of size 5-by-5 with elements H(I, j)=0.5/(n-i-j+1.5).
>> B  =  gallery('ris',5)
B =
    0.1111    0.1429    0.2000    0.3333    1.0000
    0.1429    0.2000    0.3333    1.0000   -1.0000
    0.2000    0.3333    1.0000   -1.0000   -0.3333
    0.3333    1.0000   -1.0000   -0.3333   -0.2000
    1.0000   -1.0000   -0.3333   -0.2000   -0.1429
```

```
>> % Smoke matrix  of size 3-by-3 - complex, with "smoke ring" pseudo-
spectrum.
>> SM=gallery('smoke', 3)
SM =
   -0.5000 + 0.8660i    1.0000 + 0.0000i    0.0000 + 0.0000i
    0.0000 + 0.0000i   -0.5000 - 0.8660i    1.0000 + 0.0000i
    1.0000 + 0.0000i    0.0000 + 0.0000i    1.0000 + 0.0000i
```

These standard and gallery matrices have special properties that can be of great use in various numerical simulations and analysis problems. For instance, these standard matrices—ones(), eye(), zeros(), rand(), randn()—are used often for signal processing, data analysis, and memory allocation in large computations.

Vector Spaces

In signal processing, numerical analyses, and building computer simulation models, vector spaces are very important. For instance, the logarithmic space is used for digital signal processing when frequencies go over a unit circle. There are several straightforward ways by which vectors, vector spaces, and arrays with equal spaces between their elements can be created. Let's suppose that we need to create a vector W that begins with a value $w1$ and ends with $w2$, as shown in Figure 7-28.

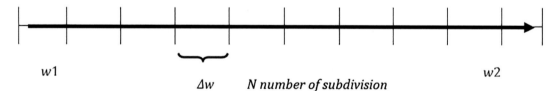

$w1$ Δw N *number of subdivision* $w2$

Figure 7-28. *Vector space*

If the size Δw is known, then the space can be expressed by $W = w1: \Delta w: w2$. For instance, the whole space can be defined in terms of $w1 = 1$, $w2 = 13$, $\Delta w = 0.1$ with the following:

```
>> w=1:0.1:13;
```

Moreover, if *N* number of points between the start and end boundaries of a space are known, the linear space function `linspace()` can be used.

```
>> % This creates a linear space of w array with equally spaced k number of
elements
>> w=linspace(1, 13, N);
```

Note   If *N* is not specified in the `linspace()` command, its default value is 100.

This simplex example generates sound waves with a sine function.

```
fs=3e4;        % Sampling frequency
% Different signal frequencies:
f1=100; f2=200; f3=300; f4=400; f5=500; f6=600;
t=0:1/fs:5;   % Time
% Signal: sum of sine waves
x=sin(2*pi*t*f1)+sin(2*pi*t*f2)+sin(2*pi*t*f3)+
sin(2*pi*t*f4)+sin(2*pi*t*f5);
[m, n]=size(x);              % Gets the size of the created vector space
sound(x, fs)                 % Plays a created sound & hear from sound cards
```

The `linspace()` command creates linearly spaced vector spaces/arrays. In MATLAB, there is another similar function, called `logspace()`, that creates logarithmic scaled vector spaces. For example, you use the following command to create a logarithmic space of the *x* array containing 130 logarithmically spaced elements (here, $N = 130$) between boundary points 0 and 13:

```
>> x=logspace(1,13,130);
```

Likewise, use this command to create 50 logarithmic spaced points between 0 and π:

```
>> s=logspace(0, pi);
```

Note   If *N* is not specified in `logspace()`, then its default value is 50.

Polynomials Represented by Vectors

For numerical simulations in MATLAB, polynomials are represented via vectors using coefficients of polynomials in descending order. For instance, a fifth-order polynomial is given as follows:

$$12x^5 + 13x^4 - 15x^2 + 17x - 13$$

That is defined as a vector space in the following manner:

```
>>  f = [12, 13, 0, -15, 17, -13];
```

Note, that MATLAB reads vector entries as a vector of length $n+1$ as an n-th order polynomial. Thus, if any of the given polynomial misses any coefficients, zero has to be entered for its coefficient. For instance, in the previous example, 0 is entered for the coefficient of x^3.

There are several functions that can be used to compute the roots of polynomials. They are as follows:

- Using the `roots()` MATLAB function

- Using the `zero()` Control System Toolbox function

- Using the `solve()` Symbolic MATH Toolbox function

You find roots of the given polynomial using the base MATLAB function, `roots()`.

```
>> x_sols=roots(f)
x_sols =
  -1.2403 + 0.9412i
  -1.2403 - 0.9412i
   0.7941 + 0.0000i
   0.3015 + 0.6869i
   0.3015 - 0.6869i
```

Note that the given polynomial has only one real value root and four complex valued roots.

The roots are computed by using the `solve()` function of MATLAB to find symbolic solutions of the polynomial, and then solutions are converted (note that conversion may be not necessary) to obtain a shorter number of decimal point numeric data using the `double()` function with the following entries in the Command window:

```
>> syms x
>> syms x
>> Sol=solve(12*x^5+13*x^4-15*x^2+17*x-13)
Sol =
root(z^5 + (13*z^4)/12 - (5*z^2)/4 + (17*z)/12 - 13/12, z, 1)
root(z^5 + (13*z^4)/12 - (5*z^2)/4 + (17*z)/12 - 13/12, z, 2)
root(z^5 + (13*z^4)/12 - (5*z^2)/4 + (17*z)/12 - 13/12, z, 3)
root(z^5 + (13*z^4)/12 - (5*z^2)/4 + (17*z)/12 - 13/12, z, 4)
root(z^5 + (13*z^4)/12 - (5*z^2)/4 + (17*z)/12 - 13/12, z, 5)
>> double(Sol)
ans =
   0.3015 - 0.6869i
   0.3015 + 0.6869i
   0.7941 + 0.0000i
  -1.2403 - 0.9412i
  -1.2403 + 0.9412i
```

Roots can be computed by using zero(), which is a function of the Control Toolbox of MATLAB:

```
>> F_tf = tf(f, 1)
F_tf =
   12 s^5 + 13 s^4 - 15 s^2 + 17 s - 13
Continuous-time transfer function.

>> x_sols = zero(F_tf)
x_sols =
  -1.2403 + 0.9412i
  -1.2403 - 0.9412i
   0.7941 + 0.0000i
   0.3015 + 0.6869i
   0.3015 - 0.6869i
```

Note in this case, a transfer function (ratio of two polynomials) with a denominator of 1 in the "s" domain is created first. Then the roots of s are computed, which would make the polynomial equal to zero.

The values of polynomials at specific input argument values can be computed using MATLAB's built-in function polyval(). Here is an example how to use this function:

```
>> f = [12, 13, 0, -15, 17, -13];        % Given polynomial
>> x = linspace(-10, 10, 500);
>> f_val = polyval(f,x);                  % Computed polynomial values
```

Simulink Model-Based Solution of Polynomials

To solve polynomials via Simulink modeling, use the MATLAB Fcn block, the Constant block to input the polynomial coefficients, and the Display block to see the computed roots. Figure 7-29 shows the complete model saved as Polynomial_Solver.slx.

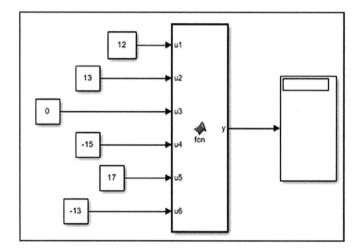

Figure 7-29. *Simulink model to solve the polynomial*
$12x^5 + 13x^4 - 15x^2 + 17x - 13 = 0$

The MATLAB Function block has the following command syntax embedded in it:

```
function y  =  fcn(u1,  u2,  u3,  u4,  u5,  u6)
y = roots([u1, u2, u3, u4, u5, u6]);
```

The MATLAB Fcn block calls the MATLAB function roots() and computes the roots of the polynomial with respect to its coefficients given by the input variables u1, u2, ... u6 since we are solving a fifth-order polynomial. As it is, this model does not run, and there are two more issues related to the size of the variables and solver type. First, the *solver type* has to be a fixed-step size type. That can be adjusted via Simulation ➤ Model

Configuration Parameters ➤ Solver Selection ➤ Fixed Step Solver. By default, the solver is a variable type.

Second, you need to change the size of the output variable y. You can do that by clicking the ⌨ icon and selecting Model Explorer ➤ [Model Hierarchy] ➤ Polynomial_Solver.slx ➤ MATLAB Function ➤ y Output ➤ Size. Set the size to 5 and click Apply. (The fifth-order polynomial has five roots.) After clicking the Run button in the menu of the Simulink model window or pressing Ctrl+T on the keyboard, you'll see the results displayed in Figure 7-30.

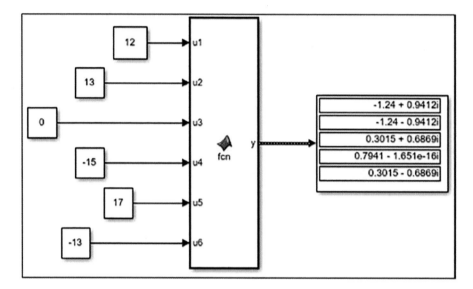

Figure 7-30. *Complete model with computed roots of the polynomial*
$12x^5 + 13x^4 - 15x^2 + 17x - 13 = 0$

The computed roots of the given polynomial match the ones computed by the MATLAB commands `roots()` and `zero()` to four decimal places.

Eigen-Values and Eigen-Vectors

Eigen-values and eigen-vectors have broad applications, not only in linear algebra but also in many engineering problems. For instance, they are used with vibrations, modal analysis, control applications, robotics, and so forth.

Definition 1. An *eigen-value* and *eigen-vector* of a square matrix A are, respectively, a scalar λ and a nonzero vector v that satisfy the following:

$$Av = \lambda v \qquad \text{(Equation 7-6)}$$

Definition 2. Given a linear transformation A (a square matrix), a nonzero vector v is defined to be an *eigen-vector* of the transformation if it satisfies the following *eigen-value* equation for some scalar λ:

$$[A]\{v\} = \lambda\{v\} \qquad \text{(Equation 7-7)}$$

In this case, the scalar λ is called an eigen-value of A corresponding to the eigen-vector $\{v\}$.

$$\underbrace{\begin{bmatrix} a & b & c \\ d & e & f \\ g & h & i \end{bmatrix}}_{[A]} * \underbrace{\begin{Bmatrix} x \\ y \\ z \end{Bmatrix}}_{[X]} = \lambda * \underbrace{\begin{Bmatrix} x \\ y \\ z \end{Bmatrix}}_{[X]}$$

$$[A]*[\boldsymbol{X}] - \lambda *[I]*[\boldsymbol{X}] = 0 \qquad \text{(Equation 7-8)}$$

Here, $[I]$ is the identity matrix. Now by rearranging, the next formulation can be written as follows:

$$\left([A]*[X] - [\lambda]*[I]\right)*[X] = 0 \qquad \text{(Equation 7-9)}$$

Let's assume that there is an inverse matrix of the coefficient of $[X]$, *i.e.*, $([A] - [\lambda] * [I])$.

$$\left([A] - [\lambda]*[I]\right)^{-1} = 0 \qquad \text{(Equation 7-10)}$$

There can be other solutions apart from a trivial solution $[X] = 0$. So, this means $([A] - [\lambda] * [I]) = 0$ is obtained via determinant of this matrix equal to 0.

$$det\left\{[A] - [\lambda]*[I]\right\} = 0 \qquad \text{(Equation 7-11)}$$

The left side of Equation 7-11 is called a *characteristic polynomial*. So, when this equation is expanded, it will lead to a polynomial equation of λ. Use the following example to compute eigen-values and eigen-vectors:

$$\begin{cases} 2.3x_1 + 3.4x_2 + 5x_3 = 0 \\ 3x_1 + 2.4x_2 - 1.5x_3 = 0 \\ 2x_1 - 0.4x_2 - 7.2x_3 = 0 \end{cases}$$

Now, the given system's equations are written in matrix form.

$$\begin{bmatrix} 2.3 & 3.4 & 5 \\ 3 & 2.4 & -1.5 \\ 2 & -0.4 & -7.2 \end{bmatrix} \begin{bmatrix} x_1 \\ x_2 \\ x_3 \end{bmatrix} = 0$$

Eigen-values of this transformation matrix are defined to be:

$$\det \begin{bmatrix} 2.3-\lambda & 3.4 & 5 \\ 3 & 2.4-\lambda & -1.5 \\ 2 & -0.4 & -7.2-\lambda \end{bmatrix} = 0$$

$$-7.884 + 49.12\,\lambda - 2.5\,\lambda^2 - \lambda^3 = 0$$

Solutions of this characteristic polynomial equation are as follows:

$$\lambda_1 = -8.434; \lambda_2 = 0.162; \lambda_3 = 5.772$$

Further, three eigen-vectors are computed by plugging in each eigen-value one by one into the equation. Hand calculations of eigen-values and eigen-vectors for larger systems are tedious and time-consuming. For very large systems of linear equations, it is infeasible to compute eigen-values and eigen-vectors with hand calculations. All of these computations can be performed with a single built-in function of MATLAB, called eig(A):

```
>> A = [2.3 3.4, 5; 3, 2.4, -1.5; 2, -0.4, -7.2]
A =
    2.3000    3.4000    5.0000
    3.0000    2.4000   -1.5000
    2.0000   -0.4000   -7.2000

>> [v, lambda]=eig(A)
v =
   -0.7649   -0.6510   -0.4725
   -0.6366    0.7276    0.2479
   -0.0983   -0.2164    0.8458
lambda =
    5.7726         0         0
         0    0.1619         0
         0         0   -8.4345
```

```
>> A*v - v*lambda    % Verify:  eigen-vectors and eigen-values;
ans =
   1.0e-14 *
   -0.0888    -0.0638    -0.1776
   -0.1776    -0.0763          0
    0.0555    -0.1783    -0.1776
```

Note that there are several different syntax forms of the eig() function to compute eigen-values and eigen-vectors of square arrays, and there is another command, called eigs(A), to compute eigen-values and eigen-vectors.

```
d = eig(A)
d = eig(A,B)
[V,D] = eig(A)
[V,D] = eig(A,'nobalance')
[V,D] = eig(A,B)
[V,D] = eig(A,B,flag)
```

To evaluate the largest eigen-values and eigen-vectors, use this:

```
d    = eigs(A)
[V,D]   = eigs(A)
[V,D,flag] = eigs(A); eigs(A,B)
eigs(A,k)
eigs(A,B,k)
eigs(A,k,sigma); eigs(A,B,k,sigma); eigs(A,K,sigma,opts);
eigs(A,B,k,sigma,opts)
```

Matrix Decomposition

The matrix decompositions have broad and valuable applications in many areas of linear algebra and engineering problem solving, for instance, solving linear equations, linear least squares, nonlinear optimization, Monte-Carlo simulation, experimental data analysis, modal analysis, circuit design, filter design, and many more. There are a few types of matrix transformations and decompositions, including QR, LU, LQ, Cholesky, Schur, singular value decomposition, and so forth. We very briefly already discussed the command syntaxes of QR, LU, LQ, chol() Cholesky, and svd() singular value decompositions while solving the systems of linear equations. This section explains how to compute matrix decompostions by using MATLAB's built-in functions.

QR Decomposition

QR decomposition is also called *orthogonal-triangular decomposition*. It's the process of factoring out a given matrix as a product of two matrices. They are traditionally called the Q and R matrices, and they are the orthogonal matrix Q and the upper triangular matrix R.

$$A = QR \qquad\qquad (\text{Equation } 7-12)$$

$$Q^T Q = I \qquad\qquad (\text{Equation } 7-13)$$

Here, Q is an orthogonal matrix, Q^T is a transpose of Q, R is an upper triangular matrix, and I is an identity matrix. The QR decomposition is based on the Gram-Schmidt method. More details of the Gram-Schmidt method can be found on Wikipedia [1]. In MATLAB for the QR decomposition computation, there is a function called qr(). It has a few different syntax methods that evaluate Q, R, and other relevant matrices.

```
[Q,R] = qr(A)   %Produces  upper  triangular  matrix  R  &  unit  matrix  Q
[Q,R] = qr(A,0)    %Produces the economy-size decomposition
[Q,R,E] = qr(A)    %Produces Q, R and permutation matrix E =>A*E = Q*R
[Q,R,E] = qr(A,0)  %Produces  economy-size  decomposition:  A(:,E) =  Q*R
X  =  qr(A)        %Produces  matrix  X.  triu(X)  is  upper  triangular
factor  R
X = qr(A,0)        % The same as X = qr(A);
R = qr(A)          % Used when A is a sparse matrix and computes a Q-less
% QR decomposition and returns R.
```

Example: Computing QR Decomposition of a 5-by-5 Matrix

Let's take matrix [A] of size 5x5 generated from a normally distributed random number generator, called randn(). Compute the QR decompositions of the [A] matrix.

```
>> format short
>> A = randn(5)
A =
     0.3335    -0.4762    -0.3349     0.6601     0.0230
     0.3914     0.8620     0.5528    -0.0679     0.0513
     0.4517    -1.3617     1.0391    -0.1952     0.8261
    -0.1303     0.4550    -1.1176    -0.2176     1.5270
     0.1837    -0.8487     1.2607    -0.3031     0.4669
>> [Q, R]=qr(A)
Q =
    -0.4629     0.0336     0.6635     0.4906     0.3219
    -0.5432    -0.7902    -0.2551    -0.1236     0.0155
    -0.6269     0.4640     0.0710    -0.4160    -0.4622
     0.1808    -0.1702     0.5228    -0.7440     0.3339
    -0.2549     0.3609    -0.4650    -0.1322     0.7557
R =
    -0.7205     0.9045    -1.3201    -0.1084    -0.3993
          0    -1.7127     0.6793    -0.0872     0.2522
          0          0    -1.4600     0.4687     0.6421
          0          0          0     0.6154    -1.5365
          0          0          0          0     0.4891
>> [Q,  R]=qr(A,  0)
Q =
    -0.4629     0.0336     0.6635     0.4906     0.3219
    -0.5432    -0.7902    -0.2551    -0.1236     0.0155
    -0.6269     0.4640     0.0710    -0.4160    -0.4622
     0.1808    -0.1702     0.5228    -0.7440     0.3339
    -0.2549     0.3609    -0.4650    -0.1322     0.7557
R =
    -0.7205     0.9045    -1.3201    -0.1084    -0.3993
          0    -1.7127     0.6793    -0.0872     0.2522
          0          0    -1.4600     0.4687     0.6421
          0          0          0     0.6154    -1.5365
          0          0          0          0     0.4891
```

```
>> [Q,R,E]=qr(A)
Q =
    -0.1608    -0.0026    -0.4424     0.8652     0.1726
     0.2655    -0.0455     0.7987     0.4982    -0.2035
     0.4990    -0.4921    -0.3661     0.0262    -0.6116
    -0.5367    -0.8164     0.1799    -0.0321     0.1099
     0.6054    -0.2988    -0.0065    -0.0386     0.7366
R =
     2.0823    -0.1148    -1.1322    -0.2883     0.4568
          0    -1.7950     0.5142     0.3657    -0.1895
          0          0     1.4850    -0.3119    -0.0250
          0          0          0     0.5510     0.4924
          0          0          0          0    -0.1773
E =
     0     0     0     0     1
     0     0     1     0     0
     1     0     0     0     0
     0     0     0     1     0
     0     1     0     0     0
>> A*E
ans =
    -0.3349     0.0230    -0.4762     0.6601     0.3335
     0.5528     0.0513     0.8620    -0.0679     0.3914
     1.0391     0.8261    -1.3617    -0.1952     0.4517
    -1.1176     1.5270     0.4550    -0.2176    -0.1303
     1.2607     0.4669    -0.8487    -0.3031     0.1837
>> Q*R
ans =
    -0.3349     0.0230    -0.4762     0.6601     0.3335
     0.5528     0.0513     0.8620    -0.0679     0.3914
     1.0391     0.8261    -1.3617    -0.1952     0.4517
    -1.1176     1.5270     0.4550    -0.2176    -0.1303
     1.2607     0.4669    -0.8487    -0.3031     0.1837
```

LU Decomposition

The LU decomposition or factorization is also called a modified form of the Gauss elimination method and was introduced by Alan Turing [2]. It is defined as follows:

$$A = LU \hspace{3cm} \text{(Equation 7-14)}$$

Here, A is a rectangular matrix, and L and U are the lower and upper triangular matrices, respectively.

For example, a 3-by-3 matrix can be LU factorized with the following expressions:

$$\begin{bmatrix} A_{11} & A_{12} & A_{13} \\ A_{21} & A_{22} & A_{23} \\ A_{31} & A_{32} & A_{33} \end{bmatrix} = \begin{bmatrix} 1 & 0 & 0 \\ L_{21} & 1 & 0 \\ L_{31} & L_{32} & L_{33} \end{bmatrix} * \begin{bmatrix} U_{11} & U_{12} & U_{13} \\ 0 & U_{22} & U_{23} \\ 0 & 0 & U_{33} \end{bmatrix}$$

In MATLAB, the LU decomposition is evaluated using the following syntax of the built-in function lu():

```
Y = lu(A)        %Produces matrix Y, for sparse  A.  Y  contains  only
                 L  [L,U]  =  lu(A)    %Produces U and L
[L,U,P] = lu(A)   %Produces U & L with a unit diagonal & permutation
                 matrix P
[L,U,P,Q]   =   lu(A)   % Produces U, L, and row permutation matrix P
                        % and column reordering matrix Q, so that
                        P*A*Q = L*U
[L,U,P,Q,R]  =  lu(A)   % Produces U,L, & permutation matrices P and Q,
                        % diagonal scaling matrix R so that   P*(R\
                          A)*Q  =  L*U
                        % for sparse non-empty A.
[...] = lu(A,'vector')  %Produces the permutation information in two %row
                        vectors p and q. A user can specify from 1 to 5
                        outputs.
[...] = lu(A,thresh)
[...] = lu(A,thresh,'vector')
```

Example: Computing LU Composition of a 3-by-3 Pascal Matrix

Let's compute L, U, and other (P, Q, R) matrices from any given rectangular matrix. For this task, you write a small script called LU_decomposition.m with MATLAB's built-in function lu(). The script takes one user entry (input), which has to be a rectangular matrix. You'll employ in this script another built-in function of MATLAB, called issparse(). It identifies whether the user-entered matrix is a sparse matrix or not.

```
% LU_decomposition.m
A=input('Enter rectangular matrix: ');
if    issparse(A)
      Y  =  lu(A)                        %#ok
      [L,U,P,Q]   =   lu(A)              %#ok
      disp('  oops more  ')
      [L,U,P,Q,R]   =  lu(A)             %#ok
      [L, U, P, Q, R] = lu(A,'vector') %#ok
else
      [L,U]  =  lu(A)    %#ok
      [L,U,P]   =   lu(A)    %#ok
      % Check evaluation results:
      ERROR=P*A-L*U    %#ok
      [L,U,P]  =  lu(A,  'vector')    %#ok
end
```

Run the script LU_decomposition.m and enter a standard matrix, called pascal(3), as an input matrix.

```
Enter rectangular matrix: pascal(3)
L =
     1.0000        0          0
     1.0000     0.5000     1.0000
     1.0000     1.0000        0
U =
     1.0000     1.0000     1.0000
        0       2.0000     5.0000
        0          0      -0.5000
```

```
L =
    1.0000         0         0
    1.0000    1.0000         0
    1.0000    0.5000    1.0000
U =
    1.0000    1.0000    1.0000
         0    2.0000    5.0000
         0         0   -0.5000
P =
    1    0    0
    0    0    1
    0    1    0
ERROR =
    0    0    0
    0    0    0
    0    0    0
L =
    1.0000         0         0
    1.0000    1.0000         0
    1.0000    0.5000    1.0000
U =
    1.0000    1.0000    1.0000
         0    2.0000    5.0000
         0         0   -0.5000
P =
    1    3    2
```

Rerun the script and use a sparse matrix of size 3-by-3 as input.

```
Enter rectangular matrix: sparse(3)
Y =
   (1,1)         3
   oops more
L =
   (1,1)         1
```

```
U =

    (1,1)          1
P =

    (1,1)          1
Q =

    (1,1)          1
R =

    (1,1)          3
L =

    (1,1)          1
U =

    (1,1)          1
P =

      1

Q =

      1

R =

    (1,1)          3
```

Example: Solving [A]{x}=[b] Using LU Composition

LU composition can be employed to solve the $[A]\{x\} = [b]$ system of linear equations using the MATLAB's mldivide() or backslash (\) operator.

$$[A]\{x\} = [b] \rightarrow [A] = [P]' * [L] * [U]$$
$$[y] = [L]([P] * [b]) \rightarrow \{x\} = [U]\backslash[y]$$

Let's take the following example:

$$\begin{cases} 3x - \dfrac{2}{3}y + z = 1 \\[2mm] 2x + y - \dfrac{1}{2}z = 2 \\[2mm] \dfrac{3}{4}x - y - z = 3 \end{cases}$$

The solution of this example is as follows:

```
A = [3 -2/3 1; 2 1 -1/2; 3/4 -1 -1];
b = [1;2;3];
[L, U, P]=lu(A);
y=mldivide(L,(P*b));
x = U\y
x =

                0.82
             -0.555
             -1.83
```

Cholesky Decomposition

The Cholesky decomposition is particularly important for Monte Carlo simulations and Kalman filter designs. This type of matrix factorization is applicable only to square matrices and to Cholesky triangles, which are decompositions of positive and definite matrixes that is decomposed into a product of a lower triangular matrix and its transpose. The Cholesky decomposition [3, 4] can be expressed via the following formulation:

$$A = U^T U \qquad \text{(Equation 7-15)}$$

Here, A is a square matrix, and U and U^T are an upper triangular matrix and its transpose, respectively. This formulation can be written with lower triangular matrix (L) and its transpose (L^T) as well.

$$A = LL^T \qquad \text{(Equation 7-16)}$$

In MATLAB, the Cholesky decompositions are evaluated using the following syntax options of the MATLAB's built-in function, chol():

```
R = chol(A) % Produces an upper triangular matrix R satisfying: R'*R=A
L = chol(A,'lower') % Produces a lower triangular matrix R satisfying:
                % L*L'=A
[R,p] = chol(A) %Produces an upper triangular matrix R and p is 0
[L,p] = chol(A,'lower') %Produces lower triangular matrix R&p is 0
[R,p,S] = chol(A) % When A is a sparse matrix, produces a permutation
                % matrices S and R, and p that can be zero or non-zero
```

```
[R,p,s] = chol(A,'vector') % Produces the permutation information %as a
vector 's'
[L,p,s] = chol(A,'lower','vector') % Produces a lower triangular matrix
                                   % L and a permutation vector 's'
```

Note   Using chol (the Cholesky decomposition operator) is preferable over the eig (eigen-value and eigen-vector) operator for determining positive definiteness.

To evaluate the Cholesky decompositions of any given matrix (a user-entered matrix), you write the next script, called Chol_decoposition.m, by considering the requirements and properties of the Cholesky decompositions to compute decompositions of any matrix with respect to the formulations in Equations 7-15 and 7-16. It takes one input, which is a user entry matrix. Note that in this script, we used disp(), size(), det(), run(), and a pop-up dialog box command, warndlg().

```
% Chol_decomposition.m
clearvars; clc
disp('Note your matrix must be square & positive definite!!!')
disp('NB: Positive means all determinants must be positive.')
disp('You can enter as matrix elements ')
disp('or define your matrix 1st, ')
disp('and then just enter your matrix name')
disp('      ')
A=input('Enter a given Matrix: ');
[rows, cols]=size(A);
for k=1:rows
    % Determinants are computed
    Det_A(k)=det(A(1:k, 1:k));
end
if rows==cols
  if Det_A>0
    if issparse(A)
            [R,p,S] = chol(A) %#ok
            [R,p,s] = chol(A,'vector');
            [L,p,s] = chol(A,'lower','vector');
```

```
else
            R = chol(A) %#ok % Upper triangular matrix R: R'*R=A
            L = chol(A,'lower') %#ok % Lower triangular matrix R.
            [R,p] = chol(A);
% Verify:
Error_up  = A-R'*R;
Error_low = A-L*L';
disp('Error is with upper triangular matrix: ')
disp(Error_up)
disp('Error is with lower triangular matrix:')
disp(Error_low)
        end
else
warndlg('Sorry your matrix is not positive and definite!')
warndlg('Try again!!!')
run('Chol_decomposition')
    end
end
```

You can test the script with different input entries (matrices). Let's use a 4-by-4 standard matrix generated with pascal().

```
Note your matrix must be square & positive definite!!!
NB: Positive means all determinants must be positive.
You can enter as matrix elements
or define your matrix 1st,
and then just enter your matrix name

Enter a given Matrix: pascal(4)
R =
     1     1     1     1
     0     1     2     3
     0     0     1     3
     0     0     0     1
```

```
L =
     1     0     0     0
     1     1     0     0
     1     2     1     0
     1     3     3     1
Error is with upper triangular matrix:
     0     0     0     0
     0     0     0     0
     0     0     0     0
     0     0     0     0
Error is with lower triangular matrix:
     0     0     0     0
     0     0     0     0
     0     0     0     0
     0     0     0     0
Now, consider a magic matrix of size 3-by-3.

>> run('Chol_decomposition')
Note your matrix must be square & positive definite!!!
NB: Positive means all determinants must be positive.
You can enter as matrix elements
or define your matrix 1st,
and then just enter your matrix name

Enter a given Matrix: magic(3)
```

After running the script with an input entry of a magic square matrix of size 3-by-3, the warning dialog boxes shown in Figure 7-31 appear.

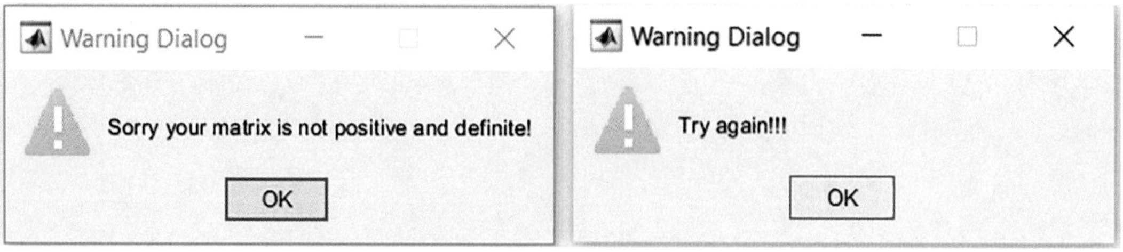

Figure 7-31. *Warnings showing that the input matrix is not positive and definite and so cannot compute the Cholesky decompositions*

Besides these two warning message boxes shown in Figure 7.31, the code keeps asking to enter a matrix. The `Chol_decoposition.m` script identifies the Cholesky decomposition properties and computes the Cholesky decomposition of a user-entered matrix. It detects a matrix type and works for given square and positive definite matrices with the MATLAB built-in function `chol()`.

Schur Decomposition

The Schur decomposition has many applications in numerical analyses, including image-processing areas in combination with other matrix decompositions or factorization tools. The Schur decomposition of a complex square matrix [A] is defined as a matrix decomposition [5]:

$$Q^H AQ = T = D + N \qquad \text{(Equation 7-17)}$$

Here, Q is a unitary matrix, Q^H is a conjugate transpose of Q, and T is an upper triangular matrix that's equal to sum of a matrix $D = diag(\lambda_1, \lambda_2, \lambda_3,..., \lambda_n)$ a diagonal matrix consisting of eigen-values λ_i of A, and strictly upper triangular matrix N. The Schur decomposition can be computed via the MATLAB's built-in function, `schur()`.

```
T  =  schur(A)    % Produces the Schur matrix of A
T  =  schur(A, flag) % Produces the Schur matrix for two cases.
%{
```

for real matrix A, returns a Schur matrix T in one of two forms depending on the value of flag:

```
'complex'   T is triangular and is complex if A has complex eigenvalues.
'real'   T has the real eigen-values on the diagonal and the complex eigen-
values in 2-by-2 blocks on the diagonal. 'real' is the default.

%}
[U,T] = schur(A,...)
```

Let's look at several examples of standard matrices and compute their Schur decompositions:

```
>> A=magic(5); B=pascal(3); C=round(randn(5,5)*10);
>> SA=schur(A)
SA =
    65.0000     0.0000    -0.0000     0.0000    -0.0000
         0    -21.2768    -2.5888     2.1871    -3.4893
         0          0    -13.1263    -3.3845    -2.8239
         0          0          0     21.2768     2.6287
         0          0          0          0     13.1263
>> SB=schur(B)
SB =
     0.1270          0          0
         0     1.0000          0
         0          0     7.8730
>> SC=schur(C)
SC =
    20.7072     7.3851    -0.2741     9.7514     1.9523
         0     -6.1453    17.4134    -5.0801   -14.3751
         0    -10.3437    -6.1453    14.7269     9.9502
         0          0          0     4.8687     2.1653
         0          0          0          0     -9.2853
>> [T,  U]=schur(A,  'complex')
T =
    -0.4472     0.0976    -0.6331     0.6145    -0.1095
    -0.4472     0.3525     0.7305     0.3760     0.0273
    -0.4472     0.5501    -0.2361    -0.6085     0.2673
    -0.4472    -0.3223     0.0793    -0.3285    -0.7628
    -0.4472    -0.6780     0.0594    -0.0535     0.5778
U =
    65.0000     0.0000    -0.0000     0.0000    -0.0000
         0    -21.2768    -2.5888     2.1871    -3.4893
         0          0    -13.1263    -3.3845    -2.8239
         0          0          0     21.2768     2.6287
         0          0          0          0     13.1263
```

```
>> [TA,  UA]=schur(B,  'real')
TA =
   -0.5438    -0.8165     0.1938
    0.7812    -0.4082     0.4722
   -0.3065     0.4082     0.8599
UA =
    0.1270         0          0
         0    1.0000          0
         0         0     7.8730
>> [T, U]=rsf2csf(U,T) %  Convert real Schur form to complex Schur form
T =
  -61.5539    20.8834    -0.0000    -0.0000     0.0000
    6.2354    18.3788     9.2270    -1.2464     3.6069
    2.0845     6.1442    -9.8039     6.8268     2.6290
    0.9854     2.9044    -4.6344   -20.6565    -1.4269
    0.0340     0.1003    -0.1601    -0.7135   -13.1055
U =
   -0.5636     0.1041     0.6400     0.4570    -0.0505
         0     0.8710    -0.2777     0.0336     0.2993
   -0.4472         0     -0.4787     0.7335    -0.3928
   -0.4472    -0.3223         0     -0.5309    -0.6067
   -0.4472    -0.6780     0.0594         0      0.6209
```

Singular Value Decomposition

The singular value decomposition (SVD) has many applications in signal processing, statistics, and image processing areas. It is formulated as a product of three matrices, which are an orthogonal matrix (U_{ij}), a diagonal matrix (D_{ij}), and the transpose of an orthogonal matrix (V_{jj}), if a given matrix A_{ij} is an i by j sized real matrix with $i > j$.

$$A = U_{ii}D_{ij}V_{jj}^T$$

(Equation 7-18)

Here, $U_{ii}^T U_{ii} = I, V_{jj}^T V_{jj} = I$. Diagonal entries of D_{ij} are known as singular values of A_{ij}.

Moreover, there are a few other important properties of the SVD.

- Left-singular vectors of *Aij* are eigen-vectors of $A_{ij}A_{ij}*$.

- Right-singular vectors of *Aij* are eigen-vectors of $A_{ij}*A_{ij}$.

- Nonzero singular values (on the diagonal entries of D_{ij}) of A_{ij} are square roots of the nonzero eigen-values of both $A_{ij}*A_{ij}$ and $A_{ij}*$.

There are a few ways to evaluate the SVD, singular values, and vectors of any given matrix. You use svd() and svds(), which are MATLAB built-in functions.

```
s = svd(A)         %Produces a vector of singular values
[U,D,V] = svd(A) %Produces a diagonal matrix D of the same dimension
%as A, with nonnegative diagonal elements in decreasing order, and
% unitary matrices U and V so that X = U*D*V'.
[U,D,V] = svd(A,0) % Produces the "economy size" decomposition. If A
%  is   m-by-n with m > n, then SVD computes only the first n columns of
%U and  D  is n-by-n. s = svds(A)
s = svds(A,k)
s = svds(A,k,sigma) s = svds(A,k,'L')
s = svds(A,k,sigma,options) [U,D,V] = svds(A,...)
[U,D,V,flag] = svds(A,...)
```

Now, take two matrices (of size 2-by-3 and 3-by-3) and evaluate their SVDs.

```
>> A=ceil(randn(2,3)*10); B=pascal(3);
>> A=ceil(randn(2,3)*10); B=pascal(3);
>> A
A =
    -2    -4     3
   -15    -1    -2
>> B
B =
     1     1     1
     1     2     3
     1     3     6
```

```
>> svd(A)
ans =
   15.2914
    5.0172
>> [U,V,D]=svd(A)
U =
   -0.1354    -0.9908
   -0.9908     0.1354
V =
   15.2914          0          0
        0     5.0172          0
D =
    0.9896    -0.0100    -0.1434
    0.1002     0.7629     0.6387
    0.1030    -0.6464     0.7560
>> [U,V,D]=svd(A, 0)
U =
   -0.1354    -0.9908
   -0.9908     0.1354
V =
   15.2914          0          0
        0     5.0172          0
D =
    0.9896    -0.0100    -0.1434
    0.1002     0.7629     0.6387
    0.1030    -0.6464     0.7560
>> SA = svds(A)
SA =
   15.2914
    5.0172
>> SB = svds(B)
SB =
    7.8730
    1.0000
    0.1270
```

```
>> SA = svds(A, 2)
SA =
   15.2914
    5.0172
>> SB = svds(B, 2)
SB =
    7.8730
    1.0000
>> SB = svds(B, 3)
SB =
    7.8730
    1.0000
    0.1270
```

Logic Operators, Indexes, and Conversions

MATLAB uses logic 1 and logic 0 for system variables to denote logic values for true and false, respectively. Variables of logical values are distinguished by a logical data type.

Table 7-2 is a list of logic operators and their operational functions used in MATLAB.

Table 7-2. *Logical Expressions and Operators in MATLAB*

Operator	Operation
true, false	Setting logical value
& (and), | (or), ~ (not), xor, any, all	Logical operations
&&, ||	Short-circuits operations
bitand, bitcmp, bitor, bitmax, bitxor, bitset, bitget, bitshift	Bitwise operations
==(eq), ~=(ne), <(lt), >(gt), <=(le), >=(ge)	Relational operations
strcmp, strncmp, strcmpi, strncmpi	String comparisons

Note   To get a complete list of relational operators, their functions, and how to use them, type >> `help relop` in the Command window.

Logical Indexing

Logic operators are one of the most central and essential keys to any programming language. Logic operators introduce another method for accessing data in MATLAB variables. For instance, given a magic matrix [A] of size 5-by-5, say you need to separate out the elements of [A] that are equal to or less than 13.

```
>> A=magic(5)
A =
    17    24     1     8    15
    23     5     7    14    16
     4     6    13    20    22
    10    12    19    21     3
    11    18    25     2     9
>> Index = A>15 | A<5    % Show which element is greater than 15 or
less than 5
Index =
  5×5 logical array
   1   1   1   0   0
   1   0   0   0   1
   1   0   0   1   1
   0   0   1   1   1
   0   1   1   1   0
>> A(Index)
ans =
    17
    23
     4
    24
    18
     1
```

```
        19
        25
        20
        21
         2
        16
        22
         3
>> A(A>15 | A<5);  % Or in a direct way
```

Let's explore the logical indexing properties further via examples to select matrix elements.

```
>> E = eye(5)  % Identity matrix
E =
     1     0     0     0     0
     0     1     0     0     0
     0     0     1     0     0
     0     0     0     1     0
     0     0     0     0     1
```

Array indices must be positive integers or logical values.

```
>>EL=loogical(E)
EL =
  5×5 logical array
   1   0   0   0   0
   0   1   0   0   0
   0   0   1   0   0
   0   0   0   1   0
   0   0   0   0   1
>> A(EL)     % Compare Ih A(E)
ans =
      17
       5
      13
```

```
     21
      9
```
A(EL) - shows all diagonal elements of A matrix.
Note

Note The previous example demonstrates the identity matrix [e] (whose elements are 1s and 0s), which is not equivalent to the logic matrix [e_L] (whose elements are also 1s and 0s).

Moreover, there are a number of functions/commands (e.g., the is*() command) that can be used to find out whether the input is of a specified type of variable, contains any elements of a particular type, or whether such a variable or file exists, and so forth. All of these functions can be used for logical indexing. Let's look at a few simple examples:

```
>>  x=13;  isnumeric(x)     % whether x is a numeric data or not?
ans = 1
>>  x=13; islogical(x)      % whether x is a logical data Or not?
ans = 0
>> x=13; islogical(x>110)   % whether the operation
(x>110)  is  logic  or  not
ans =1
>>  x  =  13;  isempty(x)    % whether x is an empty or not
ans =1
>> x = [ ]; isempty(x)       % whether x is an empty or not
ans = logical 1
>>  x  =  [1  , 2;  3, 4]; iscell(x)   % whether x is a cell array or not
ans = logical 0
>> x = [1, 0; 0, 4];
>>  X_x   =  x/0;  isnan(X_x)   % whether any elements of X_x
are  not-a-number
ans = 2×2 logical array
0   1
1   0
```

```
>>'  e'ist''X_'',  'var')   % whether the variable called
X_x  exists  or  not
ans =1
```

In the previous example, zero divided by zero (0/0) is defined to be NaN (i.e., not-a-number) in MATLAB.

Note   The logical indexing operations have particular importance in matrix/array operations, programming, data analysis, and processing since they can be used to sort out, locate, or change particular elements of matrices/arrays/data sets.

Example: Logical Indexing to Locate and Substitute Elements of [A] Matrix

Given: 3-by-3 matrix [A] with some elements equal to infinity $A = \begin{bmatrix} 17 & \infty & -6 \\ 5 & -3 & 11 \\ \infty & 13 & \infty \end{bmatrix}$

How do you substitute the elements equal to inf with 1000 and all negative-valued elements with 0? This task can be solved easily using logical indexing operations.

```
>> A = [17, Inf,   -6 ; 5 -3, 11; Inf, 13, Inf] % [A] is entered
A =
    17    Inf    -6
     5    -3     11
   Inf    13    Inf
>>   Index_inf  =  (A==1/0)   % Find out which elements of A
are  equal  to  inf
Index_inf =
  3×3 logical array
   0   1   0
   0   0   0
   1   0   1
```

```
>> A(Index_inf) =1000        % Set inf elements equal to 1000
A =
              17          1000             -6
               5            -3             11
            1000            13           1000
>> Index_neg = A<0           % Find out which elements  of  A  are  negative
Index_neg =
  3×3 logical array
   0   0   1
   0   1   0
   0   0   0
>> A(Index_neg)=0            % Set all negative elements equal to "0"
A =
              17          1000              0
               5             0             11
            1000            13           1000
```

Note   The division of any value by 0 gives the value of Inf in MATLAB.

Let's look at another example. Given a matrix [A] of size 4-by-5 with NaN (not a number) and inf (infinity) elements, how do you substitute NaN elements with 0 and inf with 100? This task can be solved easily with logical indexing similar to the previously demonstrated example.

```
>> A  =  [2, -3, -2, -3, 1; Inf, -2,   3, -1, NaN; -3, 0, Inf, 3, 2; 3,
NaN, 0, 2, Inf]
A =
     2    -3    -2    -3    1
   Inf    -2     3    -1   NaN
    -3     0   Inf     3    2
     3   NaN     0     2   Inf
>> Index_nan = isnan(A)    % Find out which elements are NaN
```

```
Index_nan =
  4×5 logical array
   0   0   0   0   0
   0   0   0   0   1
   0   0   0   0   0
   0   1   0   0   0
>> A(Index_nan)=0  % Set all NaN elements equal to "0"
A =
     2    -3    -2    -3    1
   Inf    -2     3    -1    0
    -3     0   Inf     3    2
     3     0     0     2  Inf
```

Note that this section contains rather simple and small examples to demonstrate how easily you can substitute specific (valued) elements of a matrix using logical indexing operations. This technique (logical indexing or relational operators) can be applied to matrices, arrays, and data sets of any size. Therefore, the logical indexing is particularly useful in analysis and processing of large data sets. It is fast and efficient and does not require any additional effort to program with loop (for ... end, while ... end) and conditional (if ditio. endlyit) operators.

Conversions

There are many examples in signal processing where you need to convert something. Analog to digital converters and vice versa, data processing and analysis, and programming when analog signal data format or type needs to be converted into digital or vice versa. For instance, to resolve memory issues in image processing, you might need to convert decimal (double) formatted data into binary numbers. That can be easily accomplished in MATLAB using DEC2BIN(). Conversely, BIN2DEC() is used to convert binary strings into decimal (double) type of data. DEC2BIN(D) returns the binary representation of D as a string. D must be a non-negative integer smaller than 2^{52}.

DEC2BIN(D,N) produces a binary representation with at least N bits.

Another conversion example is character conversion. You need to convert numbers into character strings and vice versa. MATLAB uses the CHAR() command to convert numbers into ASCII/ANSI formatted characters, DOUBLE() to convert characters and symbolic representations of numbers into double precision format, STR2NUM() to convert strings into binary numbers, and NUM2STR() to convert any number into a string. Let's consider several examples of employing these conversion commands:

```
>> dec2bin(11) % Converts  decimal  (integer) into  a  binary  string
ans =
    '1011'
>> dec2bin(23)
ans =
'10111'
>> dec2bin(22) ans =
'10110'
>> x=13.125/5.5;
>> dec2bin(x)
ans =
    '10'
>> dec2bin(11.11)
ans =
    '1011'
>> dec2bin(11)
ans =
    '1011'
>> bin2dec('1101') % Converts  a  binary  number  into  decimal  one
ans =
13
>> bin2dec('10110')
ans =
22
>> dec2bin(64)
ans =
10000000
>> char(bin2dec('10000000'))
ans =
@
>>   G='MatLab' G =
MatLab
```

```
>> GO=G+0 GO =
77    97    116    76    97    98
>> d2bGO=dec2bin(GO) d2bGO =
1001101
1100001
1110100
1001100
1100001
1100010
>> b2dGO=bin2dec(d2bGO) b2dGO =
77
97
116
76
97
98
>> char(b2dGO)' ans =
MatLab
>> num2str(123) ans =
'123'
>> num2str('matlab') ans =
'matlab'
>> ans+0
109    97    116    108    97    98
```

Example: Creating Character Strings with char()

Create the following letters in a progressive format by writing a script that has one input argument that has to be an integer. All the other letters need to be generated programmatically.

```
a
b c
d e f
g h i j
k l m n o
```

These characters can be generated in several ways. First, you need to determine the ASCII/ANSI numeric representation of a. Then you can generate all the other letters.

```
>> format short
>> double('a')
    97
>> char(97)
    'a'
>> double('b')
    98
```

The letter as numeric representation in ASCII/ANSI is 97, b is represented by 98, and so forth. Based on these, you can generate linear space of integers starting at 97 and convert them to character strings one row at a time. In other words, you display one character on the first row, two characters on the second, three in the third row, etc. Here is the complete script (print_character.m), which prints the letters in progressive order:

```
% print_character.m
% Part 1.
Start = 97;
for ii = 1:5
    for jj = 1:ii
        fprintf(char(Start));
        Start = Start+1;
    end
    fprintf('\n')
end
```

Here is the result of the script:

```
a
bc
def
ghij
klmno
```

Let's consider the following example, which prints a series of uppercase characters:

```
ABCDEF
GHIJK
```

LMNO

PQR

ST

U

This example is similar to the previous example with a few small differences—it requires uppercase characters, starts with six letters, and reduces in the following rows.

Again, you can determine the numerical representation of A in ANSI/ASCII with the following commands:

```
>> double('A')
ans =
    65
>> 'A' + 0     % An alternative way:
ans =
    65
```

So now you know that the numerical representation of A is 65. You can then edit the script (print_characters.m) by introducing two small changes and then write this script:

```
%% Part 2. Upper cases
Start = 65;
for ii = 1:2:9
    for jj = 1:ii
        fprintf(char(Start));
        Start = Start+1;
    end
    fprintf('\n')
end
```

When you execute this script, you obtain the following output in the Command window:

ABCDEF

GHIJK

LMNO
PQR
ST
U

Via a few examples, this section discussed logic operators, conversions, and indexing issues briefly. Applications of the issues of conversions are demonstrated via more extended examples in other chapters.

Summary

This chapter introduced linear algebra, matrix operations, vector spaces, polynomials, methods of solving linear systems of equations, and matrix decompositions and conversions. Via examples, you learned how to use MATLAB's built-in functions and commands, how to develop Simulink blocks in association with the MATLAB Command window, and how to use functions and the MATLAB Fcn block. The following MATLAB functions were discussed and explained in examples:

- Matrix operations +,-, *, and /
- Elementwise operations .*, .^, and ./
- Backslash operator (\) and mldivide()
- Solving linear equations with linsolve()
- Matrix inverse operators inv() and pinv()
- Eigen-values and eigen-vectors eig()
- Polynomial solvers roots(), solve(), and zero()
- Symbolic math equation solver solve()
- Standard matrices and gallery matrices, magic(), gallery(), and sparse()
- Vector spaces linspace() and logspace()
- Matrix operations and factorization methods, such as QR, LU, Cholesky, SVD, Schur: qr(), lu(), chol(), svd(), schur(), and decomposition()

- Logical operators (`<=`, `~=`, `>=`, `|`, `&...`, `is*()`) and indexing options

- Conversion tools and operators (`bin2dec`, `dec2bin`, `double`, and `char`)

References

[1]. Wikipedia, `http://en.wikipedia.org/wiki/Gram-Schmidt_process`, viewed on September 19, 2013.

[2]. Bunch, James R.; Hopcroft, John (1974), "Triangular Factorization and Inversion by Fast Matrix Multiplication," *Mathematics of Computation* 28: 231–236, ISSN 0025-5718.

[3]. Gentle, J. E. "Cholesky Factorization." §3.2.2 in *Numerical Linear Algebra for Applications in Statistics.* Berlin: Springer-Verlag, pp. 93-95, 1998.

[4]. Nash, J. C. "The Choleski Decomposition." Ch. 7 in *Compact Numerical Methods for Computers: Linear Algebra and Function Minimisation, 2nd ed.* Bristol, England: Adam Hilger, pp. 84-93, 1990.

[5]. Mathworld , `http://mathworld.wolfram.com/SchurDecomposition.html`, viewed on September 20, 2013.

Exercises for Self-Testing

Exercise 1

Solve the following equations for variables x, y, and z:

$$\begin{cases} 3x + 5y + 4z = -2 \\ -2x + 3y - 2z = 2 \\ x + 6\left(y - \dfrac{z}{2}\right) = 0 \end{cases}$$

1. Use the backslash (\) operator or `mldivide()` to solve the given system of equations.

2. Use the inverse matrix method `inv()` to solve the given system of equations.

3. Use the `linsolve()` function to solve the given system of equations.

4. Use the `solve()` function to solve the given system of equations.

5. Use `chol()` to solve the given system of equations.

6. Use Simulink blocks to solve the given system of equations.

7. Compute errors by computing norms for each of the methods.

Exercise 2

Solve the following equations, using the matrix inverse:

$$\begin{cases} 2q_1 + 9q_2 + 3q_3 = 15 \\ 13q_1 + 2q_2 - 5q_3 = 11 \\ q_1 - 2q_2 + 2q_3 = 9 \end{cases}$$

1. Use the inverse matrix method `inv()`.

2. Use the least squares method `lsqr()`.

3. Use the Gauss Elimination method with the `lu()`.

4. Use `rref()`.

5. Use the `solve()`.

6. Use Simulink blocks.

7. Compare the accuracy (to eight decimal places) of each solution.

Exercise 3

Solve the following equations:

$$\begin{cases} 2.5x_1 - x_2 + 3.3x_3 = 5 \\ -2.2x_1 + 2x_2 - 5x_3 = -2 \\ x_1 - 2x_2 + 2.5x_3 = 3 \end{cases}$$

1. Use the inverse matrix method qr() to solve the given system of equations.

2. Use the reduced row echelon method step-by-step by multiplying rows by scalars and adding or subtracting from each other (don't use rref()).

3. Use the reduced row echelon method rref() to solve the given system of equations.

4. Use the decomposition() function to solve the given system of equations.

5. Use the solve() function to solve the given system of equations.

6. Use Simulink blocks to solve the given system of equations.

7. Compare the accuracy (to 10 decimal places) of these four methods.

Exercise 4

Solve the following equations:

$$\begin{cases} 2.5x_1 - x_2 + 3.3x_3 - 0.3x_5 = 5 \\ -1.2x_1 + 2.5x_2 - 2x_3 - 2.2x_4 + 5.2x_5 = -3 \\ x_1 + 3x_2 - 2.5x_3 - x_5 = 1 \\ 2x_1 + x_2 - 5x_3 - 3x_4 - 4.3x_5 = -6 \\ 3x_1 - 2.4x_2 + 1.75x_5 = 13 \end{cases}$$

1. Use the inverse matrix method `inv()` to solve the given system of equations.

2. Use the singular decomposition `svd()` method to solve the given system of equations.

3. Use the `linsolve()` function to solve the given system of equations.

4. Use the `solve()` function to solve the given system of equations.

5. Use Simulink blocks to solve the given system of equations.

6. Compare the accuracy (to 13 decimal places) of these methods.

Exercise 5

Given:

$$3x + 6y - cz = 0$$

$$2x + 4y - 6z = 0$$

$$x + 2y - 3z = 0$$

-   Find for which values of c the set of equations has a trivial solution.

-   Find for which values of c the set of equations has an infinite number of solutions.

-   Find relations between x, y, and z.

Exercise 6

Find the inverse of the given matrix:

$$A = \begin{bmatrix} 3 & 6 & -12 \\ 2 & 4 & -6 \\ 1 & 2 & -3 \end{bmatrix}$$

- Explain why the given matrix does not have an inverse.

- Compute the determinant of the matrix.

- Find eigen-values and eigen-vectors of the given system by using eig().

- Find eigen-values by using roots().

Exercise 7

Find the inverse of the given matrix:

$$A = \begin{bmatrix} 3 & 6 & -2 \\ 1 & 2 & -4 \\ 0 & 1 & -3 \end{bmatrix}$$

- Compute determinant of the matrix.

- Find eigen-values and eigen-vectors of the given system using eig().

- Find eigen-values using roots().

Exercise 8

Find a solution to the following set of equations representing an underdetermined system, using the left division (\ backslash) method and the pseudo-inverse method (pinv). Compare your obtained results and discuss the differences.

$$2.5x_1 - x_2 + 3.3x_3 + 1.3x_4 - 0.3x_5 = 11$$

$$-1.2x_1 + 2x_2 - 5x_3 - 2.1x_4 + 5.2x_5 = -2$$

$$x_1 - 2x_2 + 2.5x_3 = 3$$

Exercise 9

Solve the following set of equations using the backslash (\) operator, as well as the linsolve(), inv(), lsqr(), and solve() functions:

$$2x - 3y = 5$$

$$6x + 10y = 70$$

$$10x - 4y = 53$$

Exercise 10

Show why there is no solution to the following set of equations:

$$-2x - 3y = 2$$
$$-3x - 5y = 7$$
$$5x - 2y = -4$$

Exercise 11

Solve the following equations:

$$\begin{cases} 2.5x_1 - x_2 + 3.3x_3 - 0.3x_5 = 5 \\ -1.2x_1 + 2.5x_2 - 2x_3 - 2.2x_4 + 5.2x_5 = -3 \\ x_1 + 3x_2 - 2.5x_3 - x_5 = 1 \\ 2x_1 + x_2 - 5x_3 - 3x_4 - 4.3x_5 = -6 \\ 3x_1 - 2.4x_2 + 1.75x_5 = v \end{cases}$$

$$v = \begin{bmatrix} -10, -9, -8, \ldots 9, 10 \end{bmatrix}$$

1. Use the inverse matrix method `mldivide()` to solve the given system of equations.

2. Use the singular decomposition `svd()` method to solve the given system of equations.

3. Use the `linsolve()` function to solve the given system of equations.

4. Use the `solve()` function to solve the given system of equations.

Exercise 12

Compute the eigen-values and vectors of the following set of equations:

$$\left\{ \begin{array}{l} 3x - 2y + 5z + 2u - w = 0 \\ x + y + z - w = 0 \\ -2x + 3y + 4z - 5u - 7w = 0 \\ -5y - \dfrac{3}{4}z + \dfrac{7}{13}u + 9w = 0 \\ 10x - 11y + 8u - 8w = 0 \end{array} \right.$$

Exercise 13

Create the matrix $[C]$ from the given two $[A]$ and $[B]$ matrices by using logic operators.
Explain why some of the elements of new array are zeros.

$$C = \begin{bmatrix} 0 & 0 & -3 \\ 1 & 0 & 0 \\ 0 & 0 & 1 \\ 1 & 0 & -2 \\ 2 & 1 & 0 \\ 0 & 0 & 1 \end{bmatrix}, A = \begin{bmatrix} 3 & 6 & -3 \\ 1 & 2 & -2 \\ 0 & 1 & 1 \\ 1 & 3 & -2 \\ 2 & 1 & -1 \\ 1 & 0 & 1 \end{bmatrix}, B = \begin{bmatrix} 1 & 2 & -3 \\ 1 & 1 & -3 \\ -1 & -2 & 1 \\ 1 & 1 & -1 \\ 2 & 2 & -2 \\ -2 & 0 & 1 \end{bmatrix}$$

Hints   Use logic operators ($<$, $=$) and element-wise matrix multiplication.

Exercise 14

The useful life of a machine bearing depends on its operating temperature, as the following data shows. Obtain a functional description (linear, square, and cubic polynomials) of this data. Plot the found fit functions and the data on the same plot. Estimate a bearing's life if it operates at 52.5°C.

Temperature (^{0}C)	40	45	50	55	60	65	70
Bearing life (hours x 10^3)	28	21	15	11	8	6	4

Exercise 15

The following represents pressure samples, in MPa, taken in a fuel line once every second for 10 sec:

Time (Sec)	Pressure (MPa)	Time (Sec)	Pressure (MPa)
1	2.61	6	3.06
2	2.70	7	3.11
3	2.82	8	3.13
4	2.90	9	3.10
5	2.98	10	3.05

a. Fit a *first – degree* polynomial, a *second – degree* polynomial, and a *third – degree* polynomial to this data. Plot the curve fits along with the data points.

  b. Use the results from part *a* to predict the pressure at $t = 11$ *sec*.

Exercise 16

The distance a spring stretches from its "free length" is a function of how much tension force is applied to it. The following table gives the spring length y that the given applied force F produced in a particular spring. The spring's free length is 4.7 *m*. Find functional relation between F and x, the extension from the free length ($x = y - 4.7$).

Force F(kN)	Spring Length y (m)
0	4.7
0.47	7.2
1.15	10.6
1.64	12.9

Also, plot experimental data (F versus x) and functional relation based fit (F_linear vs. x) in the same plot. Use the appropriate plot maker type, color, size, etc., options.

Exercise 17

Perform the following:

- Obtain an eye matrix of the size 5-by-5 from the magic matrix of the size of 5-by-5.

- Create a square eye matrix of the size 10-by-10 from the random square matrix of the size 10-by-10.

- Obtain a replicated square matrix of size 3-by-9 from the gallery matrix `pascal()` of size 3-by-3.

Exercise 18

Solve the following equations and discuss the solutions for two cases: $a = 13$ and $a = 29$.

$$\begin{cases} q_1 + q_2 = 1 \\ 13q_1 + 23q_2 = a \\ q_1 - 2q_2 = 9 \end{cases}$$

Write a script with logic and loop operators (`if`, `break`, `for`, and `end`) to find such value of a that gives real solutions to these equations. Consider that a has an integer value that lies within 1 to 50.

Hints   Use the `rank()` function and backslash (\\) operators.

Exercise 19

Solve the following polynomials with `roots()`, `solve()`, `zero()`, and the Simulink model.

$$2u^9 - 3u^8 + 5u^4 - 13u^2 - 131 = 0$$

$$y^7 + 5y^5 - \frac{13y^4}{4} - 11y^3 + 9y + 3 = 0$$

$$5x^5 + \frac{4x^4}{11} - \frac{3x^3}{13} + x^2 - 269x = 13$$

Exercise 20

Create a logarithmic spaced array (a row vector) B of numbers starting with 10 and ending with 100, and create BB column vector from a row vector B.

Exercise 21

Play a sound that is defined in the next expression:

$$S(t) = \cos(2\pi t f_1) + \cos(2\pi t f_2) + \cot(2\pi t f_3) + \tan(2\pi t f_4) + \tan(2\pi t f_5)$$

Here, f_s = 10000 Hz (sampling frequency); t = 13 sec. (time length); f_1 = 100 Hz (1st signal); f_2 = 200 Hz (2nd signal); f_3 = 300 Hz (3rd signal); f_4 = 600 Hz (4th signal); f_5 = 700 Hz (5th signal).

Exercise 22

Answer the following questions using MATLAB:

- What are the binary representations of decimal numbers 123, 123.123, 321, 321.123, 223, 322, 333, and 333.3?

- Why are the binary representations of 123 vs. 123.123, 321 vs. 321.123, and 333 vs. 333.3 the same?

Exercise 23

Answer the following questions using MATLAB:

- What are the decimal representations of the binary numbers 1001, 01010, 111100, 0101011?

- What are character representations of the binary numbers 1001, 01010, 111100, 0101011?

Exercise 24

Write a script that takes one input number (an integer) and prints out the following characters in the order in the Command window:

```
A
BCD
EFGHI
JKLMNOP
QRSTUVWXY
```

Exercise 25

Use numeric values of matrices [A] and [B] from Exercise 11 to evaluate the QR, LU, LQ, Cholesky, Schur, and singular value decompositions. Explain why some of the decompositions (matrix factorizations) of [A] and [B] cannot be computed.

Exercise 26

Create the Hilbert matrix of size 5-by-5 using gallery matrix functions and compute Cholesky decomposition using the Chol_decoposition.m script. Edit the script (Chol_decoposition.m) in order to make it compute only the lower triangular matrix of Cholesky decomposition.

Exercise 27

Create the Riemann matrix of size 3-by-3 using gallery matrix functions and compute its QR, LU, LQ, Cholesky, Schur, and singular value decompositions.

Exercise 28

Perform the following:

- Create the 4-by-4 random matrix with normalized columns and specified singular values using gallery matrix functions. Hint: Use randcolu.

- Compute the QR, LU, LQ, and decompositions of the matrix you just created.

Exercise 29

Perform the following:

- Create one 5-by-5 random matrix with random integer elements varying in the range of 1 to 13 and name it A_mat.

- Create one 5-by-5 Krylov matrix using a matrix gallery of Krylov and name it K_mat.

- Create logic valued 5-by-5 matrix called Logic_A by using logic operation $(A\_mat \geq K\_mat)$ and elementwise matrix multiplication from A_mat and K_mat

Exercise 30

Create the following 10-by-10 matrix:

```
AaA =
    17   24    1    8   15   17   24    1    8   15
    23    5    7   14   16   23    5    7   14   16
     4    6   13   20   22    4    6   13   20   22
    10   12   19   21    3   10   12   19   21    3
    11   18   25    2    9   11   18   25    2    9
    17   24    1    8   15   17   24    1    8   15
    23    5    7   14   16   23    5    7   14   16
     4    6   13   20   22    4    6   13   20   22
    10   12   19   21    3   10   12   19   21    3
    11   18   25    2    9   11   18   25    2    9
```

Hint   Use `magic()` and `repmat()`.

CHAPTER 8

Ordinary Differential Equations

Many modeling problems with engineering applications can be formulated using ordinary differential equations (ODEs). There are a few different definitions of differential equations. One of the simplest is "A differential equation is any equation which contains derivatives, either ordinary derivatives or partial derivatives," as given in source [1]. From this definition, we can derive two types of differential equations: ordinary differential equations (ODEs) and partial differential equations (PDEs). ODEs contain one type of derivative or one independent variable, and PDEs, on the contrary, contain two or more derivatives or independent variables. For example, first-order ODEs can be expressed as follows:

$$\frac{dy}{dx} = f(x,y) \qquad \text{(Equation 8-1)}$$

Here, $y(x)$ is a dependent variable whose values depend on the values of the independent variable of x. Another good example of ODEs is Newton's Second Law of Motion, formulated as follows:

$$ma = \frac{dp}{dt} = \frac{mdv}{dt} = f(t,v) \qquad \text{(Equation 8-2)}$$

Here, $F(t, v)$ is force, which is a function of time (t) and velocity (v). $\frac{dv}{dt}$ is a velocity change rate (acceleration) of a moving object; m is the mass of a moving object; a is an acceleration of a moving object; p is momentum; and dp/dt is its derivative. This formulation of Newton's Second Law can be also rewritten the following way:

$$\frac{md}{dt}\left(\frac{dx}{dt}\right) = \frac{md^2x}{dt^2} = F\left(t, x, \frac{dx}{dt}\right) \qquad \text{(Equation 8-3)}$$

© Sulaymon Eshkabilov 2022
S. Eshkabilov, *Beginning MATLAB and Simulink*, https://doi.org/10.1007/978-1-4842-8748-4_8

Here, the derivative $\left(\dfrac{dx}{dt}\right)$ of the displacement (x) of a moving object is the velocity (v).

In other words, the velocity is the rate of change of the displacement $x(t)$ of a moving object in time. This can be visualized with the flowchart displayed in Figure 8-1.

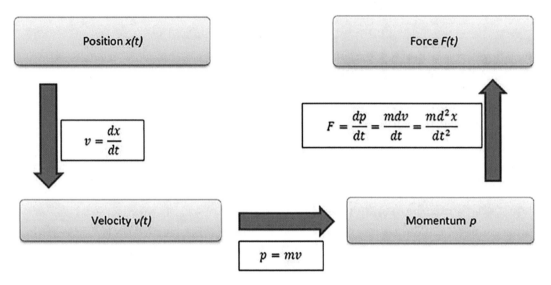

Figure 8-1. *Flowchart expressing motion and exerted force of a moving object*

Classifying ODEs

There are two classifications of ODE-related problems.

- *Initial value problems (IVPs)*: $\ddot{x} = xt - 3\dot{x}$ with initial conditions $x(0) = 3, \dot{x}(0) = 1$

- *Boundary value problems (BVPs)*: $\ddot{x} = xt - 3\dot{x}$ with boundary conditions $x(0) = 3, x(2) = 1.50$

IVPs are defined with ODEs together with a specified value, called the *initial condition*, of the unknown function at a given point in the solution domain. In the IVP of ODEs, there can be a unique solution, no solution, or many solutions. By definition, the IVP of ODEs can be explicitly or implicitly defined. Most of the IVP are explicitly defined. Let's start with explicitly defined IVPs and then move to implicitly defined ones. In addition to solution type—how solution values change over the solution search space—the IVPs are divided into stiff and nonstiff problems. Moreover, ODEs are grouped into two categories, linear and nonlinear, and divided into two groups, homogeneous and nonhomogeneous.

Here are some specific examples of different ODE types, categories, and groups:

- *Stiff ODEs*: $\dot{y} = 3 * 10^8 y,\ t \in [0, 40]$

- *Nonstiff*: $\dot{y} + 2y = 2t$

- *Linear ODEs*: $\dot{v} = 9.81 - 0.198v$

- *Nonlinear ODEs*: $\dot{v} = 9.81 - 0.198v^2$

- *Homogeneous ODEs*: $\dot{y} + 2y = 0$

- *Nonhomogeneous ODEs*: $\dot{y} + 2y = \sin(2t)$

The following are several examples of ODEs and their application areas.

Example 1: Unconstrained Growth of Biological Organisms

This is an exponential growth problem that describes the unconstrained growth of biological organisms (such as bacteria). This behavior can also describe real estate or investment values, membership increase of a popular networking site, growth in retail businesses, positive feedback of electrical systems, and generated chemical reactions. The problem is formulated by the following first-order ODE:

$$\frac{dy}{dt} = \mu y \text{ has a solution: } y(t) = y0 e^{\mu t}$$

Example 2: Radioactive Decay

This refers to exponential decay, which describes many phenomena in nature and in engineering, such as radioactive decay, washout of chemicals in a reactor, discharge of a capacitor, and decomposition of material in a river. It's expressed using this first-order ODE:

$$\frac{dy}{dt} = -\mu y \text{ has a solution: } y(t) = y0 e^{-\mu t}$$

Examples 1 and 2 are two simple examples of first-order ODEs.

Example 3: Newton's Second Law

The motion of a falling object is expressed in the following equation using Newton's Second Law:

$$\frac{md^2y}{dt^2} = mg - \frac{\gamma dy}{dt}$$

This is a second-order ODE that has a solution in the following form:

$$y(t) = C_1 e^{-\left(\frac{\gamma t}{m}\right)} - \frac{m(mg - g\gamma t)}{\gamma^2} + C_2$$

Here, m is the mass of the falling object, g is gravitational acceleration, and γ is an air-drag coefficient of a falling object. Three parameters—m, g, and γ—are constant, the solution of a falling object, and C_1 and C_2 are arbitrary numbers that are dependent on the initial conditions. In other words, they can be computed considering the initial condition of a falling object.

There are a few methods that evaluate analytical solutions of ODEs, including separation of variables, introduction of new variables, and others. We look at specific examples of these types of ODEs to see how to evaluate their analytical solutions and compute numerical solutions. We do this by employing different techniques in the MATLAB/Simulink environment and writing scripts and building models. In computing analytical solutions of ODEs, we explain via specific examples how to use built-in functions of the Symbolic Math Toolbox.

For obvious reasons, considerable effort is placed on numerical solution methods rather than analytical solution search tools. It is not always possible or is too costly to evaluate analytical solutions of ODEs. Therefore, a numerical solution search is often best. There are a number of numerical methods. They are Euler (forward, backward, modified), Heun, the midpoint rule, Runge-Kutta, Runge-Kutta-Gill, Adams-Bashforth, Milne, Adams-Moulton, Taylor series, and trapezoidal rule methods.

Some of these methods are explicit, and others are implicit. To demonstrate how to employ these methods, we first describe their formulations and then work on their implementation algorithm for writing scripts (programs) explicitly. We do not attempt to derive any of the formulations used in these numerical methods. There are many literature sources [see 2, 3, 4, 5] that explain the theoretical aspects of these methods.

In solving the IVP using numerical methods, we start at an initial point (initial conditions) and then take a step (equal step-size or varying step-size) forward in time to compute the following numerical solution. Some of the previously named numerical methods (e.g., Euler's methods) are single-step methods, and others (Runge-Kutta, Adams-Bashforth, Milne, Adams-Moulton, and the Taylor series) are multistep methods. Single-step methods refer to only one previous point and its derivative to determine the current value. Other methods, such as Runge-Kutta methods, take some intermediate steps to obtain a higher-order step and then drop off values before taking the next step. Unlike single-step methods, multistep methods keep and use values from the previous steps instead of discarding them. This way, multistep methods link a few previously obtained values (solutions) and derivative values. All of these methods, i.e., the single-step and multistep methods, are assessed based on their accuracy and efficiency in terms of the computation time and resources (e.g., machine time) spent to compute numerical solutions for specific types of IVPs of ODEs. Nevertheless, it remains true that most solutions of the first-, second-, or higher-order IVPs cannot be found by analytical means. Therefore, we need to employ various numerical methods.

Analytical Methods

The Symbolic Math Toolbox has several functions that are capable of evaluating analytical solutions of many analytically solvable ODEs. There are two commands (built-in functions) by which analytical solutions of some ODEs can be evaluated—dsolve and ilaplace/laplace.

Note that in this section we demonstrate—via a few examples of first- and second-order ODEs and systems of coupled differential equations—how to compute analytical solutions of ODEs.

DSOLVE

One ODE solver tool for computing an analytical (or general) solution of any given ODE in MATLAB is dsolve. It can be used with the following general syntax:

```
Solution = dsolve(equation)
Solution = dsolve(equation, conditions)
Solution = dsolve(equation, conditions, Name, Value)
[y1,...,yN] = dsolve(equations)
[y1,...,yN] = dsolve(equations, conditions)
[y1,...,yN] = dsolve(equations, conditions, Name, Value)
```

Example 1: Using DSOLVE

Given a first-order ODE: $y + 2ty^2 = 0$ with no known initial or boundary conditions. Let' solve it using dsolve().

```
>> y_solution=dsolve('Dy=-2*y^2*t')

Y_solution=
  -1/(C3-t^2)
```

Note that C3 is defined from the initial or boundary conditions of the given ODE. There is also an alternative command. In later versions of MATLAB (starting with MATLAB 2012), we can solve the given problem by using the following command syntax:

```
>>syms y(t); y_sol=dsolve(diff(y) == - 2*y^2*t)
 y_sol =
    0
    -1/(- t^2+C3)
```

Example 2: Plotting the Found Solution with dsolve

Given a first-order ODE: $\dot{y} + 2ty^2 = 0$ with the initial condition $y(0) = 0.50$. Let's solve it using dsolve().

```
>> Solution=dsolve('Dy=-2*y^2*t', 'y(0)=0.5')
Solution =
1/(t^2 + 2)
```

The alternative command syntax with later versions of MATLAB is as follows:

```
>> syms y(t); Solution=dsolve(diff(y) == -2*y^2*t, y(0)==0.5)
```

```
Solution = 1/(t^2 + 2)
```

The evaluated analytical solution in a symbolic formulation can be plotted with fplot. (See Figure 8-2.)

```
>> fplot(Solution, [-5, 5], 'ro-'); grid on; xlabel('t'); ylabel 'y(t)'
```

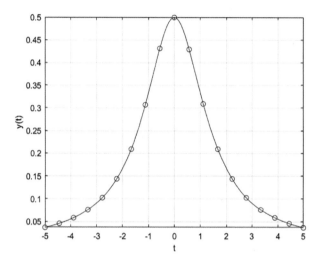

Figure 8-2. *Analytical solution of* $\dot{y} + 2ty^2 = 0$ *with the initial condition y(0) = 0.50*

Numerical values of the analytical solution (the equation) can be computed by vectorizing (parameterizing) the symbolic formulation (the solution), as shown here:

```
>> ysol=vectorize(solution)
ysol =
1./(t.^2 + 2)
>> t=(-5:.1:5); ysol_values=eval(ysol);
```

Example 3: Adding an Unspecified Parameter

Given $\dot{y} + kty^2 = 0$, $y(0) = 0.50$. Let's solve it using dsolve(). Note that this exercise has one unspecified parameter k.

```
>> syms k
>> solution=dsolve('Dy=-k*y^2*t', 'y(0)=0.5')
solution =
1/((k*t^2)/2 + 2)
```

An alternative command syntax is as follows:

```
>> syms y(t) k;
solution=dsolve(diff(y) == -k*y^2*t, y(0)==0.5)
solution =
1/((k*t^2)/2 + 2)
```

Note   Options in `dsolve` need to be set appropriately depending on the problem type. In MatlaB 2008–2010 or earlier versions, you should set `IgnoreAnalyticConstraints` to none to obtain all possible solutions.

Here's an example:

```
solution=dsolve('Dy=-k*y^2*t','y(0)=0.5','IgnoreAnalyticConstraints','none')
```

Note   For MatlaB 2012 or later versions, you should set `IgnoreAnalyticConstraints` to `false` to get all possible correct answers for all the argument values. Otherwise, `dsolve` may output an incorrect answer because of its pre-algebraic simplifications.

Here's an example:

```
solution=dsolve(diff(y)==-k*y^2*t, y(0)==0.5,'IgnoreAnalyticConstraints',
false)
```

Second-Order ODEs and a System of ODEs

There are a myriad of processes and phenomena that are expressed via second-order differential equations. Examples include simple harmonic motions of a spring-mass system, motions of objects with some acceleration (Newton's Second Law), damped vibrations, current flows in resistor-capacitor-inductance circuits, and so forth. In general, second-order ODEs are expressed in two different forms—homogeneous (Equation 8-4) and nonhomogeneous (Equation 8-5).

$$\ddot{y} + p(x)\dot{y} + q(x)y = 0 \qquad \text{(Equation 8-4)}$$

$$\ddot{y} + p(x)\dot{y} + q(x)y = p(x) \qquad \text{(Equation 8-5)}$$

Note that the homogeneous ODEs in Equation 8-4 always have one trivial solution, which is $y(x) = 0$. It satisfies the givens in Equation 8-4. With respect to the independent functions $p(x)$, $q(x)$, and $g(x)$, the ODEs can be linear or nonlinear. In some cases, the independent functions $p(x)$, $q(x)$, and $g(x)$ can be constant values or nonconstant values.

Let's consider several examples of second-order ODEs to see how to compute general and particular solutions with MATLAB's Symbolic Math Toolbox.

Example 1: dsolve with a Second-Order ODE

Given a second-order ODE: $\ddot{u} + 100u = 2.5\sin(10t), u(0) = 0, \dot{u}(0) = 0$ with no known initial or boundary conditions. Let's solve it using dsolve().

```
usol=dsolve('D2u+100*u=2.5*sin(10*t)', 'u(0)=0', 'Du(0)=0'); pretty(usol)
%% Alternative syntax
syms u(t)
Du = diff(u);
u(t) = dsolve(diff(u, 2)==2.5*sin(10*t)-100*u, u(0)==0, Du(0) == 0);
pretty(u(t))
```

The following is the output from executing the two short scripts/commands:

```
sin(10 t) 3   sin(30 t)                 / t   sin(20 t) \
----------- - --------- - cos(10 t) | - - --------- |
   320           320                 \ 8      160    /
```

Example 2: System ODEs

Given a system of ODEs: $\begin{cases} y_1' = y_2 \\ y_2' = -y_1 - 0.125y_2 \end{cases}$, $y_1(0) = 1, y_2(0) = 0.$

Let's solve it using dsolve().

557

The given problem is a system of two first-order ODEs. This problem can be solved directly with dsolve, similar to the previous examples.

```
%% System of two 1st-order ODEs solved with dsolve
yt=dsolve('Dy1=y2', 'Dy2=-y1-0.125*y2','y1(0)=1', 'y2(0)=0');
pretty(yt.y1) pretty(yt.y2)
%% Alternative  syntax
syms y1(t) y2(t)
z=dsolve(diff(y1,1)==y2, diff(y2,1)==(-y1-0.125*y2), y1(0)==1, y2(0)==0);
pretty(z.y1), pretty(z.y2)
```

The computed analytical solutions of the problem displayed in the command window are not shown here. These are the computed solutions:

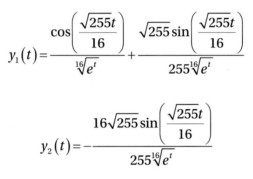

$$y_1(t) = \frac{\cos\left(\dfrac{\sqrt{255}t}{16}\right)}{\sqrt[16]{e^t}} + \frac{\sqrt{255}\sin\left(\dfrac{\sqrt{255}t}{16}\right)}{255\sqrt[16]{e^t}}$$

$$y_2(t) = -\frac{16\sqrt{255}\sin\left(\dfrac{\sqrt{255}t}{16}\right)}{255\sqrt[16]{e^t}}$$

Example 3: Unsolvable Solutions Using dsolve

Given a second-order ODE: $2\ddot{y}+3\dot{y}-|y|\cos(100t)=2$, $y(0)=1$, $\dot{y}(0)=2$.
Let' solve it using dsolve().

```
>> syms y(t); Dy = diff(y,t); D2y = diff(y, t,2);
>> Solution=dsolve(2*D2y+3*(Dy^3)-cos(100*t)*abs(y)- 2==0, y(0)==1,
Dy(0)==2)
Warning: Unable to find symbolic solution.
 Solution =
   [ empty sym ]
```

When the analytical solutions cannot be found with dsolve,the only option will be numerical solution.

Example 4: Computing an Analytical Solution

Say we have this: $\dot{y}-|y|e^t=2,\ y(0)=2$. Here are the commands used to compute an analytical solution of the given exercise:

```
syms y(t)
 Dy = diff(y,t);
 Solution = dsolve(Dy==2+abs(y)*exp(t), y(0)==2);
 fplot(Solution, [0, 5], 'r--o'), grid on
 xlabel 't'
 ylabel('y(t)')
```

Figure 8-3 shows the plot of the found analytical solution.

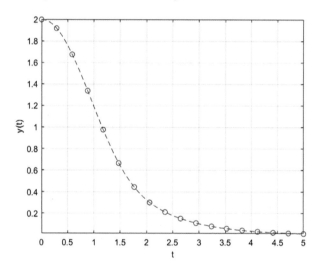

Figure 8-3. *The found analytical solution $\dot{y}-|y|e^t=2,\ y(0)=2$*

Example 5: An Interesting ODE

Let's consider a second-order ODE, as follows: $\ddot{u} + \dot{u} = \sin(t)$, $u(0) = 1$, $\dot{u}(0) = 2$.

```
syms u(t)
Du = diff(u,t);
D2u = diff(u,t, 2);
Solution = dsolve(D2u==sin(t)-Du, u(0)==1, Du(0)==2);
fplot(Solution, [0, 4*pi], 'b-.s'), grid on
xlabel ('t'); ylabel('u(t)')
```

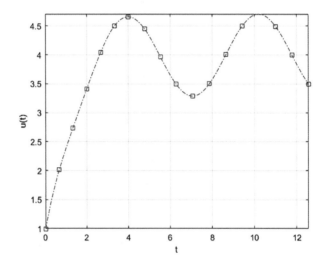

Figure 8-4. *Analytical solution plot of $\ddot{u} + \dot{u} = \sin(t)$, $u(0) = 1$, $\dot{u}(0) = 2$*

Laplace Transforms

Solutions of linear ordinary differential equations with constant coefficients can be evaluated by using the Laplace transformation. One of the most important features of the Laplace transforms in solving a differential equation is that the transformed equation is an algebraic equation. It will be used to define a solution to the given differential equation. In general, the Laplace transform application to solving differential equations can be formulated in the following way.

Let's consider the nth order derivative of $y^n(x) = f(t)$. The Laplace transform of $y^n(x)$ is as follows:

$$\mathcal{L}\left\{\frac{d^n u}{dt^n} = f(t)\right\} \Longrightarrow s^n U(s) - s^{n-1} u^{n-1}(0) - \ldots - s\dot{u}(0) - su(0) = F(s) \qquad \text{(Equation 8-6)}$$

Or

$$s^n U(s) - \sum_{i=1}^{n} s^{n-1} u^{i-1}(0) = F(s) \qquad \text{(Equation 8-7)}$$

In Equation 8-6 or 8-7, if you substitute constant values of initial conditions at $t = 0$ given as $y(0) = a_0$; $\dot{y}(0) = a_1$; $\ddot{y}(0) = a_2$; ...; $y^{n-2}(0) = a_{n-2}$; $y^{n-1}(0) = a_{n-1}$, you can rewrite the expression (Equation 8-6 or 8-7) as follows:

$$s^n U(s) - s^{n-1} a_0 - \ldots - sa_n - sa_{n-1} = F(s) \qquad \text{(Equation 8-8)}$$

Subsequently, we first solve for Y(s), take the inverse Laplace transform from Y(s), and obtain the solution y(t) of the nth order differential equation.

The general procedure for applying the Laplace and the inverse Laplace transforms to determine the solution of differential equations with constant coefficients is as follows:

[1]. Take the Laplace transforms from both sides of the given equation.

[2]. Solve for Y(s) in terms of F(s) and other unknowns.

[3]. Take the inverse Laplace transform of the found expression to obtain the final solution of the problem.

Note that in step 3, you should also break the expression from step 2 into partial fractions in order to use tables of the inverse Laplace transform correspondences.

A schematic view of the Laplace and inverse Laplace transforms is given in the flowchart shown in Figure 8-5.

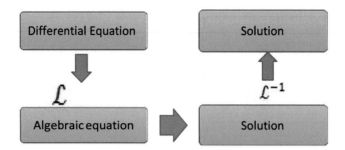

Figure 8-5. *Flowchart of solving ODE with Laplace transform and its inverse*

Example 1: First Laplace Transform

Let's consider a second-order nonhomogeneous differential equation.

$$\frac{d^2y}{dt} + \frac{Ady}{dt} + C = e^{nt}, y(0) = k, dy(0) = m$$

Now, by applying the steps depicted in the flowchart of the Laplace and inverse transforms from the flowchart, you write the Laplace transform of the given problem in explicit steps.

$$\mathcal{L}\left\{\frac{d^2y}{dt} + \frac{Ady}{dt} + C = e^{nt}\right\} = \mathcal{L}\left\{\frac{d^2y}{dt^2}\right\} + \mathcal{L}\left\{\frac{Ady}{dt}\right\} + \mathcal{L}\{C\} = \mathcal{L}\{e^{nt}\} \qquad \text{(Equation 8-9)}$$

$$\mathcal{L}\{e^{nt}\} = \frac{1}{s-n} \qquad \text{(Equation 8-10)}$$

Now, you can work on the left-hand side of Equation 8-9 starting from the highest order.

$$\mathcal{L}\left\{\frac{d^2y}{dt^2}\right\} = s^2Y(s) - dy(0) - s*y(0) = s^2Y(s) - k - s*m \qquad \text{(Equation 8-11)}$$

$$\mathcal{L}\left\{\frac{Ady}{dt}\right\} = A*\mathcal{L}\left\{\frac{dy}{dt}\right\} = A*(s*Y(s) - y(0)) = A*s*Y(s) - m \qquad \text{(Equation 8-12)}$$

$$\mathcal{L}\{C\} = C \qquad \text{(Equation 8-13)}$$

By plugging Equations 8-10, 8-11, 8-12, and 8-13 back into Equation 8-9, you obtain the assembled expression given in Equation 8-14.

$$s^2Y(s)-k-sm+AsY(s)-m+C=\frac{1}{s-n} \qquad \text{(Equation 8-14)}$$

You solve Equation 8-15 for Y(s).

$$Y(s)=\frac{1+(sm+(k+m-C))(s-n)}{(s-n)(s^2+As)}=\frac{Cn-Cs-kn-mn+ks+ms+ms^2}{s(A+s)(n-s)} \qquad \text{(Equation 8-15)}$$

From the expression of Y(s) in Equation 8-15, you can split this into partial fractions and take the inverse Laplace transform of both sides. You obtain Equation 8-16, which is the y(t) of the given differential equation:

$$y(t)=\frac{e^{nt}}{An+n^2}-\frac{Cn-kn+mn+1}{An}+\frac{Cn-kn-mn-A(k-mn-C)+1}{Ae^{At}(A+n)} \qquad \text{(Equation 8-16)}$$

The built-in laplace() function of the Symbolic Math Toolbox is used to evaluate the Laplace transform of any algebraic expression or differential equation. Likewise, the ilaplace() function of the Symbolic Math Toolbox is used to compute the inverse of the evaluated Laplace transformed s domain expression. These two functions handle all transformations by breaking up the partial fraction procedures automatically and then compute an analytical solution of a given ODE exercise.

LAPLACE/ILAPLACE

As mentioned, laplace/ilaplace are based on the Laplace and inverse Laplace transforms, which are built-in function tools of the Symbolic Math Toolbox. The general syntax of laplace/ilaplace is as follows:

```
F=laplace(f)
F=laplace(f, t)
F=laplace(f, var1, var2)
```

and as follows:

```
f=ilaplace(F)
f=ilaplace(F, s) f=ilaplace(F, var1, var2)
```

Example 2: Using LAPLACE

Given $x(t) = \sin(2t)$, the Laplace transform of $x(t)$ is computed with the following command syntax:

```
>> syms t
>> xt=sin(2*t); Xs=laplace(xt)
2/(s^2 + 4)
```

Given $y(t) = \sin(Kt)$. Let's compute its Laplace transform with laplace().

```
>> syms t K
>> yt=sin(K*t); Ys=laplace(yt)
K/(K^2 + s^2)
```

Example 3: A Final LAPLACE

Compute the Laplace transform of $y(x) = ax^3 + b$.

```
>> syms x a b y(x)
>> y(x)=a*x^3+b; Ys=laplace(y(x))
Ys =
(6*a)/s^4 + b/s
```

You can also obtain the t variable domain instead of s.

```
>> syms x t a b
>> y=a*x^3+b; Yt=laplace(y, x, t)
Yt =
(6*a)/t^4 + b/t
```

The `ilaplace()` function syntax and implementation are exactly the same for `laplace`. Let's look at several ODE exercises to see how to use `laplace/ilaplace` and compare their evaluated solutions to the ones obtained with `dsolve`.

Example 4: Comparing LAPLACE/ILAPLACE with DSOLVE

Let's solve $\dot{y} + 2y = 0$, $y(0) = 0.5$ with `laplace/ilaplace` and `dsolve`. The following script (`ODE_Laplace.m`) shows the solution:

```
% ODE_Laplace.m
clearvars; clc; close all
% Step #1. Define symbolic variables' names
syms t s y(t) Y
Dy = diff(y(t),t);
ODE1=Dy==-2*y(t);
% Step #2. Laplace Transforms
LT_A=laplace(ODE1, t, s);
% Step #3. Substitute ICs and initiate an unknown Y
LT_A=subs(LT_A,{laplace(y(t),t, s),y(0)},{Y,0.5});
% Step #4. Solve for Y (unknown)
Y=solve(LT_A, Y);
display('Laplace Transforms of the given ODE with ICs'); disp(Y)
% Step #5. Evaluate Inverse Laplace Transform
Solution_Laplace=ilaplace(Y);
display('Solution found using Laplace Transforms: ')
pretty(Solution_Laplace)
% Step #6. Compute numerical values and plot them
t=0:.01:2.5; LTsol=eval(vectorize(Solution_Laplace));
figure, semilogx(t, LTsol, 'ko')
xlabel('t'), ylabel('solution values')
title('laplace/ilaplace vs dsolve ')
grid on; hold on
%% Compare with dsolve solution method
clearvars;
syms y(t)
Dy = diff(y, t);
Y_d=dsolve(Dy==-2*y, y(0)==0.5);
disp('Solution with dsolve')
t=0:.01:2.5;
pretty(Y_d); Y_sol=eval(vectorize(Y_d));
plot(t,Y_sol, 'b-', 'linewidth', 2), grid minor
legend('laplace+ilaplace', 'dsolve')
hold off; axis tight
```

After executing the ODE_Laplace.m script, the following output is obtained:

```
Laplace Transforms of the given ODE with ICs
1/(2*s + 4)
Solution found using Laplace Transforms: exp(-2 t)
--------
2
Solution with dsolve exp(-2 t)
--------
2
```

The plot of solutions shown in Figure 8-6 clearly displays a perfect convergence of solutions found with laplace/ilaplace and dsolve.

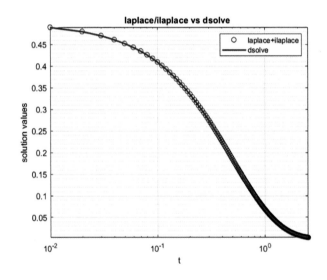

Figure 8-6. *The problem $\dot{y} + 2y = 0$, $y\,(0) = 0.5$ solved with laplace/ilaplace and dsolve*

Example 5: Convergent Answers

Given $\ddot{y} + \dot{y} = \sin(t)$, $y(0) = 1$, $\dot{y}(0) = 2$, here is the solution script (Laplace_vs_Dsolve.m) of this second-order nonhomogeneous ODE with laplace, ilaplace, and dsolve:

```
clearvars, clc, close all
syms t s y(t) Y(s)
Dy = diff(y(t), t);
D2Y = diff(y(t), t, 2);
ODE2nd=D2Y== sin(t)-Dy;
% Step 1. Laplace Transforms
LT_A=laplace(ODE2nd, t, s);
% Step 2. Substitute ICs and initiate an unknown Y
LT_A=subs(LT_A,{laplace(y(t),t, s),subs(diff(y(t),t),  t,
0),y(0)},{Y(s),2,1});
% Step 3. Solve for Y unknown
Y=isolate(LT_A, Y);
%disp('Laplace Transforms of the given ODE with ICs');
disp(Y)
Solution_Laplace=ilaplace(rhs(Y));
disp('Solution found using Laplace Transforms: ')
pretty(Solution_Laplace);
t=0:.01:13; LTsol=eval(vectorize(Solution_Laplace));
figure, plot(t, LTsol, 'ro-'); xlabel('t'), ylabel('solution values')
title('laplace/ilaplace vs. dsolve: ddy+dy=sin(t)'); hold on
% dsolve solution method
Y=dsolve('D2y+Dy=sin(t)', 'y(0)=1, Dy(0)=2', 't');
disp('Solution with dsolve:   ');
pretty(Y)
fplot(Y, [0, 13], 'b-', 'linewidth', 2); grid minor
legend('laplace+ilaplace', 'dsolve', 'location', 'SE'); hold off
```

567

The computed analytical solutions are as follows:

```
Y(s) == (s + 1/(s^2 + 1) + 3)/(s^2 + s)
Solution found using Laplace Transforms:
    cos(t)   sin(t)   5 exp(-t)
4 - ------ - ------ - ---------
       2        2         2
Solution with dsolve:
               /      pi \
    sqrt(2) cos| t - -- |
               \      4 /   5 exp(-t)
4 - --------------------- - ---------
              2                 2
```

From the plot displayed in Figure 8-7, it is clear that the solutions found via the Laplace transforms (laplace/ilaplace) and the dsolve functions converge perfectly well. Both functions evaluate the same analytical solution of a given ODE.

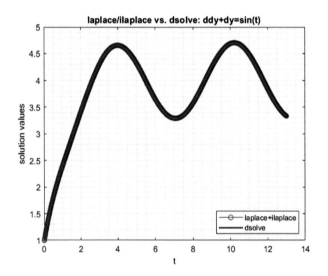

Figure 8-7. *Analytical solutions of* $\ddot{y}+\dot{y}=\sin(t)$, $y(0)=1, \dot{y}(0)=2$

Example 6: No Analytical Solution

Given the following second-order nonhomogeneous and nonlinear ODE, let's try to solve it using syms, diff(), and laplace().

$$2\ddot{y} + 3\dot{y} - |y|\cos(100t) = 2, y(0) = 1, \dot{y}(0) = 2.$$

Let's solve Here is the solution script (Lap_inv_Lap.m) with the Laplace and inverse Laplace transforms:

```
% Lap_inv_Lap.m
clearvars, clc, close all
syms t s y(t) Y(s)
Dy=diff(y(t),t);
D2y=diff(y(t),t,2);
ODE2nd=D2y==0.5*(-3*(Dy)^3+cos(100*t)*abs(y(t))+2);
% Step 1. Laplace Transforms
LT_A=laplace(ODE2nd, t, s);
% Step 2. Substitute ICs and initiate an unknown Y
LT_A=subs(LT_A,{laplace(y(t),t, s),subs(diff(y(t),t),  t,
0),y(0)},{Y(s),0,0});
% Step 3. Solve for Y unknown
Y=isolate(LT_A, Y);
% Step 3.   Solve for Y unknown Y=solve(LT_A, Y);
disp('Laplace Transforms of the given ODE with ICs');
disp(Y)
Solution_Laplace=ilaplace(Y);
disp('Solution found using Laplace Transforms: ')
pretty(Solution_Laplace)
```

The Lap_inv_Lap.m script produces the following output in the Command window:

```
Laplace Transforms of the given ODE with ICs
Y(s) == (4 + s*laplace(abs(y(t)), t, s - 100i) + s*laplace(abs(y(t)), t, s
+ 100i) - 6*s*laplace(diff(y(t), t)^3, t, s))/(4*s^3)
Solution found using Laplace Transforms:
                              t
                              /
                              |                  /   d           \3
                          3   |  (t - u23) |  ---- y(u23) |  du23
                      2       /                \ du23        /
                      t       0
ilaplace(Y(s), s, t) == -- -  -----------------------------------------
                      2                              2
            / laplace(|y(t)|, t, s - 100i)        \
      ilaplace|  ---------------------------, s, t |
            |                2                     |
            \               s                      /
   + ---------------------------------------------------
                              4
            / laplace(|y(t)|, t, s + 100i)        \
      ilaplace|  ---------------------------, s, t |
            |                2                     |
            \               s                      /
   + ---------------------------------------------
            4
```

This output means that no analytical solution is computed explicitly with laplace/ilaplace, just like with the dsolve function tools.

Example 7: Demonstrating Efficiency and Effortlessness

Given a second-order nonhomogeneous ODE where $g(t)$ is a forcing function that is discontinuous and defined by the following expression, let's solve it by applying the Laplace transform.

$$g(t) = u_2(t) - u_{10}(t) = \begin{cases} 5, & 2 \le t \le 10 \\ 0, & 0 \le t < 2 \text{ and } t \ge 10 \end{cases}$$

The Laplace transform of the given equation is as follows:

$$\mathcal{L}\{2y+3y-2y\} = \mathcal{L}\{u_2(t)-u_{10}(t)\}$$

$$2s^2Y(s)-2sy(0)-2\dot{y}(0)+3sY(s)-y(0)-2*Y(s) = \frac{5\left(e^{-2s}-e^{-10s}\right)}{s}$$

$$Y(s) = \frac{5\left(e^{-2s}-e^{-10s}\right)}{s\left(2s^2+3s-2\right)}$$

Note that e^{-2s} and e^{-10s} are explained with time delays in the system output signals; in other words, -2 and -10 mean 2 and 10 seconds of time delays. 5 is the magnitude of the Heaviside (step) function.

The formulation Y(s) is the solution of the differential equation in the s domain, but we need it in the time domain. Thus, you need to compute its inverse Laplace transform: $\mathcal{L}^{-1}\{Y(s)\} = y(t)$. By employing ilaplace(), the next short script (Lap_4_non_homog.m) is created. It solves the given problem and computes its analytical and numerical solutions.

```
% Lap_4_non_homog.m
syms t s
F=5*(exp(-2*s)-exp(-10*s))/s;
Y=2*s^2+s+2;
TF=F/Y; TFt=ilaplace(TF);
pretty(TFt);
Sol=vectorize(TFt);
t=linspace(0, 20, 400);
S=eval(Sol); plot(t, S, 'bo-'); grid minor
title('Differential Equation with Discontinuous Forcing Fcn')
grid on, xlabel('time'), ylabel('y(t) solution'), shg
```

After executing the script, the next solution plot is obtained along with the solution formulation, as shown in Figure 8-8.

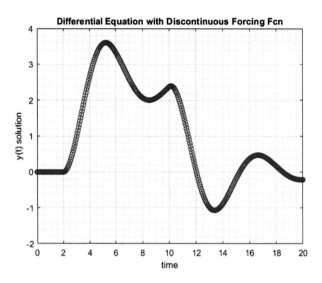

Figure 8-8. *Simulation of the second-order nonhomogeneous ODE subject to the discontinuous forcing function*

```
>> pretty(TFt)
                    /    / 5   t \ / 		   sqrt(15) sin(#1) \ 	    \
                    | exp| - - - | | cos(#1) + ---------------- | 	    |
                    |    \ 2   4 / \			      15		  / 1 |
heaviside(t - 10) | --------------------------------------------- - - |
              \                            2                          2 /

                    /    / 1   t \ / 		   sqrt(15) sin(#2) \ 	    \
                    | exp| - - - | | cos(#2) + ---------------- | 	    |
                    |    \ 2   4 / \			      15		  / 1 |
  5 - heaviside(t - 2) | --------------------------------------------- - - | 5
              \                            2                          2 /

where

        sqrt(15) (t - 10)
  #1 == -----------------
                4

        sqrt(15) (t - 2)
  #2 == ----------------
                4
```

From the previous exercise, the following points can be drawn. Using the Laplace transforms (laplace/ilaplace) to compute analytical solutions of nonhomogeneous ODEs subject to external forcing functions, which are discontinuous, is relatively easy, fast, and effortless. Such exercises are found often within control engineering problems. Moreover, note that the Laplace and inverse Laplace transforms (laplace/ilaplace) are straightforward to implement in solving ODEs. The solutions of ODEs found with them match the ones found by dsolve() perfectly well. As mentioned, many ODEs cannot be solved analytically with the laplace/ilaplace and dsolve functions. Thus, numerical methods are often the only option.

MATLAB Built-in ODEx Solvers

In MATLAB, there are a few built-in ODE solvers, namely, ode15s, ode15i, ode23, ode23s, ode23t, ode23tb, ode45, and ode113, which are efficient in finding numerical solutions of many different types of initial value problems. These solvers are based on explicit Runge-Kutta and implicit Adams-Bashforth-Moulton methods with different implementation algorithms and ODE solver methods, namely, Dormand-Prince (ode45), Bogacki-Shampine (ode23), Rosenbrock (ode23s), trapezoidal rule (ode23t), Adams-Bashforth-Moulton (ode113), Gear's method (ode15s), and so forth. Using MATLAB's built-in ODE solvers is relatively simple, and these are the following general syntaxes of the ODE solvers:

```
solver(odefun,tspan,y0)
[T,Y] = solver(odefun,tspan,y0)
[T,Y] = solver(odefun,tspan,y0,options)
[T,Y,TE,YE,IE] = solver(odefun,tspan,y0,options)
sol = solver(odefun,[t0 tf],y0...)
```

Any of ode15s, ode15i, ode23, ode23s, ode23t, ode23tb, ode45, and ode113 can be chosen depending on the given problem type, for instance, whether the given problem is stiff (how far stiff, e.g., very stiff or moderately stiff) or nonstiff, explicit or implicit.

It is worth noting that an ODE solver type needs to be selected carefully. In selecting a solver type, the recommendations given in Table 8-1 should be considered. These are taken from the help library of the MATLAB package.

Table 8-1. *MATLAB's Built-in ODEx Solvers*

Solver Type	Problem Type	Accuracy	When to Apply
ode15i	Fully Implicit	Medium	For only fully implicit IVP
ode15s	Stiff	Low to Medium	If ode45 is too slow in finding solutions of the problem due to its stiffness
ode23	Nonstiff	Low	For moderately stiff problems with crude error tolerances
ode23s	Stiff	Low	For stiff problems with crude error tolerances
ode23t	Moderately stiff	Low	For moderately stiff problems
ode23tb	Stiff	Low	For stiff problems with crude error tolerances
ode45	**Nonstiff**	**Medium**	**Recommended for most problems; must be the first ODE solver to try**
ode113	Nonstiff	Low to high	For problems with tight error tolerances

Moreover, the efficiency of these solvers depends on the chosen step type (fixed or variable), the size, and the relative and absolute error tolerances that directly affect the accuracy of simulation results and efficiency of computation processes. While using built-in ODE solvers, the step size can be chosen as variable (automatically chosen) or fixed/specified by a user. All built-in ODE solvers by default will take variable step sizes automatically depending on the type of a given IVP (e.g., a stiffness level) and a solution search space. Error tolerance can be controlled in ODE solvers via their setting options. Hereafter, we study in real exercises all these key aspects and settings of ODE built-in solvers.

ODEFUN for the ODE solvers can be defined by using the following:

1. Anonymous function with function (@)

2. Function file (*.m file)

3. matlabFunction: function file (*.m file) by employing the Symbolic Math Toolbox

4. Inline function (in the future MATLAB versions will be removed)

Note   You need to be careful while recalling the function (name) ODEFUN. If it is defined via anonymous function (@) or inline function, then you should use the following command syntax:

```
[T Y]=ODEx(my_Function, t, y0);
```

If you define a given problem (function/expression) via a function file, then you need to use one of the following command syntaxes:

```
[T Y]=ODEx(@Fun_File, t, y0);
[T Y]=ODEx('Fun_File', t, y0);
```

Time space can be predefined as a row or column vector of time values or with two elements, namely, starting and end values, e.g., t = linspace(0, 13, 1000); t = (0:0.001:13).'; t = [0, 13].

ODEx solvers will automatically take different number of steps or step size with respect to the nature of the given ODE (stiff or nonstiff, linear or nonlinear, etc.).

Example 8: Demonstrating MATLAB Built-in ODEx Solvers

Here is the example problem: $\dot{y} + 2ty^2 = 0$, $y_0 = 0.5$. In this case, our function file called Fun_File.m is defined via the next function file:

```
function F=Fun_File(t, y)
F=(-2*y^2*t);
```

We will look at several different problems of how to implement these built-in tools and their options in defining ODEFUN. In the first example, we show how to use the anonymous function (@) to simulate a first-order ODE: $\dot{y} + 2ty^2 = 0$, $y_0 = 0.5$.

The following script (Example_8.m) shows the implementation of ode45, ode23, and ode113 solvers with an anonymous function (@) with a fixed step size, $h = 0.1$:

```
%% Example_8.m
% Part 1
% dy/dt=-2*t*(y^2); with ICs: y(0)=0.5
clearvars
F=@(t,y)(-2*y^2*t); % Anonymous function (@)
% matlabFunction creates a function file called: Fun_F.m
syms tt u; % tt and u are used instead of t and y not to overlap.
F=-2*u^2*tt;
matlabFunction(f, 'file', 'Fun_F');
t0 = 0;             % Start of simulation
tend=10;            % End of simulation
h = 0.1;            % Time step
t=t0:h:tend;        % Time space
y0=0.5;             % Ics: y0 at t0
[t1, Yode45]=ode45(F, t, y0);  % F is anonymous function (@)
[t2, Yode23]=ode23(@Fun_File,t,y0); % Fun_File.m - function file
[t3, Yode113]=ode113('Fun_F',t,y0); % Fun_F.m - matlabFunction
plot(t1, Yode45, 'ks-', t2, Yode23, 'ro-.',t3, Yode113,'bx--'),
grid on;
title('\it Solutions of: $$\frac{dy}{dt}+2*t^2=0, y_0=0.5$$',
'interpreter', 'latex')
legend ('ode45','ode23','ode113')
xlabel('Time, t'), ylabel('Solution, y(t)'), shg
```

Figure 8-9 shows the output plot of the script. You can conclude that for the given problem, ode23, ode45, and ode113 performs very well with the fixed step size of $h = 0.1$.

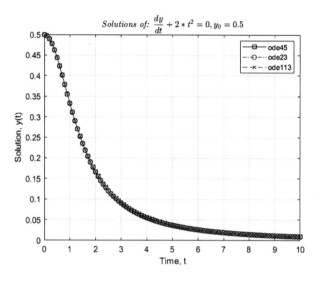

Figure 8-9. *Simulation results of ODE23, ODE45, and ODE113*

Note   If you do not specify the output variable names, e.g., `ode45(F, t, y0)`, then the chosen solver displays computation results in a plot figure and no numerical outputs are saved in the workspace.

Let's look at the issue of how MATLAB built-in solvers take variable steps in solving a given problem (Example 1. $\dot{y} + 2ty^2 = 0$, $y_0 = 0.5$) and how the step size will influence the accuracy of simulations and computation (elapsed) time costs.

```
%%  Example_8.m
%% Part 2
t0 = 0;                % Start of simulation
tend=100;              % End of simulation
t=[t0, tend];          % Time space
y0=0.5;                % ICs: y0 at t0
F=@(t,y)(-2*y^2*t);
tic
[t1, Yode45]=ode45(F, t, y0);
Tode45=toc; fprintf('Tode45 = %2.6f  \n', Tode45)
clearvars -except t y0
```

```
tic
[t2, Yode23]=ode23(@Fun_File, t, y0);
Tode23=toc; fprintf('Tode23 = %2.6f \n', Tode23)
clearvars -except t y0
tic
[t3, Yode113]=ode113('Fun_F', t, y0);
Tode113=toc; fprintf('Tode113 = %2.6f \n', Tode113)
```

In Part 2 of the script, the time space ($[t_0, t_{end}]$) is defined by the initial and end time values. Thus, in this case, each solver has taken variable steps while performing simulations. The simulations are performed on a laptop computer with these specs: Windows 10, Intel Core i7 – 9750 CPU @ 2.60 GHz, 16 GB RAM. The script outputs the following data that are computational time of the solvers: ode45, ode23, and ode113.

```
Tode45 = 0.018722
Tode23 = 0.007884
Tode113 = 0.010358
```

Note that the computation time (Tode45) of ode45 (in seconds) is the shortest.

Example 9: MATLAB Built-in ODEx Solvers for Second-Order ODEs

When solving second- or higher-order ODEs, you need to rewrite a given problem as a system of first-order ODEs.

Here's the nonhomogenous and nonlinear second-order ODE problem:

$2\ddot{y}+3\dot{y}-|y|\cos(100t)=2$, $y(0)=1$, $\dot{y}(0)=2$.

Note that this exercise can't be solved analytically using dsolve or laplace/ilaplace (see Example 6 given earlier).

Before writing a script of commands for MATLAB built-in ODE solvers, you need to rewrite the given second-order ODE as a system of two first-order ODEs by introducing new variables.

$\ddot{y}=\dfrac{1}{2}\left(-3\dot{y}+|y|\cos(100t)+2\right)$ is re-written: $\begin{cases} \dot{y}_1 = y_2 \\ \dot{y}_2 = \dfrac{1}{2}\left(-3y_2 +|y_1|\cos(100t)+2\right) \end{cases}$

Note that $y_1 = y$ and $\dot{y}_2 = \ddot{y}$.

The previously written system of first-order ODEs can be expressed by matlabFunction, anonymous function (@), function file, and inline function (be removed in the future MATLAB releases) in scripts.

Note   ODE45 is a recommended solver to try when solving the IVPs if the given problem is not stiff or implicitly defined.

The script (Example_9.m) embeds command syntaxes of the ODE solvers, namely, ode45, ode23, and ode113, to compute numerical solutions of the given problem.

```
% Example_9.m
clearvars; close all
t0=0;          % Start of simulations
tend=2;        % End of simulations
t=[t0, tend];
y(1,:)=[1; 2];         % Initial Conditions
% ode45 - RUNGGE-KUTTA 4/5 Order
Fun = @(t, y)([y(2); (1/2)*(-3*y(2)+abs(y(1))*cos(100*t)+2)]);
[T1, U1]=ode45(Fun, t, y, []);
plot(T1, U1(:,1), 'rp', 'markersize', 9); grid on; hold on
% ode23 - RUNGGE-KUTTA 2/3 Order
[T2, U2]=ode23(Fun, t, y);
plot(T2, U2(:,1), 'b:o', 'markersize', 9)
% ode113 - ADAMS Higher Order
[T3, U3]=ode113(Fun, t, y);
plot(T3, U3(:,1), 'k-', 'linewidth', 2)
legend('ode45', 'ode23', 'ode113', 'location', 'SE')
title('Simulation of: $$\frac{2d^2y}{dt^2}+\frac{3dy}{dt}-
|y|cos(100t)=2$$', 'interpreter', 'latex')
xlabel('Time, $$t$$', 'interpreter', 'latex'),
ylabel('Solution, $$y(t)$$', 'interpreter', 'latex')
axis tight
```

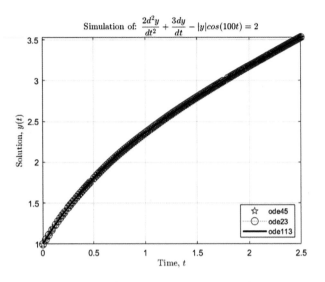

Figure 8-10. *Simulation results of ODE23, ODE45, and ODE113*

This exercise shows that the employed *ODE45*, *ODE23*, and *ODE113* built-in solvers have found well-converged numerical solutions of the given nonhomogeneous and nonlinear second-order ODE problem.

Note   There are some exercises that have a nonzero starting time of IVPs. In solving such problems, the simulation has to start at a given initial time (value). For example, for $u\left(\dfrac{\pi}{2}\right)=\dfrac{2}{3}$, the simulation has to start at $t=\dfrac{\pi}{2}$. This is applicable for all built-in ODEx solvers, scripts, and Simulink models.

Example 10: Simulink Modeling

Solving second or higher-order ODEs with Simulink modeling should be started with the Integrator block to obtain a sought solution from second- or higher-order derivative variable. For example, if you are solving a first-order ODE, you need one integrator block, and similarly, if you are solving second- or third-order ODE, you need two or three Integrator cblocks.

Let's consider the following second-order ODE example to demonstrate how to build a Simulink model.

$$\frac{1}{2}\ddot{u}+\frac{2}{5}\dot{u}+u=t, \; u(0)=1 \text{ and } \dot{u}(0)=2.$$

You first rewrite the given second-order ODE before starting to model it.

$$\ddot{u} = 2t - 0.8\dot{u} - 2u.$$

Note that to obtain $u(t)$ from $\ddot{u}$ that must be integrated twice, as shown in Figure 8-11, you need two Integrator blocks to build a sought model.

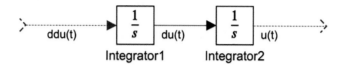

Figure 8-11. *Double Integration with Integrator blocks*

The initial conditions of the given ODE exercise are set up in the Integrator1 and Integrator block parameters by double-clicking each integrator block shown in Figure 8-11 in a sequential order. The Integrator1 block parameters, including the Initial condition entry window, are shown in Figure 8-11. Similarly, by double-clicking Integrator2, the block parameters are accessed and set up. Alternatively, the integrator block parameters can be accessed via one click and using the right mouse option of Block Parameters (Integrator). Note in this case that the initial condition source is chosen to be internal but can be also chosen to be external.

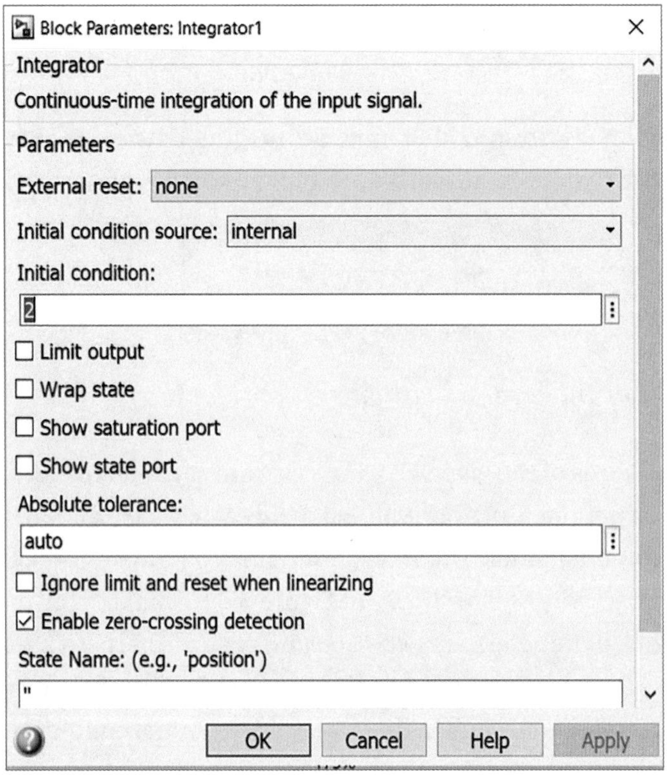

Block Parameters: Integrator1   ×

Integrator

Continuous-time integration of the input signal.

Parameters

External reset: none   ▾

Initial condition source: internal   ▾

Initial condition:

2

☐ Limit output

☐ Wrap state

☐ Show saturation port

☐ Show state port

Absolute tolerance:

auto

☐ Ignore limit and reset when linearizing

☑ Enable zero-crossing detection

State Name: (e.g., 'position')

"

       ?   OK   Cancel   Help   Apply

Figure 8-12.   *Setting up the initial condition using Block Parameters: Integrator*

The complete model of this exercise is `Example_10.slx`, which is shown in
Figure 8-13.

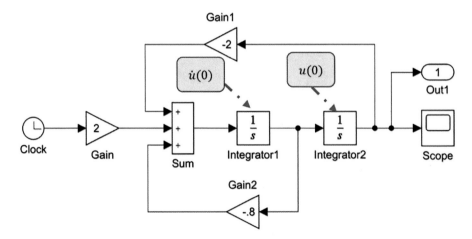

Figure 8-13.   *Simulink model of the problem:* $\frac{1}{2}\ddot{u}+\frac{2}{5}\dot{u}+u=t$, $u(0)=1,\dot{u}(0)=2$

In the Simulink model in Figure 8-12, the Integrator1 block has an internal initial condition value of 2.0, and the other one has an internal initial condition value of 1.0. By executing the model (Figure 8-13), the simulation results obtained via the Scope block shown in Figure 8-14 are obtained.

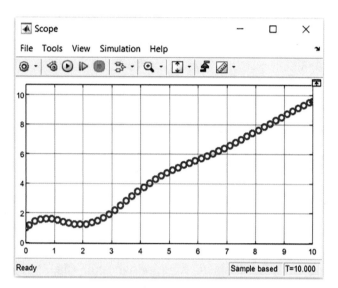

Figure 8-14. *Simulation results shown in the Scope block of the Simulink model, Figure 8-13*

The simulation results of $\frac{1}{2}\ddot{u}+\frac{2}{5}\dot{u}+u=t$, $u(0)=1, \dot{u}(0)=2$ are displayed in the Scope block, as shown in Figure 8-14. Note that in the Scope block shown earlier, we have made some adjustments, e.g.,, by adjusting/selecting its background color and plotting data points from parameters ( ) of the block, that are a marker and line type and the color of plotted data points, which are similar to plot tools of MATLAB. Note that the Out1 block is optional to include in the model. By including this model, you obtain two outputs (tout and yout) in the MATLAB workspace, which are plotted data points shown in Figure 8-13. This Simulink model, called Example_10.slx, can be executed without opening it from MATLAB using the sim() command, and the simulation data points (tout and yout) can be also plotted in MATLAB. Here's an example:

```
[t, u]=sim('Example_10.slx');
plot(t, u(:,1), 'bo'), grid on
xlabel('time, [s]')
ylabel('Solution, u(t)')
```

Note that in the simulation results, t represents the time taken from tout, and u represents the solution results taken from (yout) two integrator blocks, which are the displacement and velocity values.

Note   The options in the Scope block parameters to change the background color, the plotted data's line type, and the marker type and axis color, as well as add legends, are only available starting from MATLAB 2012/Simulink 8.0.

The accuracy of the found numerical solutions from Simulink models depends on the solver type (variable or fixed step solver) and solver (ode45, ode113, ode23, ode1, ode2, odeN, etc.), relative and absolute error tolerances, zero-crossings, step size (if a fixed step solver type), and other settings.

Note   By default, the variable-step solver with ode45 is chosen that can be switched to a fixed step solver. Moreover, solver settings can be adjusted from the Simulink model window's GUI tools via the Modeling tab. Click Model Settings or use the simset() function from MATLAB.

The solver settings can be adjusted using GUI tools from the Simulink model window via the Modeling tab. Clicking Model Settings opens the Configuration Parameters: Solver, Data Import/Export, Math and Data Types, and so forth. By default, Solver is selected, and this shows all Solver details and setting options. All Solver settings can be also accessed and changed from MATLAB using the simset() function.

Let's test the previous example by changing the solver settings, such as solver type and relative and absolute error tolerances, and switching off the zero-crossing option using a MATLAB script. Here is a complete script (SimSet_Simulate.m) to simulate the Simulink model (Example_10.slx):

```
% SimSet_Simulate.m
% Part 1. Variable step solver
% Solver 1 (Variable-step solver): ode45;
Time = [0, 25];
OPTIONS = simset('solver', 'ode45', 'zerocross', 'on');
[t1, u1]=sim('Example_10.slx', Time, OPTIONS);
% Solver 2 (Variable-step solver): ode113;
OPTIONS = simset('solver', 'ode113', 'zerocross', 'on');
[t2, u2]=sim('Example_10.slx', Time, OPTIONS);
% Solver 3 (Variable-step solver): ode23s;
OPTIONS = simset('solver', 'ode23s', 'zerocross', 'on');
[t3, u3]=sim('Example_10.slx', Time, OPTIONS);
plot(t1, u1(:,1), 'bo', t2, u2(:,1), 'r*', t3, u3(:,1), 'kp'), grid on
L=legend('ode45', 'ode113', 'ode23s', 'location', 'SE');
title(L,'Solver type: Variable-step')

title('Solution: $$\frac{d^2u}{2dt^2}+\frac{2du}{5dt}+u=t$$',
'interpreter', 'latex')
xlabel('Time, $$t$$', 'interpreter', 'latex')
ylabel('Solution, $$u(t)$$', 'interpreter', 'latex')
axis tight

%% Part 2. Fixed-step solver
% Solver 1: ode1 (Euler); reltol=1e-3; abstol=1e-5;
OPTIONS = simset('solver', 'ode1', 'reltol', '1e-3', 'abstol', '1e-5',
'zerocross', 'off');
[t1, u1]=sim('Example_10.slx', Time, OPTIONS);
% Solver 2: ode3 (Bogacki-Shampine); reltol=1e-3; abstol=1e-5;
OPTIONS = simset('solver', 'ode3', 'reltol', '1e-3', 'abstol', '1e-5',
'zerocross', 'off');
[t2, u2]=sim('Example_10.slx', Time, OPTIONS);
% Solver 3: ode14x (Extrapolation); reltol=1e-3; abstol=1e-5;
OPTIONS = simset('solver', 'ode14x', 'reltol', '1e-3', 'abstol', '1e-5',
'zerocross', 'off');
[t3, u3]=sim('Example_10.slx', Time, OPTIONS);
plot(t1, u1(:,1), 'bo', t2, u2(:,1), 'r*', t3, u3(:,1), 'kp'), grid on
```

```
L=legend('ode1', 'ode3', 'ode14x', 'location', 'SE');
title(L,'Solver type: Fixed-step')
title('Solution: $$\frac{d^2u}{2dt^2}+\frac{2du}{5dt}+u=t$$',
'interpreter', 'latex')
xlabel('Time, $$t$$', 'interpreter', 'latex')
ylabel('Solution, $$u(t)$$', 'interpreter', 'latex')
```

After simulating the script (SimSet_Simulate.m), we get the following simulation results from the variable step solvers—ode45, ode113, ode23s, and fixed step solvers—ode1, ode3, ode14x. From the simulation of the variable step solvers shown in Figure 8-15, the found numerical solutions from the three solvers are well converged. On the other hand, the results from the fixed step-solvers shown in Figure 8-16 show that not all fixed step solvers can compute accurate numerical solutions despite the same error tolerances. The solver ode1 (Euler method) exhibits significantly inaccurate solutions of the problem. This is a good example that shows the importance of selecting a right solver type and solver with respect to a given ODE problem nature and its stiffness level. Another important observation in this example is that the variable-step solver takes a varying step size, and fixed step-solvers with the same error tolerance settings take the same number of steps to compute numerical solutions.

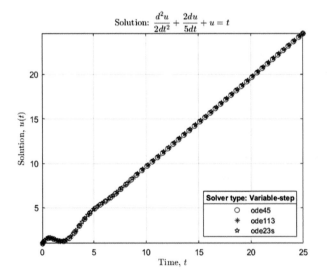

Figure 8-15. *Simulation results from the variable-step type solvers of the Simulink model, Example_10.slx*

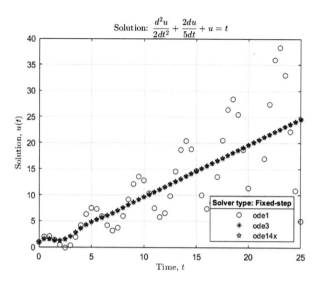

Figure 8-16. *Simulation results from the fixed-step type solvers of the Simulink model, Example_10.slx*

Summary

This chapter covered briefly analytical solution functions (`dsolve`, `laplace/ilaplace`) of MATLAB to solve ODE exercises. Not all ODE problems can be solved analytically using `dsolve` and `laplace/ilaplace` functions. On the other hand, the Laplace transforms (`laplace/ilaplace`) can be employed to solve ODEs with discontinuous forcing functions, which have broad engineering applications.

The chapter introduced key steps of using MATLAB's ODEx numerical solvers, such as `ODE23`, `ODE45`, and `ODE113`, for first and second-order ODEs. Moreover, you learned how to use Simulink modeling aspects to solve IVPs. The chapter demonstrated how to adjust the Simulink solver type and solver settings using the `simset()` function from MATLAB.

References

[1]. `http://tutorial.math.lamar.edu/Classes/DE/Definitions.aspx`, viewed September, 2013.

[2]. Gear, C.W., Numerical Initial-Value problems in Ordinary Differential Equations, Prentice-Hall, Englewood Cliffs, N.J. (1971).

[3]. Potter M. C., Goldberg J.L., Aboufadel E.F., *Advanced Engineering Mathematics,* 3rd Edition, Oxford University Press, (2005).

[4]. Boyce W.E., Diprima R.C., *Elementary Differential Equations and Boundary Value Problems,* 7th Edition, John Wiley & Son, Inc, (2003).

[5]. Hairer E., Norsett S. P., Wanner G., *Solving ordinary differential equations I: Non-stiff problems,* 2nd Edition, Berlin: Springer Verlag, ISBN 978-3-540-56670-0, (1993).

Self-Study Exercises

Exercise 1

The following are IVPs of second-order nonhomogeneous ODEs:

- $\ddot{y} + 9y^2 = \sin(2t)$, $y(0) = 0$ and $\dot{y}(0) = 6$ for $t \, \epsilon \, [0, \ 3\pi]$

- $\ddot{y} + 4\dot{y} + 104y = 2\cos(10t)$, $y(0) = 0$ and $\dot{y}(0) = 0$ for $t \, \epsilon \, [0, \ 5\pi]$

- $\ddot{y} + y = e^x + 2$, $y(0) = 0$ and $\dot{y}(0) = 0$ for $x \, \epsilon \, [0, \ 13]$

- $\ddot{y} + 2\dot{y} + y = 2x$, $\dot{y}(0) = 0$ and $\dot{y}(0) = 6$ for $x \, \epsilon \, [0, \ 15]$

- $\ddot{y} + 2\dot{y} + 101y = 5\sin(10t)$, $y(0) = 0$ and $\dot{y}(0) = 20$ for $t \, \epsilon \, [0, \ 5\pi]$

Solve each of the second-order ODEs with the following methods:

a) Using MATLAB built-in ODE solvers ode23, ode45, and ode113 and adjusting their settings, namely, relative and absolute error tolerances

b) Building a Simulink model (see Chapters 5 and 8 for Simulink modeling) and using a solver ode3

Next compare the solutions found from (a) and (b) and figure out which approach is the most efficient and accurate (correct and has smallest error margins) one.

Finally, is it possible to compute an analytical solution of given problems by using dsolve and Laplace transforms (laplace and ilaplace)? If yes, plot analytical solutions against numerical solutions found from (a) and (b).

Exercise 2

First, solve the second-order nonhomogeneous ODE: $x^2\ddot{y} - 5x\dot{y} + 8y = e^{3x}$, $y(1) = 0$ and $\dot{y}(1) = 24$ for $x \in [1, 15]$. Note that the initial point is at $x = 1$. Use the following methods:

a) Using MATLAB built-in ODE solvers: ode23, ode45, ode113

b) Building a Simulink model (see Chapter 5 for Simulink modeling) and employing a solver ode2

Next, is it possible to compute an analytical solution of the given problem by using dsolve and Laplace transforms (laplace and ilaplace)? If yes, plot analytical solutions against numerical solutions found from (a) and (b).

Exercise 3

First, solve the following second-order nonhomogeneous and nonlinear ODE:
$\ddot{y} + 16|y|\dot{y} + 12y = 3t^3 + 12\cos(3t)$, $y(0) = -1$ and $\dot{y}(0) = 0$ for $x \in [0, 13]$. Use the following methods:

a) Using MATLAB built-in ODE solvers: ode23, ode45, ode113

b) Building a Simulink model (see Chapters 5 and 8 for Simulink modeling) and employing a solver ode4

Next, compare the solutions found from (a) and (b) by plotting t versus $y(t)$ and t versus $\dot{y}(t)$, and find out which approach is the most adequate (meaning it's correct and has the smallest error margins) and efficient.

Exercise 4

First, solve the given IVP of this second-order nonhomogeneous and nonlinear ODE:
$\ddot{y} + 2y^2\dot{y}^2 + 101y = e^{5t} + 2t^2 + 5\sin(10t)$, $y(0) = 0$ and $\dot{y}(0) = 20$ for $t \in [0, 6\pi]$. Use the following methods:

a) Using MATLAB built-in ODE solvers: ode23, ode45, ode113

b) Building a Simulink model (see Chapter 5 and 8 for Simulink modelling) and employing a solver ode14x

Next, compare the solutions found from (a) and (b) and find out which approach is the most efficient. Take smaller time steps if necessary.

Finally, is it possible to compute an analytical solution of the given problem by using dsolve and Laplace transforms (laplace and ilaplace)?

Exercise 5

First, solve the following second-order nonhomogeneous and nonlinear ODE:

$\ddot{y}^2 - 5y^2|\dot{y}| = e^{2t} + 2t$, $y(0) = 1$ and $\dot{y}(0) = 0$ for $t\,\epsilon\,[0,\ \ 13]$. Use the following methods:

a) Using MATLAB built-in ODE solvers: ode23, ode45, ode113

b) Building a Simulink model (see Chapters 5 and 8 for Simulink modeling) and employing a solver ode1

Next, compare the solutions found from (a) and (b) and find out which approach is the most efficient.

Exercise 6

Given an equation of charge in resistor-inductance-capacitor (RLC) circuit shown in the

below given figure in a series by Kirchhoff's law: $L\ddot{q} + R\dot{q} + \dfrac{q}{C} = \mathcal{E}_{max}\cos\omega t$

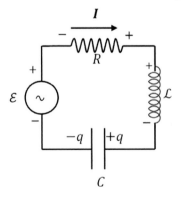

with $q(0) = \dot{q}(0) = 0$ for $t\,\epsilon\,[0,\ \ 4\pi]$.

EMF: $\mathcal{E}_{max} = 110$ $[V]$

Resistance: $R = 7.17$ $[\Omega]$

Capacitor: $C = 50 * 10^{-3}[F]$

Armature inductance: $\mathcal{L} = 9.53 * 10^{-4} \, [H]$

Frequency: $\omega = 60 \, [Hz]$.

First do the following:

a) Find numerical solutions of $q(t)$ using MATLAB built-in ODE solvers: ode23, ode45, ode113.

b) Find numerical solutions of $q(t)$ by building a Simulink model (see Chapters 5 and 8 for Simulink modeling) and employing a solver ode8.

c) Compare the solutions found from (a) and (b) and find out which approach is the most efficient and correct/appropriate. If necessary, take reasonably smaller time steps and specify the (appropriate) initial step size, as well as relative and absolute tolerances.

Then, is it possible to compute analytical solution of the problem using dsolve and Laplace transforms (laplace and ilaplace)? If yes, plot analytical solutions against numerical solutions found from (a) and (b).

Exercise 7

First, solve the given IVP of the fourth-order nonhomogeneous

ODE: $y^{iv} + 3\ddot{y}^3 - \cos(100t)\dot{y} + 8y = t^2 + 10\sin(100t)$ with

$y(0) = 0, \dot{y}(0) = 1, \ddot{y}(0) = 2,$ and $\dddot{y}(0) = -3$. For $t \, \epsilon \, [0, \; 3\pi]$. Use the following methods:

a) Solve the problem by using MATLAB built-in ODE solvers (ode23, ode45, ode113) and adequately setting up relative and absolute error tolerances.

b) Solve the problem by using MATLAB built-in ODE solvers (ode23s, ode15s, ode23tb) and obtain the numerical solution of the problem in plot only (hints: set up OutputFcn for @odeplot with odeset).

c) Solve the problem by building a Simulink model (see Chapters 5 and 8 for Simulink modeling) with a solver ode2.

Finally, compare all the solutions found from (a) to (c) and find out which approach is the most efficient and adequate.

Exercise 8

Solve the given IVP of the fourth-order nonhomogeneous ODE:
$y^{iv} + 2\ddot{y} + \ddot{y} + 8\dot{y} - 12y = 12\,sin\,(25t) - e^{-5t}$, $y(0) = 3, \dot{y}(0) = 0, \ddot{y}(0) = -1$ and $\dddot{y}(0) = 2$, for $t \in [0,\ 5\pi]$. Use the following methods:

a)   Solve the problem by using MATLAB built-in ODE solvers (ode23s, ode15s, ode113) by setting up relative and absolute tolerances.

b)   Solve the problem by using MATLAB built-in ODE solvers (ode23, ode45, ode23tb) and obtain the numerical solutions of the problem in plot only (hints: set up 'OutputFcn' for @odeplot with odeset).

c)   Solve the problem by building a Simulink model (see Chapters 5 and 8 for Simulink modeling) with the solver ode8.

Then, compare all the solutions found from (a) to (c) and find out which approach is the most efficient and appropriate.

Exercise 9

Find numerical solutions of the following systems of coupled ODEs defined by the following:

1. $\begin{cases} \dfrac{dx_1}{dt} = -x_2 + \cos(t) \\ \dfrac{dx_2}{dt} = x_1 + \sin(t) \end{cases}$ with ICs: $x_1(1) = 2.5,\ \ x_2(1) = 3.5,\ \ \ t \in [1, 13]$.

2. $\begin{cases} \dfrac{dx}{dt} = -3x + 5y + 2ye^x \\ \dfrac{dy}{dt} = -13x - x^2 - y^2 \end{cases}$ with ICs: $x(0.5) = 2,\ \ y(0.5) = -2,\ \ \ t \leq 5.55$.

3. $\begin{cases} \dfrac{dx}{dt} = (1+x)\sin(y) \\ \dfrac{dy}{dt} = 1 - x - \cos(y) \end{cases}$ with ICs: $x\left(\dfrac{\pi}{4}\right) = 1.25,\ y\left(\dfrac{\pi}{4}\right) = 0.75,\ t\epsilon\left[\dfrac{\pi}{4}, \dfrac{7\pi}{2}\right].$

For each of the systems, perform the following tasks:

- Write an anonymous function of the coupled system.

- Create a Function file called, e.g., CoupleODE.m.

- Solve the problem by building a Simulink model (see Chapters 5 and 8 for Simulink modeling) called, e.g., CoupledODEsim.mdl, with a fixed step solver ode3.

- Find the numerical solutions of the problem by employing ode23, ode45, and ode113. Compare the solutions from ODEx solvers and the Simulink model and check the efficiency of each approach. Take smaller time steps, adjust the relative and absolute tolerances, and simulate your created Simulink model (CoupledODEsim.mdl) from an M-file (hint: use sim() and simset()).

Exercise 10

By using the Laplace transforms (laplace, ilaplace), solve the following second-order nonhomogeneous ODEs subject to discontinuous forcing function:

1. $\ddot{y} + 5y = h(t),\ y(0) = 0, \dot{y}(0) = 0,\ h(t) = \begin{cases} 0, & 0 \leq t < 3 \\ (t-3)/3 & 3 \leq t < 11 \\ 13 & t \geq 11 \end{cases}$

2. $\ddot{y} + 5\dot{y} + 5y = g(t),\ y(0) = 0, \dot{y}(0) = 2,\ g(t) = \begin{cases} 5 & \pi \leq t < 3\pi \\ 0 & 0 \leq t < \pi \ \text{and}\ t \geq 3\pi \end{cases}$

3. $\ddot{x} + 5\dot{x} + \dfrac{5}{6}x = u(t),\ x(0) = 0, \dot{x}(0) = 0,\ u(t) = \begin{cases} \sin(t) & 0 \leq t < 2\pi \\ 0 & t \geq 2\pi \end{cases}$

Plot numerical values of the analytical solutions for a sufficient time.

Index

© Sulaymon Eshkabilov 2022
S. Eshkabilov, *Beginning MATLAB and Simulink*, https://doi.org/10.1007/978-1-4842-8748-4

Index

A

Acceleration equation, 114
Acell cell array, 58
Additional plot-related commands, 431
addpath(), 10, 367
Analytical methods, ODEs
    dsolve, 553, 554
    plotted with fplot, 554
    unspecified parameter, 555
Animated plot commands, 431
Animated plots
    with drawnow function, 428, 429
    with getframe(), 428, 429
Anonymous function, 236, 237
App Designer, 258–261, 275
Archimedes, 255
ASCII/ANSI character symbols, 40
Astr and Bstr structure arrays, 64
axis command, 385
Axis labels, 72, 196–197, 385, 391, 392, 411

B

Blank Model, 322, 324
Blank Model window, 325, 327, 349, 359
Boolean logical arrays, 47
Boundary value problems (BVPs), 550

C

Callback functions, 257, 258,
    268–272, 279–283

C/C++
    MATLAB, 300
    MEX files, 312
    M-file, 302
    source code type selection
        generation, 304
Cell arrays, 54, 55, 57, 58, 60, 65, 88, 196
Characteristic polynomial, 505, 506
Cholesky decomposition, 456,
    515–519, 549
Code generation, 60, 201, 300, 303, 358,
    369, 371
colon (:) operator, 30, 34
Color specifiers, 389
Command History window, 18
Command window, 3, 4, 6, 7, 13, 14,
    19, 24, 321
% Comments, 71, 72, 75
Computing values of math functions
    adjustments to the model, 340
    blocks, 338
    Configuration properties of Scope,
        341, 342
    Ex3_Function_Compute_Simple.
        slx, 345
    Ex3_Function_Compute.slx, 338
    fixed-step size, 339
    input/output signals
        from/to the MATLAB
        workspace, 345–348
    modeling process, 338
    Scope block, 342

L